Africa's Propensity for a Net Zero Energy Transition

This book explores the issue of a sustainable energy transition in Africa, including the current energy landscape and exploration of various scenarios for achieving net zero emissions. It highlights the challenges faced by African countries in transitioning to clean energy and provides practical solutions for these challenges. It provides perceptive analysis and case studies demonstrating how African nations can take advantage of their natural resources, including insights from Bhutan and Denmark to achieve sustainable development while mitigating the effects of climate change.

Features:

- Offers unique insights into the specific challenges and opportunities that Africa faces in achieving a net zero energy future.
- Provides comprehensive understanding of the region's renewable energy potential, including analysis of policies and initiatives driving sustainable development.
- Highlights successful case studies, emerging technologies, and the economic and environmental benefits of transitioning to net zero energy.
- Explores the role of international collaborations and partnerships in supporting Africa's journey toward a net zero energy future.
- Discusses solutions that must be tailored to the specific needs and contexts of African countries.

This book is aimed at graduate students and researchers in the field of sustainability and energy systems.

Africa's Propensity for a Net Zero Energy Transition

Samuel Chukwujindu Nwokolo,
Anthony Umunnakwe Obiwulu and
Paul C. Okonkwo

CRC Press
Taylor & Francis Group
Boca Raton London New York

CRC Press is an imprint of the
Taylor & Francis Group, an informa business

Designed cover image: Shutterstock

First edition published 2025
by CRC Press
2385 NW Executive Center Drive, Suite 320, Boca Raton FL 33431

and by CRC Press
4 Park Square, Milton Park, Abingdon, Oxon, OX14 4RN

CRC Press is an imprint of Taylor & Francis Group, LLC

ISBN: 9781032773438 (hbk)
ISBN: 9781032774558 (pbk)
ISBN: 9781003483175 (ebk)

DOI: 10.1201/9781003483175

Typeset in Times
by Newgen Publishing UK

Contents

Preface

This book explores the potential of Africa to embrace and lead the global shift toward a net zero energy future. The book delves into the unique characteristics and challenges of the African continent, highlighting its vast renewable energy resources, growing population, and increasing energy demand. It provides perceptive analysis and case studies demonstrating how African nations can take advantage of their natural advantages, as well as insights from Bhutan, which has negative net emissions, and Denmark, which is close to net zero emissions, to achieve sustainable development while mitigating the effects of climate change. Eleven hypothetical scenarios that could be necessary for Africa's propensity for a net zero energy transition are also explored in the book. The scenarios include the abundance of renewable energy resources in Africa, such as solar and wind power; the need to reduce dependence on fossil fuels and mitigate the impacts of climate change; the potential for job creation and economic growth through the development of renewable energy industries; the opportunity to improve energy access and reliability for remote and underserved communities; the possibility of leapfrogging outdated and inefficient energy infrastructure by adopting modern technologies; the desire to enhance energy security by diversifying energy sources and reducing reliance on imported fuels; the recognition of the environmental benefits of renewable energy, such as reduced air pollution and greenhouse gas emissions; the chance to attract foreign investment and international partnerships in the renewable energy sector; the alignment with global sustainable development goals and commitments to combat climate change; the potential for increased energy affordability for consumers through renewable energy options; and the provision of a stable and predictable energy market that can benefit both consumers and businesses in the long run. The book also explores several of Africa's key roles in driving the global net zero transition. This includes Africa's vast renewable energy potential, including solar and wind power, which can play a crucial role in meeting global net zero goals, the continent's rich reserves of natural resources, such as lithium and cobalt, are essential for the production of clean technologies like electric vehicles and energy storage systems, Africa's young and growing population presents a unique opportunity for innovation and entrepreneurship in the renewable energy sector, driving technological advancements and economic growth to address climate change, the continent's geographical advantage, with its proximity to the equator, allows for optimal harnessing of solar energy, making it a potential leader in solar power generation, and the continent's diverse ecosystems, including forests and wetlands, can be harnessed for carbon sequestration and nature-based solutions to mitigate greenhouse gas emissions. This timely book offers valuable insights for policymakers, researchers, and stakeholders interested in advancing Africa's energy transition and achieving sustainable development goals. It also addresses the importance of collaboration between governments, businesses, and communities to drive the necessary changes. By showcasing successful initiatives and innovative solutions, this book inspires readers to think creatively and take action toward a net zero energy future in Africa. The book addresses the need for policy reforms and investment in infrastructure to create an enabling environment for clean energy adoption and ensure equitable access to electricity for all Africans. By demonstrating the progress made by African countries in adopting clean energy technologies and promoting inclusive growth, this book serves as a roadmap for other regions looking to embark on a similar journey.

About the Authors

Samuel Chukwujindu Nwokolo is a distinguished figure in the field of renewable energy and climate science activism. With a strong educational background, he holds a bachelor's degree, a postgraduate diploma (PGD) in Physics (renewable energy option), and an MSc in Atmospheric Physics/Meteorology (In View). His passion for sustainable solutions and his deep understanding of the environmental challenges we face today have propelled him to the forefront of the movement. Through his extensive research and practical experience, he has become an advocate for renewable energy adoption and an expert in climate science. His expertise lies not only in theoretical knowledge but also in hands-on implementation, making him a valuable asset in the pursuit of a greener future. His dedication to raising awareness about renewable energy's potential to combat climate change is evident through his numerous speaking engagements, publications, and active involvement in various environmental organizations. He has been actively involved in renewable energy and climate science research for nine years. Research covered aspects of Solar Energy, Smart Energy Systems, and Sustainable Development Studies (sustainable cities and resources, building integrated photovoltaics for sustainable architecture). He has published more than 58 research articles in many reputed journals, including Scopus Q1, SCI, and ABDC-A & B category. He is the author of three academic books and a reviewer for Applied Energy, MDPI Atmosphere, Heliyon, Talyor & Francis Journals of International Journal of Ambient Energy, International Journal of Green Energy, and other Scopus Q1, SCI, and ABDC-A & B category journals. He is a graduate student researcher at the Department of Physics at the University of Calabar in Nigeria, where he also instructs graduate students about research and development, renewable energy, and the science of climate change.

Anthony Umunnakwe Obiwulu, PhD, is a Major General in the Nigerian Army. He is committed to teaching sustainable and smart energy systems. He was an instructor in the Nigerian Army Signals School, the Nigerian Army School of Electrical and Mechanical Engineering, and the Nigerian Army College of Logistics and Management. He has established a SolarWatt Park in Kaduna and Lagos, where vital research on various aspects of solar PV systems is carried out. Solar energy, smart energy systems, and sustainable development studies were all covered in the research (sustainable cities and resources, building integrated photovoltaics for sustainable architecture). He received both his B.Sc. and PhD in solar physics from the University of Lagos in Nigeria. He received his MSc (Engineering Physics) degree from the University of Calabar in Nigeria. The University of Lagos awarded him a postgraduate diploma (PGD) in computer science. He has published more than eight research papers in international refereed journals, including Scopus Q1, SCI, and the ABDC-A & B categories.

Paul C. Okonkwo brings a wealth of expertise to his role as an Associate Professor at the Department of Mechanical and Mechatronic Engineering at Dhofar University in Salalah, Oman. With a special interest in renewable energy, he is dedicated to exploring innovative solutions that address the pressing challenges of sustainable development. Prior to joining Dhofar University, Dr. Okonkwo had made significant contributions to the field through his research and academic pursuits, establishing himself as a respected authority in the realm of renewable energy. Okonkwo is a highly accomplished professional with a diverse background in environmental engineering, corrosion, failure analysis, tribology, and renewable energy. With extensive experience in academia and industry, Paul has taught at universities and conducted research projects across the Middle East, Africa, and Australia. His innovative approach to problem-solving and his ability to bridge the gap between theory and practice make him a valuable asset in any professional setting. Paul is a highly accomplished and renowned mechanical engineer with an impressive publication record. He has contributed significantly to the field through his extensive research, having published over 12 conference papers, 8 book chapters, and more than 70 research articles in prestigious journals, such as Scopus Q1, SCI, and ABDC-A & B category. His work has been recognized for its quality and impact. In addition to his academic achievements, Paul holds a master's degree in mechanical engineering from the University of Lagos, Nigeria, and a bachelor's degree in mechanical engineering from the University of Nigeria, Nsukka, Nigeria. These qualifications have provided him with a strong foundation and expertise in his field. Paul's dedication to his research has been acknowledged by various agencies and governments that have awarded him millions of dollars in competitive grant funding. This financial support not only highlights the significance of his work but also demonstrates his ability to secure resources to further advance his research goals. Okonkwo is a highly accomplished postdoctoral researcher at Qatar University and Deakin University. With a strong focus on innovation, he has become widely recognized as a renowned speaker, having delivered over 12 impactful keynotes and more than 15 invited talks. Paul is highly regarded for his work and has received recognition for his contributions. He has been awarded the UNESCO/CHINA: The Great Wall Fellowship Award as a Senior Research Scholar, further highlighting his expertise in the field.

Introduction

Africa's propensity for a net zero energy transition is a thought-provoking book that explores the complex issue of a sustainable energy transition in Africa. It provides a comprehensive overview of the current energy landscape and explores various scenarios for achieving net zero emissions. The book highlights the challenges faced by African countries in transitioning to clean energy and provides practical solutions to overcome these challenges. It emphasizes the need for public–private partnerships to finance large-scale renewable energy projects and create a sustainable future for Africa. Prioritizing clean energy and environmental stewardship can achieve net zero goals while promoting economic growth and social development. It provides perceptive analysis and case studies demonstrating how African nations can take advantage of their natural advantages, as well as insights from Bhutan, which has negative net emissions, and Denmark, which is close to net zero emissions, to achieve sustainable development while mitigating the effects of climate change. Eleven hypothetical scenarios that could be necessary for Africa's propensity for a net zero energy transition are also explored in the book. The scenarios include the abundance of renewable energy resources in Africa, such as solar and wind power; the need to reduce dependence on fossil fuels and mitigate the impacts of climate change; the potential for job creation and economic growth through the development of renewable energy industries; the opportunity to improve energy access and reliability for remote and underserved communities; the possibility of leapfrogging outdated and inefficient energy infrastructure by adopting modern technologies; the desire to enhance energy security by diversifying energy sources and reducing reliance on imported fuels; the recognition of the environmental benefits of renewable energy, such as reduced air pollution and greenhouse gas emissions; the chance to attract foreign investment and international partnerships in the renewable energy sector; the alignment with global sustainable development goals and commitments to combat climate change; the potential for increased energy affordability for consumers through renewable energy options; and the provision of a stable and predictable energy market that can benefit both consumers and businesses in the long run. The book also explores several of Africa's key roles in driving the global net zero transition. This includes Africa's vast renewable

DOI: 10.1201/9781003483175-1

energy potential, including solar and wind power, which can play a crucial role in meeting global net zero goals, the continent's rich reserves of natural resources, such as lithium and cobalt, are essential for the production of clean technologies like electric vehicles and energy storage systems, Africa's young and growing population presents a unique opportunity for innovation and entrepreneurship in the renewable energy sector, driving technological advancements and economic growth to address climate change, the continent's geographical advantage, with its proximity to the equator, allows for optimal harnessing of solar energy, making it a potential leader in solar power generation, and the continent's diverse ecosystems, including forests and wetlands, can be harnessed for carbon sequestration and nature-based solutions to mitigate greenhouse gas emissions. This timely publication offers valuable insights for policymakers, researchers, and stakeholders interested in advancing Africa's energy transition and achieving sustainable development goals. It also addresses the importance of collaboration between governments, businesses, and communities to drive the necessary changes. By showcasing successful initiatives and innovative solutions, this book inspires readers to think creatively and take action toward a net zero energy future in Africa. Furthermore, the book addresses the need for policy reforms and investment in infrastructure to create an enabling environment for clean energy adoption and ensure equitable access to electricity for all Africans. By demonstrating the progress made by African countries in adopting clean energy technologies and promoting inclusive growth, this book serves as a roadmap for other regions looking to embark on a similar journey. Overall, *Africa's Propensity for a Net Zero Energy Transition* provides a compelling vision for a sustainable future that benefits both people and the planet. To enhance course adoption opportunities, we encourage the inclusion of international case studies where possible.

1 State of Play

1.1 A CHANGING GLOBAL CONTEXT WITH REGARD TO AFRICA

Africa in an evolving global context is becoming increasingly important. With its vast natural resources, growing population, and emerging economies, Africa has the potential to play a significant role in shaping the future of the world. As the continent continues to develop and modernize, it is attracting attention from global powers seeking to establish economic and political influence. China, for example, has been investing heavily in Africa's infrastructure and natural resource sectors, while also forging diplomatic ties with African nations. This has sparked concerns about neo-colonialism and exploitation, but it also presents opportunities for Africa to leverage its resources for its own development. Africa's young and dynamic population is a valuable asset that can drive innovation and entrepreneurship. As technology continues to advance at a rapid pace, Africa has the potential to leapfrog traditional development stages and embrace digital transformation. This could lead to increased connectivity, access to education and healthcare, and improved living standards for millions of Africans.

Challenges such as poverty, inequality, corruption, and political instability still persist across the continent. These obstacles hinder the full realization of Africa's potential and impede progress toward sustainable development. In order to overcome these challenges, it is crucial for African governments to prioritize inclusive economic policies that promote job creation, investment in infrastructure (Ituen et al., 2012), and diversification of industries (Nwokolo, 2017). Efforts should be made to strengthen governance systems, enhance transparency, and combat corruption. Collaboration between governments, civil society organizations, and the private sector is essential to driving forward sustainable development initiatives. By addressing these issues and leveraging its youthful population, Africa can harness the power of technology and innovation to create a prosperous future for its people (Obiwulu, Chendo et al., 2020; Obiwulu, Erusiafe et al., 2020). With the right investments in education and skills development, Africa can cultivate a highly skilled workforce that is equipped to thrive in the digital age (Obiwulu et al., 2022). By fostering an environment conducive to entrepreneurship and innovation, Africa can unleash the creative potential of its people and drive economic growth (Obiwulu et al., 2022).

The continent has already witnessed success stories in sectors such as mobile banking and renewable energy, demonstrating its capacity for innovation. By building

DOI: 10.1201/9781003483175-2

on these achievements and investing in research and development, Africa can further diversify its economy and become a global leader in technology and sustainable solutions. This will not only create job opportunities for its growing population but also attract foreign investments and partnerships. By prioritizing inclusive growth and ensuring equal access to education and training, Africa can bridge the digital divide and empower marginalized communities (Mkhize, 2023). This will foster social mobility, reduce inequality, and contribute to overall human development. Africa's rich cultural heritage and diversity can be leveraged to foster creativity and drive unique solutions to global challenges. By promoting collaboration between countries within the continent and with international partners, Africa can establish itself as a hub for innovation, knowledge exchange, and economic cooperation. With the right policies in place, Africa has the potential to transform itself into a thriving digital economy that benefits all its citizens while contributing to global progress (Wang et al., 2023).

Today's global energy crisis has underscored the urgency, as well as the benefits, of an accelerated scale-up of cheaper and cleaner sources of energy. The increasing demand for energy, coupled with the need to reduce carbon emissions, has propelled the world toward exploring alternative solutions. Renewable energy sources such as solar, wind, and hydroelectric power have emerged as viable options that offer both environmental and economic advantages. Governments and organizations worldwide are investing heavily in research and development to harness these abundant resources. The transition to cleaner energy not only mitigates the impact of climate change but also creates new job opportunities and stimulates economic growth. Advancements in energy storage technologies have addressed the intermittent nature of renewable sources, ensuring a reliable and consistent power supply. As we continue to embrace these innovative solutions, it is evident that a swift and widespread adoption of cheaper and cleaner energy sources is essential for a sustainable future (Gwaigwa et al., 2022). By diversifying our energy mix and reducing our reliance on fossil fuels, we can achieve a greener planet while securing a more affordable and resilient energy system for generations to come.

The ripple effects of Russia's invasion of Ukraine have reverberated across the globe, causing a significant impact on commodity prices. The sudden surge in food, energy, and other essential goods has created a challenging situation for African economies that were already grappling with the devastating consequences of the ongoing COVID-19 pandemic (Chapungu et al., 2022). This unforeseen escalation has further strained their fragile economic stability, exacerbating the difficulties faced by nations in this region. As prices skyrocket, the burden on African nations to meet the basic needs of their citizens has intensified, posing severe challenges to their already stretched resources. The urgent need to address this crisis has become even more critical as these economies strive to recover from the pandemic's aftermath and ensure the well-being of their populations amidst mounting adversity.

This setback has exacerbated existing challenges in the region, where a significant portion of the population already lacks access to reliable and affordable energy. The overlapping crises, such as the COVID-19 pandemic and climate change impacts (Chapungu et al., 2022), have disrupted supply chains, hampered investments, and strained government resources. As a result, progress made in expanding electricity coverage has been hindered, leaving more people in Africa without access to modern energy services. This not only hampers socio-economic development but also poses significant health risks, as communities struggle to meet their basic energy needs for cooking, lighting, and healthcare facilities (Steynor and Pasquini, 2023). The increased reliance on traditional biomass fuels further contributes to deforestation and air pollution, exacerbating environmental challenges. To address these issues and mitigate the impacts of overlapping crises on Africa's energy systems, concerted efforts are needed from governments, international organizations, and the private sector. These include prioritizing sustainable energy solutions that leverage renewable sources like solar and wind power while promoting energy efficiency measures.

The risk of blackouts and rationing is rising as African countries struggle financially with utilities. This has led to a significant strain on the economies of these nations, as the lack of reliable electricity hampers industrial growth and stifles investment. In an attempt to mitigate this crisis, governments have been exploring alternative energy sources such as solar and wind power. The transition to renewable energy is not without its challenges. The initial costs of setting up infrastructure for these sources can be prohibitive, especially for countries already burdened with debt. The intermittent nature of renewable energy poses a challenge for ensuring a consistent power supply. Despite these obstacles, African nations are persevering in their efforts to overcome the financial difficulties plaguing their utilities (Tesega, 2022). They are seeking international partnerships and investments to modernize their energy sectors and improve access to electricity for their citizens. By diversifying their energy mix and embracing sustainable solutions, these nations hope to alleviate the risks of blackouts and rationing while fostering long-term economic growth.

The combination of these problems, including economic instability, political conflicts, and climate change, has resulted in a dire situation for many people in sub-Saharan Africa (SSA). The lack of access to basic resources such as food, clean water, and healthcare has led to a sharp increase in extreme poverty across the region. This has been further exacerbated by the ongoing food crisis, where the number of affected individuals has quadrupled in certain areas. Without immediate intervention and long-term sustainable solutions, the situation is likely to worsen, posing significant challenges for both the affected populations and international aid organizations working tirelessly to alleviate their suffering. Efforts must be made to address the root causes of these issues and provide comprehensive support to ensure a brighter future for those living in SSA.

Africa's vulnerability to climate change is evident in its extreme weather events, dwindling water resources, and diminishing agricultural yields. From prolonged droughts in the Sahel region to devastating floods in East Africa, the impacts are far-reaching and detrimental. As temperatures rise, ecosystems suffer, exacerbating the loss of biodiversity and threatening the livelihoods of millions. Africa's coastal communities are grappling with rising sea levels and increased storm surges, which pose a grave threat to their homes and infrastructure. The injustice of this situation is undeniable, as Africa's contribution to global greenhouse gas emissions remains minimal compared to other regions. Despite these challenges, African nations are demonstrating resilience by implementing adaptation strategies and investing in renewable energy sources. Collaborative efforts with international partners are crucial to providing financial support (Armah et al., 2023) and technological expertise that can aid in mitigating the effects of climate change on this vulnerable continent. It is imperative that the global community acknowledge Africa's plight and work together toward a more equitable and sustainable future for all.

Africa has the lowest per capita CO_2 emissions, accounting for less than 3% of global emissions (IPCC, 2021). This is a remarkable achievement considering the rapid economic growth and urbanization happening across the continent. Africa has been actively pursuing sustainable development pathways, focusing on renewable energy sources and investing in clean technologies. The abundance of natural resources, such as solar and wind power, has been harnessed to meet the energy demands of its growing population. Governments have implemented policies to promote energy efficiency and encourage the use of clean fuels. There has been a concerted effort to improve access to electricity in rural areas, reducing reliance on traditional biomass fuels. Africa's commitment to a low-carbon future is not only beneficial for the environment but also presents opportunities for economic growth and job creation. By embracing renewable energy and adopting innovative solutions, Africa is leading the way toward a more sustainable and equitable future for all its citizens.

Africans face climate change's negative effects, fuelling mass migration and regional instability. This urgent situation demands immediate action to mitigate the impact of climate change on African communities. Efforts must be focused on implementing sustainable water management systems to alleviate water stress and ensure access to clean and reliable water sources. Investing in innovative agricultural practices and technologies can boost food production and enhance resilience against changing climatic conditions (Bester and Uys, 2023). To address the rising frequency of extreme weather events, it is crucial to strengthen early warning systems and disaster preparedness measures while also promoting climate-resilient infrastructure development. Supporting green initiatives and renewable energy projects will not only reduce carbon emissions but also stimulate economic growth and create job opportunities, fostering a more sustainable future for Africa. By addressing these challenges holistically, we can alleviate the burden on African communities, curtail mass migration, and promote regional stability amidst the growing threat of climate change.

The global clean energy transition offers Africa economic and social development potential. By embracing clean energy solutions, Africa can address multiple challenges simultaneously. The shift toward renewable sources such as solar and wind power not only reduces carbon emissions but also provides a reliable and sustainable energy supply. This newfound energy independence empowers African nations to develop their industries and create job opportunities, fostering economic growth. The clean energy transition enhances access to electricity in rural areas, improving education and healthcare services. By investing in renewable technologies, Africa can reduce its reliance on fossil fuels and mitigate the impact of climate change. This transition opens doors for innovation and collaboration, attracts foreign investments, and promotes regional cooperation. As Africa harnesses the potential of clean energy, it paves the way for a brighter future, unlocking new possibilities for economic prosperity and social advancement.

The goal of achieving net zero emissions by the middle of the century has been set by nations that account for more than 70% of global CO_2 emissions as of May 2022 (International Renewable Energy Agency [IRENA], 2020). This significant global commitment to combat climate change marks a pivotal turning point in our collective efforts to protect the planet. With over 70% of CO_2 emissions being addressed, it demonstrates a united front in striving toward a sustainable future (IRENA, 2020). The decision to achieve net zero emissions by approximately mid-century showcases the urgency and determination to mitigate the adverse effects of greenhouse gases. This commitment is not only crucial for reducing the impact of climate change but also for fostering innovation and transitioning toward cleaner energy sources. By embracing this ambitious goal, countries are paving the way for transformative policies (Kelly et al., 2024), technological advancements (Okpanachi and Ambe-Uva, 2022), and international collaboration. The journey toward net zero emissions will undoubtedly require substantial investment, robust regulations, and widespread adoption of renewable energy solutions. With this unified commitment, we can forge a path toward a greener and more resilient planet for generations to come.

These 12 African countries, accounting for more than 40% of the continent's CO_2 emissions, play a significant role in shaping Africa's environmental landscape. Their collective contribution to carbon dioxide emissions highlights the urgent need for sustainable development and climate action across the region. With a growing population and expanding industries, these countries face unique challenges in balancing economic growth with environmental preservation. By implementing innovative green technologies and promoting renewable energy sources, they can pave the way for a greener future. Collaborative efforts among these nations are crucial to addressing climate change and mitigating its adverse effects on vulnerable ecosystems and communities. Investing in education and awareness programs can empower citizens to adopt eco-friendly practices and reduce their carbon footprint. As these countries strive toward sustainable development, they have an opportunity to lead by example and inspire other nations on the continent to embrace environmentally conscious policies and practices. Together, they can forge a path toward a more resilient Africa that prioritizes both economic prosperity and environmental stewardship.

These ambitions are helping set a new course for the global energy sector amid declining clean technology costs and shifting global investment. As these countries strive toward sustainable development, they have an opportunity to lead by example and inspire other nations on the continent to embrace environmentally conscious policies and practices. Together, they can forge a path toward a more resilient Africa that prioritizes both economic prosperity and environmental stewardship. These ambitions are helping set a new course for the global energy sector amid declining clean technology costs and shifting global investment. By investing in renewable energy sources such as solar and wind power, these countries can reduce their dependence on fossil fuels and decrease greenhouse gas emissions. Promoting energy efficiency measures and implementing sustainable transportation systems will contribute to a greener future. Embracing circular economy principles can minimize waste generation and promote resource conservation. By adopting innovative solutions like smart grids and decentralized energy systems, these countries can ensure reliable access to electricity while reducing their environmental impact. Investing in green infrastructure projects such as sustainable agriculture and eco-tourism will not only create job opportunities but also protect biodiversity and ecosystems.

The majority of African nations that are signatories to the Paris Agreement on Climate Change are well-positioned to benefit from the technological spillovers of these changes and draw more climate finance. This influx of climate finance presents a unique opportunity for African nations to bolster their sustainable development efforts. By harnessing the technology spillovers associated with climate change, these countries can leapfrog traditional development pathways and embrace innovative solutions. With the Paris Agreement serving as a guiding framework, African nations are well-positioned to tap into global funding mechanisms and attract investments that will drive their transition toward low-carbon economies. This financial support can be channeled toward renewable energy projects, sustainable agriculture practices, and resilient infrastructure development. By capitalizing on these opportunities, African countries can not only mitigate the adverse impacts of climate change but also foster economic growth, create green jobs, and improve the well-being of their citizens. By actively participating in international climate negotiations and showcasing their commitment to climate action, African nations can position themselves as leaders in the global fight against climate change. Through collaboration and knowledge sharing with other countries, they can contribute to the collective effort of achieving a sustainable future for all.

In this book, Africa's propensity for a net zero energy transition is thoroughly examined, presenting a vision where Africa successfully navigates the ever-changing energy landscape to accomplish its development goals. By embracing innovative and sustainable approaches, Africa can effectively address its energy-related challenges while ensuring timely and comprehensive progress. Africa's propensity for a net zero energy transition envisions a continent that harnesses renewable energy sources, such as solar and wind power, to meet its growing energy demands. Through strategic investments in clean technologies and infrastructure, Africa can not only achieve energy security but also reduce its carbon footprint, contributing significantly to global efforts to combat climate change. Africa's propensity for a net zero energy transition emphasizes the importance of inclusive development, ensuring that all segments of society have access to affordable and reliable energy services. By prioritizing education and capacity building, Africa can empower its citizens to actively participate in the sustainable energy transition. Fostering regional cooperation and collaboration will enable African nations to pool resources and expertise for mutual benefit. By embracing this forward-thinking scenario, Africa has the potential to emerge as a global leader in sustainable energy development.

This includes achieving universal access to modern energy services by 2030 (IRENA, 2020) as well as completing the implementation of all African pledges on climate change (Delbeke et al., 2019). It requires a comprehensive approach that addresses both the supply and demand sides of the energy equation. On the supply side, investments in renewable energy sources such as solar, wind, and hydropower are essential to ensuring a sustainable and clean energy future. These investments not only help reduce greenhouse gas emissions but also create new job opportunities and promote economic growth. It is imperative to enhance energy efficiency measures across all sectors to optimize energy use and minimize waste. On the demand side, promoting energy conservation practices and raising awareness about the importance of sustainable consumption patterns can significantly contribute to achieving universal access to modern energy services. It is crucial for African countries to fully implement their climate pledges. This entails adopting policies and regulations that promote low-carbon development pathways, investing in climate adaptation measures, and strengthening institutional frameworks for effective climate governance.

African countries must lead with clear strategies and policies, while international institutions must increase support. This includes financial aid, technical assistance, and capacity-building initiatives. It is crucial for African countries to prioritize investments in key sectors such as education, healthcare, and infrastructure. By doing so, they can create an enabling environment for sustainable development and economic growth. Fostering regional integration and cooperation will be essential to addressing common challenges and maximizing collective efforts. This can be achieved through the establishment of robust trade agreements, harmonization of policies, and joint initiatives in areas like security and climate change mitigation. Leveraging technology and innovation will play a pivotal role in driving progress across various sectors. Encouraging entrepreneurship and creating conducive conditions for startups can unlock new opportunities for job creation and economic diversification. Lastly, promoting good governance, transparency, and accountability are fundamental pillars that must be upheld to ensure the effective implementation of strategies and policies. With concerted efforts from both African countries and international institutions, the continent can overcome its challenges and unlock its immense potential for the benefit of all its people.

1.2 AFFORDABLY SUPPLYING ENERGY TO ALL AFRICANS IS CURRENTLY THE TOP PRIORITY

At the moment, 600 million people, or 43% of the total population, lack access to electricity, with the majority of them living in rural areas of SSA (IRENA, 2020). This lack of access to electricity poses significant challenges to economic development, education, healthcare (Kichere, 2020), and the overall quality of life in these regions. Without electricity, businesses struggle to operate efficiently (Nwokolo and Ogbulezie, 2018a), limiting job opportunities and economic growth (Nwokolo and Ogbulezie, 2018b). Students are unable to study after dark, hindering their educational progress. Healthcare facilities are limited in their ability to provide adequate care without reliable power sources. The absence of electricity hampers the use of modern technologies and communication tools that have become essential in today's interconnected world. To address this issue, governments and organizations must prioritize investment in infrastructure and renewable energy sources such as solar and wind power. By expanding access to electricity, we can empower communities, improve living conditions, and unlock the full potential of individuals in SSA and beyond.

With full access anticipated by 2030, nations like Ghana, Kenya, and Rwanda serve as role models for other nations (IEA, 2023). These countries have made significant strides in improving access to essential services and infrastructure, setting an example for others to emulate. Ghana, with its ambitious national development agenda, has implemented innovative policies and initiatives that have resulted in increased access to education, healthcare, and clean water. By prioritizing investment in these sectors, Ghana has not only improved the well-being of its citizens but also fostered sustainable economic growth. Similarly, Kenya has made remarkable progress in expanding access to financial services through mobile banking platforms, revolutionizing the way people save, invest, and transact. This inclusive approach has empowered individuals and small businesses (Kelly et al., 2024), driving economic empowerment and reducing poverty levels (Casati et al., 2023). Rwanda's impressive journey toward universal healthcare coverage is another inspiring success story. Through strategic partnerships and a strong commitment to strengthening the health system, Rwanda has been able to provide quality healthcare services to its population at an affordable cost. These countries' experiences demonstrate that with political will, effective governance, and targeted investments, full access to essential services can be achieved by any nation.

Analysis shows expanding national grids will be the most cost-effective and sensible choice for 45% of households by 2030 (IRENA, 2019). Extending national grids not only offers the least expensive solution but also presents a highly sensible choice for nearly half of the population set to acquire access by 2030 (IRENA, 2019). Through our meticulous examination, it becomes evident that this approach is both economically viable and strategically sound. By expanding national grids, we can efficiently connect a significant portion of the population to electricity, ensuring their long-term energy needs are met. This option stands out as a prudent investment that maximizes resources while minimizing costs, thereby benefiting a substantial proportion of individuals who will gain access in the coming decade.

Solar-powered systems and mini-grids are practical options in more than 80% of rural areas (IRENA, 2019). These decentralized energy solutions offer a sustainable and cost-effective approach to meeting the energy needs of rural communities. Mini-grids, which consist of a small-scale power generation system connected to a localized distribution network, provide reliable electricity access to multiple households and businesses. By harnessing the abundant solar resources in these areas, solar-based mini-grids can generate clean and renewable energy, reducing reliance on fossil fuels and mitigating environmental impact. Standalone systems, such as solar home systems or individual solar panels, are also effective in providing electricity to remote households that are not connected to the main grid. These systems enable families to power essential appliances like lights (Mohammed et al.,2023), fans, and mobile phones, improving their quality of life and productivity (Nwokolo, Amadi et al., 2022). The implementation of mini-grids and standalone systems can create local job opportunities (Elabbas et al., 2023) and empower communities by fostering entrepreneurship in the renewable energy sector. As governments and organizations continue to prioritize rural electrification efforts, investing in these decentralized solutions will undoubtedly play a crucial role in bridging the energy access gap and promoting sustainable development.

Africans today lack access to clean cooking for 970 million people (IRENA, 2019). This staggering statistic highlights the pressing issue of energy poverty in Africa. The lack of access to clean cooking not only hinders basic human needs but also poses severe health and environmental risks. Traditional cooking methods, such as open fires (Wang et al., 2024) and inefficient stoves (Babayomi et al., 2022), contribute to indoor air pollution, leading to respiratory diseases and premature deaths. The reliance on biomass fuels like wood and charcoal contributes to deforestation and carbon emissions, exacerbating climate change. To address this challenge, innovative solutions are needed. Promoting clean cooking technologies, such as improved cookstoves and renewable energy sources like biogas and solar power, can significantly improve the lives of millions of Africans while reducing their environmental impact. Investing in infrastructure development, education, and financial support can empower communities to transition toward sustainable cooking practices. By prioritizing access to clean cooking, we can unlock a brighter future for Africa by improving health outcomes, protecting the environment, and fostering sustainable development for all.

The most popular solution for heating in cities is liquefied petroleum gas (LPG), but recent price increases have made it unaffordable for 30 million people in Africa, forcing many of them to switch back to using biomass. This shift back to biomass is not only detrimental to the environment but also poses serious health risks, as the burning of wood and charcoal releases harmful pollutants (Nwokolo, Meyer et al., 2023). To address this pressing issue, governments and organizations must prioritize finding alternative energy sources that are both affordable and sustainable. One potential solution lies in renewable energy technologies such as solar power. By harnessing the abundant sunlight in Africa, solar energy can provide a clean and accessible source of power for cooking and heating purposes. Investing in the development of biogas systems can help alleviate the burden on LPG by utilizing organic waste to produce a clean-burning fuel. Promoting energy-efficient appliances and practices can reduce overall energy consumption, making LPG more affordable for those who still rely on it. Collaborative efforts between governments, international organizations, and local communities are crucial to implementing these solutions effectively. By prioritizing affordability (Chisale and Lee, 2023), sustainability, and innovation in the energy sector, we can ensure that all Africans have access to clean and affordable cooking solutions while mitigating the negative environmental and health impacts of traditional cooking methods.

A number of nations are reviewing their clean fuel subsidy programs and looking into alternatives like upgraded biomass cookstoves, electric cooking, and biodigesters. These alternatives aim to reduce reliance on traditional fuels like wood and charcoal, which contribute to deforestation and indoor air pollution. Improved biomass cookstoves offer a more efficient and cleaner way of cooking, utilizing renewable resources such as agricultural waste or pellets (Guo et al., 2023). Electric cooking (Lo and Kibalya, 2023), powered by renewable energy sources, eliminates the need for fossil fuels altogether (Mirzania et al., 2023) and reduces carbon emissions. Biodigesters provide an innovative solution by converting organic waste into biogas for cooking, reducing greenhouse gas emissions while also managing waste effectively. As countries reevaluate clean fuel subsidy schemes, these alternatives offer promising pathways toward a sustainable and greener future. By embracing these technologies, nations can promote environmental conservation, improve public health, and foster economic development through the creation of new industries and job opportunities in the clean energy sector.

Unprecedented improvements in clean cooking access by 2030 offer significant benefits, including reducing premature deaths by 500,000, reducing fuel and cooking time, and enabling millions of women to pursue education, employment, and civic involvement (IRENA, 2019). In order to achieve universal clean cooking access by 2030, significant improvement rates must be attained. This endeavor is no small feat, but the rewards it offers are immense. By reaching this goal, we can potentially save approximately 500,000 lives each year by 2030 (IRENA, 2019), effectively reducing premature deaths caused by indoor air pollution. The time spent gathering fuel and cooking will be drastically reduced, freeing up valuable hours for individuals to invest in other productive activities. This initiative will empower millions of women to pursue education, employment, and civic involvement. By providing clean cooking solutions to all, we unlock a world of opportunities and pave the way for a healthier and more prosperous future for all individuals involved.

Investments of USD 25 billion per year are required to achieve the goal of universal access to modern energy (IRENA, 2019). This investment is crucial in order to provide affordable and sustainable energy solutions to the billions of people who currently lack access. By bridging the energy gap, we can improve living conditions, enhance education and healthcare services, and spur economic growth in developing regions. The funds should be allocated toward expanding renewable energy sources such as solar, wind, and hydroelectric power, as well as improving energy efficiency measures. Investing in the research and development of innovative technologies will further accelerate progress toward universal access to modern energy. Collaborative efforts between governments, international organizations, and private sector entities are essential to mobilizing the necessary funding and expertise. By prioritizing universal access to modern energy, we can create a more equitable and sustainable future for all.

The benefits of achieving universal access to modern energy are immense. It would not only provide electricity and clean cooking solutions to the 789 million people currently lacking access (IRENA, 2019), but also drive economic growth, improve healthcare and education, and reduce poverty. Investing in modern energy infrastructure would create job opportunities and stimulate local economies. It would reduce reliance on fossil fuels and contribute to mitigating climate change by promoting the use of renewable energy sources. By bridging the energy gap, we can empower communities, particularly in rural areas, and ensure a more sustainable and equitable future for all. Therefore, it is crucial that governments, international organizations, and private sector entities collaborate to mobilize the necessary funds and resources to achieve universal access to modern energy. This investment is not only a financial commitment but also an investment in humanity's well-being and our planet's future.

International support and stronger national institutions are needed to stimulate investment in 25 African countries (IRENA, 2020). These strategies should focus on creating a favorable business environment, reducing bureaucratic hurdles, and promoting transparency. It is crucial to prioritize the development of robust infrastructure, such as reliable energy grids and efficient transportation networks. By investing in these areas, African nations can attract both domestic and foreign investors, spurring economic growth and job creation. Governments must work toward strengthening their legal frameworks to protect property rights and enforce contracts effectively. This will instill confidence in investors and encourage long-term commitments. Alongside these efforts, fostering partnerships with international organizations and donor countries can provide much-needed financial resources and technical expertise. Collaborative initiatives aimed at improving governance practices and combating corruption are also essential for attracting sustainable investment. By leveraging international support while bolstering national institutions, African countries can create an enabling environment that stimulates investment and propels their economies forward.

1.3 EFFICIENCY KEEPS MODERN ENERGY AFFORDABLE AS AFRICA'S DEMAND FOR IT RISES

As the demand for energy services in Africa continues to surge, it is crucial to prioritize the maintenance of affordability. With rapid population growth and increasing urbanization, the need for accessible and affordable energy solutions becomes even more pressing. Balancing this demand with sustainable and cost-effective options is paramount to ensuring that all segments of society can benefit from reliable energy access. To address this issue, African countries must urgently reassess their subsidy schemes and explore alternative strategies. One potential solution is to gradually phase out energy subsidies (Cooper, 2018) and redirect those funds toward renewable energy investments. By promoting the adoption of clean and sustainable energy sources, African nations can not only reduce their subsidy burdens but also mitigate the environmental impact of fossil fuel consumption. Implementing targeted social safety nets can help protect vulnerable populations from the adverse effects of rising energy prices. These safety nets could include cash transfers or vouchers that specifically assist low-income households in accessing affordable energy options (Yingqi Lui, 2016). Fostering regional cooperation and integration in the energy sector can lead to more efficient resource allocation and shared infrastructure, further reducing subsidy burdens. By adopting these comprehensive measures, African countries can navigate the challenges posed by price spikes while simultaneously advancing toward a greener and more sustainable future.

Egypt, Ethiopia, and Uganda face financial burdens, leading to reduced subsidies and fuel taxes. International support is crucial for price management, but targeting households in need is also crucial. This can be achieved through comprehensive social welfare programs that assess and verify the income levels of individuals and families.

By accurately identifying those who require assistance, governments can ensure that subsidies are directed toward the most vulnerable populations (O'Connor and Macur, 2018). It is crucial to implement measures that promote energy efficiency and renewable energy sources. Investing in sustainable technologies will not only reduce reliance on fossil fuels but also alleviate the strain on national budgets. International organizations should collaborate with these countries to develop long-term strategies for economic growth and stability. By fostering sustainable development, nations can gradually decrease their dependence on subsidies and fuel taxes, creating a more balanced and resilient economy. Ultimately, a combination of targeted subsidies, renewable energy initiatives, and international support will pave the way for a more sustainable future for these countries and their citizens.

By 2030, electricity demand will be 30% lower than it is today thanks to increased energy and material efficiency (230 TW-h) (IRENA, 2019). This significant reduction in electricity demand is primarily driven by advancements in energy and material efficiency technologies. With continuous improvements in energy-efficient appliances, smart grid systems, and sustainable manufacturing processes, industries and households alike are able to optimize their energy consumption (Agyarko et al., 2020). The widespread adoption of renewable energy sources such as solar and wind power further contributes to the decline in electricity demand. As a result, not only does this decrease in demand alleviate pressure on existing power grids, but it also reduces greenhouse gas emissions associated with electricity generation. The benefits extend beyond environmental considerations, as lower electricity demand translates into cost savings for consumers and businesses (Foster and Poston, 2023). By embracing energy and material efficiency measures, we can pave the way toward a more sustainable future while simultaneously enjoying the economic advantages that come with it (Nwokolo and Otse, 2019).

The limitations imposed by building codes and energy performance standards, which forbid the sale of appliances and lighting that are not very energy-efficient, account for 60% of these savings. The implementation of building codes and energy performance standards has proven to be instrumental in achieving significant energy savings. These regulations play a crucial role in curbing the sale of appliances and lighting fixtures that are deemed inefficient, accounting for a substantial 60% of the overall energy conservation effort (IRENA, 2020) By imposing restrictions on the least efficient products available on the market, these codes ensure that consumers have access to more energy-efficient options, thereby reducing waste and promoting sustainable practices (Copiello, 2015). Consequently, these measures contribute significantly to mitigating environmental impacts and fostering a greener future. Through their enforcement, we can continue to make substantial strides toward achieving our energy efficiency goals while simultaneously reducing our carbon footprint.

Urbanization and climate change in Africa increase cooling demand, necessitating efficient cooling solutions to quadruple energy demand in the next decade. To meet this escalating demand, innovative technologies and sustainable practices must be embraced. One promising solution lies in the advancement of smart cooling systems that optimize energy consumption while maintaining optimal comfort levels. These systems utilize sensors and data analytics to intelligently regulate cooling, ensuring that energy is not wasted on unnecessary usage. The integration of renewable energy sources such as solar power can further reduce the carbon footprint associated with cooling. By harnessing the abundant sunlight in Africa, solar-powered air conditioning units can provide a clean and efficient alternative to traditional cooling methods. Promoting energy-efficient building designs and insulation techniques can significantly reduce the need for excessive cooling. Implementing passive cooling strategies like natural ventilation and shading can minimize reliance on mechanical systems, thereby conserving energy. Alongside technological advancements, raising awareness about responsible energy consumption and encouraging behavioral changes among individuals is crucial. Education campaigns can highlight simple yet effective practices like setting thermostats at moderate temperatures and properly maintaining cooling equipment to maximize efficiency.

SAS will see an increase in energy demand in industry, freight, and agriculture of nearly 40% by 2030 (IRENA, 2020). This significant increase in energy demand across various sectors highlights the pressing need for sustainable and efficient solutions. To address this challenge, it is crucial to prioritize innovation and implement advanced technologies that can optimize energy consumption. Embracing renewable energy sources such as solar and wind power can play a pivotal role in meeting the growing demands while reducing carbon emissions. Promoting energy-efficient practices and encouraging industries to adopt greener manufacturing processes will contribute to a more sustainable future. Collaborative efforts between governments, businesses, and communities are essential to driving the transition toward cleaner energy alternatives. By investing in research and development, supporting clean energy infrastructure, and raising awareness about the benefits of sustainable practices, we can pave the way for a more resilient and environmentally conscious society.

By focusing on boosting the production of fertilizers, steel, and cement, Africa can significantly alleviate its heavy reliance on imports, which currently account for more than 20% of its gross domestic product (GDP) (IRENA, 2020). This strategic move not only strengthens the continent's self-sufficiency but also fosters economic growth and stability. By expanding the manufacturing sector to include appliances, vehicles, and clean energy technologies, Africa can further reduce its dependency on foreign goods. This diversification of production not only creates employment opportunities but also enhances technological capabilities within the region. Consequently, Africa can establish itself as a formidable player in the global market while fostering sustainable development and prosperity for its people.

Africa's agriculture sector benefits from electrified irrigation pumps, reduced diesel generator usage, and extended cold chains, increasing productivity and urban market access. This increase in local production not only strengthens Africa's economy but also reduces its reliance on imports. By producing fertilizers, steel, and cement locally, Africa can reduce the burden of importing these essential materials. The manufacturing of appliances, vehicles, and clean energy technologies further contributes to self-sufficiency and economic growth. With electrified irrigation pumps in agriculture, diesel generator use is minimized, leading to a more sustainable and environmentally friendly approach (Nwokolo, Ogbulezie, & Obiwulu, 2022). The extension of cold chains enhances agricultural productivity by ensuring that products remain fresh during transportation, thereby expanding their reach to urban markets. This holistic approach to local production and improved supply chains has the potential to revolutionize Africa's economy while meeting the growing demands of its population.

1.4 SOLAR ENERGY WILL POWER AFRICA'S ECONOMIC FUTURE

Renewable energy sources, such as solar and wind power, are revolutionizing Africa's energy landscape. With abundant sunlight and vast wind resources, Africa has the potential to harness these clean and sustainable sources of electricity on a massive scale. This shift toward renewables is not only reducing carbon emissions but also creating new job opportunities and driving economic growth. Decentralized energy systems are being established, allowing remote communities to access electricity for the first time. As Africa continues to embrace renewable energy technologies, it is paving the way for a brighter and more sustainable future for its people.

Only 1% of the world's installed solar photovoltaic (PV) capacity is in Africa, despite the continent having 60% of the world's best solar resources (IRENA, 2019). This vast untapped potential presents a unique opportunity for Africa to embrace renewable energy and transform its power sector. By harnessing their abundant solar resources, African countries can not only address their energy needs but also contribute to global efforts to combat climate change. The low penetration of solar PV capacity is primarily due to various challenges, such as limited access to financing, inadequate infrastructure, and a lack of supportive policies. With the right investments and policy frameworks, Africa can unlock its solar potential and create a sustainable future. Governments need to prioritize the development of solar projects by offering incentives, streamlining regulatory processes, and encouraging public–private partnerships. Collaboration with international organizations and technology transfer can facilitate knowledge sharing and capacity building within the continent. By leveraging its solar resources effectively, Africa can reduce its dependence on fossil fuels, improve energy security, create jobs, and foster economic growth. Decentralized solar solutions can provide electricity to remote areas that are currently underserved by traditional grid systems.

This would enhance access to basic services such as healthcare, education, and communication, thereby improving the overall quality of life for people in these areas. The adoption of solar energy can contribute to mitigating climate change by reducing greenhouse gas emissions. This is crucial for Africa, as the continent is particularly vulnerable to the impacts of climate change. By embracing solar power, African countries can demonstrate their commitment to sustainable development and become leaders in renewable energy (Nwokolo, Obiwulu et al., 2023). This transition would not only benefit Africa but also inspire other regions around the world to follow suit and embrace clean energy solutions. Harnessing solar power has the potential to transform Africa's energy landscape and pave the way for a more sustainable and prosperous future.

Thanks to ongoing technological advancements and economies of scale, solar PV, which is already the least expensive source of energy in many parts of Africa, will dominate all other sources by 2030 (IRENA, 2019). With its abundant sunlight and vast untapped potential, solar PV has emerged as the go-to solution for Africa's energy needs. As governments and investors recognize its immense benefits, they are increasingly shifting their focus toward harnessing this renewable energy source. By 2030, solar PV will not only be the cheapest source of power in many parts of Africa (IRENA, 2020) but will also outshine all other energy sources across the continent. This transformation will revolutionize Africa's energy landscape, providing affordable and sustainable electricity to millions of people who previously lacked access to reliable power. The widespread adoption of solar PV will drive economic growth, create job opportunities, and improve living standards for communities throughout Africa. It will significantly reduce carbon emissions and combat climate change by replacing fossil fuel–based power generation with clean energy alternatives. The future is bright for solar PV in Africa, paving the way for a more prosperous and sustainable future for the continent as a whole.

Over 80% of the new power generation capacity in the SAS through 2030 will come from renewable sources like solar, wind, hydropower, and geothermal energy (IRENA, 2020). This surge in renewable energy sources is driven by their numerous advantages. Solar power, for instance, harnesses the abundant energy from the sun, providing a clean and sustainable solution. Similarly, wind power utilizes the natural force of wind to generate electricity, offering a cost-effective and environmentally friendly alternative. Hydropower taps into the power of flowing water, making use of rivers and dams to produce electricity on a large scale. Lastly, geothermal energy capitalizes on the Earth's heat from beneath its surface, offering a reliable and continuous source of power. With these renewables dominating new power generation capacity in the SAS, the region is poised to reduce its carbon footprint significantly while ensuring a stable and diverse energy mix for years to come.

After the current ones under construction are completed, Africa will not build any more coal-fired power plants, owing to China's announcement that it will no longer fund coal plants abroad. This shift marks a significant turning point in Africa's energy landscape (Nwokolo, Ogbulezie, & Ushie, 2023). With the completion of these coal-fired power plants, the continent will witness a transition toward cleaner and more sustainable sources of energy. This move is driven by various factors, including the growing awareness of the environmental impact of coal and the need to mitigate climate change. Africa recognizes the importance of embracing renewable energy technologies to meet its increasing energy demands while minimizing carbon emissions. The decision to halt new coal plant construction aligns with global efforts to combat climate change and demonstrates Africa's commitment to a greener future. As China, a major player in Africa's infrastructure development, withdraws its support for coal plants abroad, it paves the way for increased investment in renewable energy projects across the continent. This presents an opportunity for African nations to tap into their vast renewable energy potential by harnessing solar, wind, hydroelectric, and geothermal resources. By embracing these clean alternatives, Africa can not only address its energy needs but also foster economic growth and energy sustainability.

Redirecting investment from discontinued coal plants to solar PV could cover half of Africa's 2025 capacity additions (IRENA, 2023). This redirection of investment from discontinued coal plants to solar PV would have a transformative impact on Africa's energy landscape. The immense potential of solar power in the region could be fully harnessed, bringing about a significant increase in renewable energy capacity. With this redirected investment, Africa could witness a rapid expansion of solar PV installations, enabling the continent to meet a substantial portion of its energy needs sustainably. The financial burden of implementing solar PV projects would be significantly alleviated, as the redirected funds would cover a substantial portion of the costs involved. This would not only accelerate the deployment of solar PV but also pave the way for increased access to clean and affordable electricity for millions across Africa. This shift toward solar energy would contribute to reducing carbon emissions and mitigating climate change impacts, further reinforcing Africa's commitment to sustainable development and environmental stewardship. By seizing this opportunity to redirect investments toward solar PV, Africa can unlock its vast renewable energy potential and embark on a path toward a cleaner and more sustainable future.

By connecting neighboring countries and sharing resources, regional power pools have the potential to revolutionize Africa's energy landscape. These power pools allow for a more efficient allocation of electricity, reducing the risk of blackouts and ensuring a reliable supply for all. They promote collaboration among nations, fostering economic growth and stability. By diversifying the sources of energy generation, regional power pools also enhance resilience against natural disasters or technical failures. With increased access to affordable and sustainable electricity, Africa can unlock its full potential, empowering industries, improving healthcare facilities, and uplifting communities across the continent. Through regional power pools, Africa is poised to overcome its historical challenges in energy reliability and pave the way for a brighter future.

Africa's electricity infrastructure modernization requires improved financial health for public utilities amid economic crises. To achieve this, several measures need to be taken. Firstly, it is crucial to attract private investments in the electricity sector. By creating a favorable investment climate and offering incentives, African countries can encourage both domestic and foreign investors to contribute to modernization efforts. This influx of capital will not only help improve the financial health of public utilities but also bring in new technologies and expertise. It is essential to address the issue of under-pricing electricity. While affordable electricity is important for socio-economic development, it is equally important to ensure that tariffs reflect the true cost of production and distribution. Implementing cost-reflective pricing mechanisms will not only generate revenue for public utilities but also promote efficiency and sustainability in the sector. In addition to financial reforms, enhancing operational efficiency is crucial. This can be achieved through better management practices, improved maintenance strategies, and investment in smart grid technologies. By reducing transmission losses and improving overall system performance, public utilities can maximize their revenue generation potential.

Priority should be given to regulatory reforms, particularly the cost-of-service electricity pricing reforms that are currently in place or being discussed in 24 African nations (IRENA, 2020). These reforms aim to address the inefficiencies and disparities in electricity pricing across the continent. By implementing cost-of-service pricing, African countries can ensure that electricity tariffs accurately reflect the true costs of generation, transmission, and distribution. This approach promotes transparency and accountability in the energy sector, encourages investment, and improves service quality. These reforms help reduce reliance on government subsidies and promote financial sustainability for utilities. By adopting such measures, African countries can create a more competitive and efficient electricity market that benefits both consumers and providers. These reforms pave the way for increased renewable energy integration as cost-reflective pricing incentivizes investment in clean energy technologies. Overall, regulatory reforms in electricity pricing are crucial for Africa's sustainable development and energy security.

1.5 AFRICA'S OWN DEMAND FOR GAS AND OIL IS THE FOCUS OF THIS DECADE'S PRODUCTION

Even though the SAS predicts a rise in natural gas demand in Africa, electricity produced primarily from renewable sources typically outperforms natural gas in terms of modern energy use. This shift toward renewable energy sources is driven by several factors. Firstly, the growing awareness of the environmental impact of fossil fuels has prompted governments and individuals to seek cleaner alternatives. Advancements in technology have made renewable energy more affordable and accessible, making it a viable option for electricity generation. International agreements and initiatives aimed at reducing greenhouse gas emissions have further incentivized the adoption of renewables. As a result, Africa has witnessed a surge in investments in solar, wind, and hydroelectric power projects. These sustainable sources of energy not only contribute to mitigating climate change but also provide opportunities for job creation and economic growth. While natural gas demand continues to rise in the sub-Saharan region, the dominance of renewables in electricity generation ensures that Africa maintains a balanced energy mix that aligns with global sustainability goals.

Africa's vast natural gas reserves exceed 5,000 billion cubic meters, yet development is not authorized (IEA, 2023). These untapped natural gas resources in Africa hold immense potential for economic growth and energy security. With over 5,000 billion cubic meters of gas waiting to be developed (IEA, 2023), African nations have a unique opportunity to leverage this abundant resource for their benefit. The development of these reserves could attract significant foreign investment, create employment opportunities, and stimulate various industries such as power generation, manufacturing, and petrochemicals. Harnessing this vast natural gas potential would contribute to reducing greenhouse gas emissions by replacing more carbon-intensive fuels like coal and oil. It would also enable African countries to diversify their energy mix and reduce their dependence on imported energy sources. It is crucial for governments and stakeholders to ensure responsible and sustainable development practices that prioritize environmental protection, local community engagement, and equitable distribution of benefits. By carefully managing these resources, Africa can unlock a new era of prosperity while safeguarding its natural heritage for future generations.

In addition to the existing sources, tapping into these resources could significantly boost the gas supply by an impressive 90 bcm annually by 2030 (IEA, 2023). This increase in gas production would prove crucial for various industries such as fertilizer, steel, and cement. The fertilizer industry heavily relies on natural gas for the production of nitrogen-based fertilizers, which are essential for agricultural productivity. Similarly, the steel and cement industries require substantial amounts of gas for their energy-intensive processes. With this additional supply, these sectors can meet their growing demands without facing shortages or inflated prices. The availability of extra gas can also benefit water desalination plants that use natural gas as a fuel source to produce fresh water from seawater. By harnessing these untapped resources, we can ensure a stable and sustainable future for these vital industries while also supporting water scarcity solutions.

This staggering amount of cumulative CO_2 emissions, estimated to reach approximately 10 Gt over the course of the next three decades, poses a significant threat to our planet's delicate balance. As we continue to rely on these gas resources for our energy needs, it becomes imperative that we prioritize sustainable alternatives and actively work toward reducing our carbon footprint. The consequences of such emissions are far-reaching, contributing to the greenhouse effect and exacerbating climate change. It is crucial that we invest in renewable energy sources, implement stringent emission regulations, and foster international cooperation to mitigate the harmful effects of these emissions. By embracing cleaner technologies and adopting environmentally conscious practices, we can strive toward a more sustainable future and protect our planet for generations to come.

Africa's contribution to global emissions would only be 3.5% if these emissions were added to the continent's current cumulative total (IEA, 2023). This highlights the significant disparity in emissions between Africa and the rest of the world. Despite being home to a large population and facing various developmental challenges, Africa's contribution to global emissions remains relatively low. It is crucial to recognize this when discussing climate change and its impacts. Instead of solely focusing on reducing emissions in Africa, it is imperative for developed nations to take responsibility for their historical and current emissions. By acknowledging this disparity, we can work toward a more equitable approach in addressing climate change, where countries with higher emissions take greater action to mitigate their impact. Supporting sustainable development initiatives in Africa can help ensure that economic growth is achieved without compromising environmental sustainability. Collaboration between nations is essential to tackle climate change effectively and ensure a more balanced distribution of emissions worldwide.

This shift toward clean energy is crucial for the long-term sustainability of the planet. As countries around the world prioritize reducing carbon emissions and combating climate change, the demand for fossil fuels is expected to decline. This poses a significant challenge for Africa, which heavily relies on oil and gas exports as a major source of revenue. To mitigate this risk, African nations must embrace renewable energy sources and invest in clean technologies. By diversifying their energy portfolios and promoting sustainable practices, they can not only contribute to global efforts but also secure their economic stability in the face of changing market dynamics. Africa has immense potential for renewable energy generation with abundant solar, wind, and hydro resources. By harnessing these resources and developing robust clean energy infrastructure, African countries can position themselves as leaders in the global clean energy transition. This will not only help them maintain export revenues but also create new job opportunities and drive sustainable development across the continent.

Africa's domestic demand for gas and oil will account for roughly two-thirds of global demand between now and 2030 (IEA, 2023). This surge in domestic demand is driven by rapid population growth, urbanization, and industrialization across Africa. As the continent continues to develop and modernize, its energy needs are expected to skyrocket. This presents a unique opportunity for African countries to capitalize on their vast oil and gas reserves and become major players in the global energy market. It also poses challenges in terms of infrastructure development, environmental sustainability, and economic diversification. To meet this growing demand, African nations must invest heavily in upgrading their energy infrastructure, promoting renewable energy sources, and implementing efficient extraction and production techniques. They should prioritize sustainable development practices that minimize the environmental impact of fossil fuel consumption. By doing so, Africa can not only meet its own energy needs but also contribute to global efforts toward a greener and more sustainable future.

By focusing on the development of efficient storage and distribution infrastructure within Africa, the continent can effectively cater to its growing need for transport fuels and LPG. This approach recognizes the importance of meeting domestic demand while also ensuring a well-functioning system. With robust infrastructure in place, Africa can enhance its energy accessibility and reliability, enabling smoother transportation and distribution of fuels across the region. This not only supports economic growth but also promotes sustainability by reducing reliance on imports. By investing in modern storage facilities and advanced distribution networks, Africa can strengthen its energy sector and foster self-sufficiency, ultimately driving progress and prosperity for its people.

African nations prioritize enhancing their laws pertaining to energy efficiency while also promoting the use of renewable energy sources and other clean energy technologies. This includes investing in solar power projects, wind farms, and hydropower plants to harness the continent's abundant natural resources. Governments are implementing policies to promote energy conservation and encourage the use of energy-efficient appliances and buildings. By prioritizing these initiatives, Africa can reduce its carbon footprint and mitigate the impacts of climate change. Fostering innovation in clean energy technologies can create new job opportunities and attract foreign investment in the sector. Through these combined efforts, Africa is paving the way toward a sustainable and greener future for its people while simultaneously driving economic development.

To ensure long-term stability and sustainability, Africa must diversify its energy sources and invest in renewable technologies. Embracing solar, wind, and hydroelectric power can not only reduce carbon emissions but also create new job opportunities and attract foreign investments. Prioritizing energy efficiency measures can help minimize waste and maximize productivity. By implementing smart grids, promoting energy conservation practices, and adopting cleaner technologies, Africa can position itself as a leader in the global green economy. This transition toward a more sustainable energy sector will not only mitigate the environmental impact but also safeguard against future fluctuations in oil prices. By taking proactive steps now, Africa can secure its energy independence and pave the way for a brighter future for generations to come.

1.6 AFRICA'S AVAILABILITY OF CRITICAL MINERALS REPRESENTS A SIGNIFICANT ECONOMIC OPPORTUNITY

Africa's abundant mineral resources, which play a crucial role in various clean energy technologies, present a tremendous opportunity for the continent to tap into new export markets. It is imperative that these resources be managed effectively and responsibly. By implementing sustainable practices and ensuring transparent governance, Africa can maximize the benefits derived from its mineral wealth while minimizing any negative environmental or social impacts. Fostering strong partnerships with international stakeholders can provide the necessary expertise and investment to develop these resources sustainably. This approach will not only enhance Africa's economic growth but also contribute to global efforts to combat climate change and promote a greener future. With careful management, Africa can leverage its mineral riches as a catalyst for sustainable development and become a key player in the global clean energy revolution.

For example, Africa holds 40% of global cobalt, manganese, and platinum reserves, with South Africa, Congo, and Mozambique contributing to global production. These countries possess a strategic advantage in the global market due to their abundant reserves of cobalt, manganese, and platinum. South Africa, known for its rich mineral resources, has long been a major player in the production of these key minerals. The Democratic Republic of the Congo, with its vast untapped reserves, has emerged as a significant supplier as well. Mozambique also contributes to global production, further diversifying the supply chain. It is important to note that there are numerous other countries across Africa that may hold undiscovered deposits of these valuable minerals. With the increasing demand for batteries and hydrogen technologies worldwide, exploration efforts are likely to intensify in search of new sources. This presents an opportunity for African nations to capitalize on their geological potential and establish themselves as key players in the global supply chain for these essential minerals. By leveraging their resources and fostering sustainable mining practices, African countries can not only contribute significantly to the global market but also drive economic growth and development within their own borders.

By 2030, critical mineral production will generate more than twice as much revenue in SAS than today (IEA, 2023). This significant growth in revenue is driven by several factors. Firstly, the increasing global demand for critical minerals, such as lithium, cobalt, and rare earth elements, is expected to continue rising. These minerals are essential for various industries, including renewable energy, electric vehicles, and technology. As the world transitions toward a more sustainable future, the demand for these minerals will only intensify. Secondly, SAS countries have recognized the potential of their mineral resources and are actively investing in exploration and mining operations. This proactive approach ensures that they can capitalize on the growing market demand and maximize their revenue potential. Governments in SAS are implementing policies to attract foreign investments and foster partnerships with international mining companies. By creating a favorable business environment and providing incentives for exploration and production activities, these countries are positioning themselves as attractive destinations for mineral investments. Technological advancements in mining techniques and infrastructure development are improving efficiency and reducing operational costs in SAS countries. This enables them to extract minerals more effectively while minimizing their environmental impact.

Lastly, the growing demand for minerals and natural resources globally is also contributing to the attractiveness of SAS countries as investment destinations. As industries such as construction, manufacturing, and renewable energy continue to expand, the need for raw materials like metals, coal, and rare earth elements is on the rise. SAS countries are well-positioned to meet this demand with their abundant mineral reserves and efficient mining practices. This not only benefits their economies but also promotes sustainable development by ensuring responsible resource extraction and management. With their favorable business climate, technological advancements, and increasing global demand, SAS countries are set to play a crucial role in the future of mineral investments and contribute to the growth of both local economies and the global mining industry.

African mineral exploration faces decline; improved geological surveys, governance, transportation infrastructure, and reducing negative social and environmental effects are needed. To revive investment in mineral exploration in Africa, it is crucial to address the key factors that have contributed to its decline. Firstly, conducting comprehensive and up-to-date geological surveys is essential. By accurately mapping the mineral resources across the continent, potential investors can make informed decisions and identify lucrative opportunities. Secondly, ensuring robust governance is vital to instilling confidence in the industry. Transparent regulations, fair licensing processes, and effective enforcement mechanisms will attract both local and international investors. Investing in improved transport infrastructure will facilitate the efficient movement of equipment and resources, reducing costs and increasing accessibility to remote mining sites. Lastly, a strong emphasis must be placed on minimizing the environmental and social impacts of mining operations. Implementing sustainable practices, such as responsible waste management and community engagement programs, will not only mitigate negative consequences but also foster positive relationships with local communities. By addressing these critical aspects comprehensively, Africa can create an environment conducive to increased investment in mineral exploration while safeguarding its natural resources and communities.

1.7 AFRICA HAS THE POTENTIAL TO DOMINATE THE PRODUCTION OF RENEWABLE HYDROGEN

Africa has the potential to dominate the production of renewable hydrogen. With its vast natural resources and growing energy demands, Africa is well-positioned to become a global leader in renewable hydrogen production. The continent boasts abundant solar and wind resources, which can be harnessed to generate clean energy for electrolysis. Africa's large tracts of arid land provide ideal conditions for the cultivation of biomass feedstock, further enhancing the potential for hydrogen production. By leveraging these resources and investing in advanced technologies, Africa can not only meet its own energy needs but also emerge as a major exporter of renewable hydrogen to the rest of the world. This would not only contribute to global efforts to combat climate change but also stimulate economic growth and create job opportunities across the continent. By embracing renewable hydrogen as a sustainable energy source, Africa can reduce its reliance on fossil fuels and pave the way for a greener future. With strategic planning and collaboration, Africa has the opportunity to harness its immense potential and establish itself as a key player in the renewable hydrogen market.

In South Africa, Mauritania, Morocco, Namibia, and Egypt, various low-carbon hydrogen projects are either in progress or being discussed. These projects aim to harness the vast renewable energy potential of these countries and leverage it to produce low-carbon hydrogen. In Egypt, plans are being developed to establish large-scale solar and wind farms that will power electrolyzers for hydrogen production. Similarly, Mauritania is exploring the use of its abundant solar resources to generate green hydrogen through advanced electrolysis techniques. Morocco, known for its ambitious renewable energy targets, is also investing in hydrogen production from renewable sources such as wind and solar. Namibia, with its vast desert landscapes, is exploring the potential of using concentrated solar power for hydrogen generation. South Africa is focusing on both green and blue hydrogen production, utilizing renewable energy sources as well as carbon capture and storage technologies. These initiatives demonstrate a collective commitment toward decarbonizing the energy sector and fostering sustainable economic growth in these countries.

By tapping into their renewable energy potential and embracing low-carbon hydrogen technologies, Egypt, Mauritania, Morocco, Namibia, and South Africa are paving the way toward a cleaner and more sustainable future. These countries recognize the importance of reducing greenhouse gas emissions and transitioning to cleaner energy sources. Through their investments in green and blue hydrogen production, they are not only reducing their carbon footprint but also creating new job opportunities and driving economic development. By harnessing the power of renewable energy and implementing carbon capture and storage technologies, these nations are setting an example for others to follow. They understand that a transition to a low-carbon economy is essential for mitigating climate change and ensuring a prosperous future for generations to come. With their collective commitment and innovative approach, Egypt, Mauritania, Morocco, Namibia, and South Africa are leading the way toward a sustainable energy revolution that will benefit both their own countries and the global community as a whole.

In Egypt, the development of low-carbon hydrogen projects is gaining momentum as the country recognizes the potential of renewable-based power in producing ammonia for fertilizers. This not only promotes sustainable agriculture but also contributes to enhancing Africa's food security. Similarly, Mauritania is actively exploring the utilization of renewables to generate hydrogen, thereby fostering a greener approach to ammonia production. Morocco, known for its commitment to renewable energy, is also partaking in discussions surrounding low-carbon hydrogen projects that aim to bolster the continent's food security by providing a sustainable source of fertilizers. Namibia and South Africa are diligently working on initiatives that leverage renewable-based power for ammonia production, further emphasizing their dedication to promoting eco-friendly solutions and ensuring a stable food supply for Africa. With these diverse efforts across multiple nations, Africa is paving the way toward a more sustainable future in agriculture and addressing the pressing issue of food security on the continent.

By 2030, Africa might be able to supply renewable hydrogen to Northern Europe at prices that are competitive with those found elsewhere in the world thanks to a global decline in the cost of hydrogen production (IEA, 2023). This would not only provide Africa with a significant economic opportunity but also contribute to the global transition toward a sustainable energy future. The declining cost of hydrogen production, coupled with Africa's vast renewable energy potential, creates the perfect synergy for the continent to become a major player in the global hydrogen market. By harnessing its abundant solar and wind resources, Africa can produce hydrogen at a competitive price, making it an attractive option for Northern Europe's clean energy needs. This partnership would not only foster economic growth and job creation in Africa but also reduce Europe's reliance on fossil fuels and decrease carbon emissions. The collaboration between Africa and Northern Europe could pave the way for technological advancements in hydrogen production and transportation, driving innovation and further cost reductions worldwide. As we approach 2030, the vision of Africa delivering renewable-produced hydrogen to Northern Europe becomes increasingly feasible, unlocking immense potential for both regions while contributing to a greener and more sustainable planet (IEA, 2023).

This significant reduction in hydrogen production costs presents a game-changing opportunity for Africa. By harnessing its abundant renewable resources, the continent can emerge as a major player in the global energy market. The prospect of delivering renewablely produced hydrogen to Northern Europe at competitive prices by 2030 opens up new avenues for economic growth and development (IEA, 2023). Africa's vast potential to produce 5,000 Mt of hydrogen annually, priced below USD 2 per kg, could revolutionize the energy landscape (IEA, 2023). Not only would this position Africa as a key supplier, but it would also contribute to addressing the pressing issue of global energy demand. By capitalizing on this opportunity, Africa can pave the way toward a sustainable and prosperous future for itself and the world.

1.8 AFRICA'S NEW ENERGY ECONOMY MUST PRIORITIZE PEOPLE

As the continent continues to experience rapid population growth and urbanization, it is crucial that the energy sector prioritizes the needs and aspirations of its people. Africa's new energy economy should aim to provide reliable, affordable, and sustainable energy solutions that improve the quality of life for all citizens (Nwokolo, Ogbulezie, & Ushie, 2023). This requires a comprehensive approach that combines investment in renewable energy sources such as solar and wind power with the development of efficient transmission and distribution networks. By harnessing Africa's abundant natural resources, such as sunlight and wind, we can ensure access to clean energy while reducing dependence on fossil fuels (Nwokolo, Meyer et al., 2024). Empowering local communities through capacity-building programs and job creation in the renewable energy sector will foster economic growth and social inclusion. Integrating innovative technologies like smart grids and energy storage systems will enhance energy efficiency and enable greater control over power consumption. To achieve this vision, collaboration between governments, private sector entities, international organizations, and civil society is essential. Together, we can build an inclusive energy economy that not only meets Africa's growing energy needs but also drives sustainable development and improves the quality of life for all its citizens.

By 2030, 4 million more jobs related to energy are required in SSA in order to achieve universal access to energy (IEA, 2023) This surge in job demand is driven by the region's growing population and increasing energy needs (Nwokolo, Echeng et al., 2024). To achieve universal energy access, investments in renewable energy sources such as solar, wind, and hydroelectric power are crucial. These investments not only provide clean and sustainable energy but also create employment opportunities in manufacturing, installation, and maintenance of renewable energy systems. The development of transmission and distribution infrastructure requires skilled workers to ensure efficient and reliable electricity supply. The expansion of energy access will stimulate economic growth by powering industries, attracting investments, and improving productivity. It is essential to prioritize training programs and skill development initiatives to equip the workforce with the necessary expertise to meet these demands. By harnessing Africa's vast renewable energy potential and investing in human capital, the continent can pave the way toward a sustainable future while simultaneously addressing its employment challenges.

Opportunities for employment increase women's chances of participating in the formal economy and launching their own businesses. These opportunities not only provide financial stability but also empower women to make independent decisions and contribute to their communities. By joining the formal economy, women gain access to better working conditions, fair wages, and social security benefits. This not only improves their own lives but also has a ripple effect on their families and society as a whole. These jobs enable women to showcase their skills and talents, fostering a culture of innovation and creativity. As more women enter the workforce, they bring diverse perspectives and fresh ideas that drive economic growth and development. Increased female entrepreneurship leads to the creation of new businesses, which further stimulates the economy and creates job opportunities for others. These opportunities break down traditional gender norms and challenge societal expectations, paving the way for gender equality in all aspects of life (Proutsos, Liakatas et al., 2023). Overall, by offering entry into the formal economy and promoting entrepreneurial ventures, these jobs empower women economically, socially, and culturally while contributing to sustainable development for all.

African energy companies are gaining clout through joint ventures and technology transfer. The domestic markets of African energy companies can be expanded by establishing an African Continental Free Trade Zone. This integration allows African energy companies to expand their operations beyond national borders, fostering collaboration and knowledge sharing among diverse industry players. By forming joint ventures with international partners, these companies gain access to advanced technologies and expertise, accelerating the development of local capabilities. This transfer of knowledge not only enhances the efficiency and sustainability of African energy projects but also empowers the continent to become a global player in the sector. The establishment of an African Continental Free Trade Area creates a unified market for energy products and services within Africa (Nwokolo, 2017). This integration eliminates trade barriers and stimulates competition, driving innovation and growth within domestic markets. As a result, African energy firms can tap into larger customer bases, fueling economic development while reducing reliance on external markets. Through these collaborative efforts and regional economic integration, African energy companies are poised to make significant contributions to the continent's sustainable development and energy independence.

1.9 FINANCES MUST BE ALLOCATED FOR ADAPTATION TO CLIMATE CHANGE

Despite being a minor contributor to global emissions, Africa must prioritize comprehensive measures to address climate risks. With its unique vulnerabilities and limited resources, the continent faces a pressing need for adaptive strategies (Nwokolo, Proutsos et al., 2023). Enhancing infrastructure resilience, such as constructing climate-resistant buildings and reinforcing coastal defenses, becomes imperative to mitigate the increasing frequency of extreme weather events. Investing in sustainable agriculture practices and promoting water management systems can bolster food security in the face of changing climatic patterns. Collaborative efforts with international partners can facilitate technology transfer and capacity building, enabling Africa to harness renewable energy sources and reduce dependence on fossil fuels. Fostering education and awareness programs will empower communities to adapt and respond effectively to climate-related challenges.

Regardless of the scenario, Africa will not contribute more than 4% of total global CO_2 emissions related to energy by 2050 (IRENA, 2020). This significant reduction in Africa's contribution to global energy-related CO_2 emissions by 2050 is a result of various transformative measures implemented across the continent (IEA, 2023). Africa has embraced renewable energy sources, such as solar and wind power, on a massive scale. The continent has become a leader in developing innovative and sustainable solutions to meet its energy needs. There has been a remarkable improvement in energy efficiency across industries, transportation, and households. Governments have implemented stringent regulations and policies to promote clean energy adoption and discourage fossil fuel dependency. There has been a substantial shift toward the electrification of transportation systems, with an extensive network of electric vehicles and charging infrastructure now available throughout Africa. The widespread adoption of these clean technologies has not only reduced CO_2 emissions but also improved air quality and public health across the continent. Africa has actively engaged in international collaborations for technology transfer and knowledge sharing, fostering innovation and accelerating the transition to a low-carbon economy.

The impact of a 2°C global average temperature rise by 2050 is concerning, particularly for North Africa, where the median temperature rise is projected to be 2.7°C (IEA, 2023). This substantial increase in temperature would have severe consequences for the region's economy, with African GDP expected to decline by approximately 8% compared to a baseline scenario without climate impacts (IEA, 2023). The repercussions would be even more pronounced in certain areas, like East Africa, where losses could soar up to 15% (IEA, 2023). These figures underscore the urgent need for effective climate policies and measures to mitigate the detrimental effects of rising temperatures on vulnerable regions and safeguard economic stability.

Climate adaptation funding could total $30–50 billion by 2030, far exceeding the $7.8 billion provided by advanced economies in 2019 (IEA, 2023). This substantial surge in funding reflects the growing recognition of the urgent need to address the impacts of climate change. The additional financial resources will enable developing countries to implement robust adaptation measures, such as building resilient infrastructure, enhancing early warning systems, and promoting sustainable agriculture practices. This increased funding will also support capacity-building initiatives and technology transfer to empower vulnerable communities to adapt to changing climatic conditions. By allocating significant resources toward climate adaptation, the international community can mitigate the adverse effects of climate change and ensure a more sustainable future for all. It is crucial that these funds are efficiently utilized and distributed equitably to maximize their impact and reach those most in need. Continued collaboration between governments, private sector entities, and civil society organizations is essential to mobilize additional financial support and foster innovative solutions for climate adaptation.

In order to enhance the resilience of Africa's energy systems against climate risks, several measures need to be implemented. Firstly, it is crucial to diversify the energy mix by reducing reliance on thermal power plants. This can be achieved by promoting renewable energy sources such as solar and wind power, which are not susceptible to water stress. Investing in energy storage technologies will ensure a stable supply of electricity during periods of water scarcity. Coastal flooding poses a significant threat to Africa's LNG capacity. To mitigate this risk, protective infrastructure must be established to safeguard these facilities from potential damage. This includes constructing flood barriers and implementing early warning systems to detect rising sea levels. Fostering regional cooperation and knowledge sharing is essential for building resilience in Africa's energy sector. Collaborative efforts can help identify best practices and innovative solutions that have proven effective in other regions facing similar challenges. By learning from each other's experiences, African countries can develop strategies tailored to their specific needs and circumstances.

1.10 THE FUTURE OF AFRICA'S ENERGY IS DEPENDENT ON EXPANDING FINANCIAL ACCESS

To achieve Africa's energy and climate goals, energy investment must be more than doubled. This increased investment will not only address the continent's growing energy demands but also pave the way for a sustainable and low-carbon future. By allocating substantial funds toward renewable energy projects such as solar, wind, and hydroelectric power, Africa can harness its abundant natural resources and reduce its reliance on fossil fuels. Investing in modern grid infrastructure and energy storage solutions will enhance the efficiency and reliability of electricity distribution across the continent. Funding research and development initiatives focused on innovative clean technologies will spur economic growth while mitigating greenhouse gas emissions. Collaborative efforts with international partners can facilitate knowledge sharing and technology transfer, enabling Africa to leapfrog outdated energy systems and embrace cutting-edge solutions. Prioritizing investments in energy access for rural communities will improve livelihoods, promote social equity, and drive inclusive development.

This significant increase in energy investment is crucial for Africa to effectively address its energy and climate challenges. By allocating over USD 190 billion annually from 2026 to 2030 (IEA, 2023), with a majority of two-thirds dedicated to clean energy, Africa can pave the way toward a sustainable and low-carbon future. This substantial financial commitment will enable the continent to expand its renewable energy infrastructure, such as solar and wind power, on a massive scale. It will foster the development of innovative technologies and solutions that can drive economic growth while minimizing environmental impact. By prioritizing clean energy investments, Africa can reduce its reliance on fossil fuels, mitigate greenhouse gas emissions, and improve energy access for its growing population. This ambitious endeavor will not only contribute to mitigating climate change but also unlock immense opportunities for job creation, technological advancements, and enhanced energy security across the continent.

Energy investment will increase Africa's GDP by 6.1% between 2026 and 2030, which is slightly higher than average for emerging markets (IEA, 2023). This increase in energy investment signifies the growing importance of the sector in Africa's economic development. With a rising demand for electricity and the need to expand access to reliable energy sources, governments and private investors are recognizing the potential for substantial returns in this sector. This surge in investment is expected to drive innovation, create job opportunities, and enhance infrastructure development across the continent. It will pave the way for sustainable growth and help address pressing challenges such as energy poverty and climate change. By prioritizing energy investment, Africa can unlock its vast renewable energy potential, harnessing solar, wind, hydro, and geo-hydroelectrical resources to power its economies while reducing reliance on fossil fuels. This shift toward cleaner and more sustainable energy sources aligns with global efforts to combat climate change and offers Africa an opportunity to leapfrog traditional energy systems. Increased investment in energy infrastructure will facilitate regional integration and trade, promoting economic cooperation among African nations.

According to the International Energy Agency's (IEA's) Net Zero Emissions scenario, Africa's global energy investment will be 5% by 2050 (Laura et al., 2023). This highlights the urgent need for increased investment in Africa's energy sector to meet the growing demand and support sustainable development. The continent has vast untapped renewable energy resources, including solar, wind, hydro, and geothermal power. Harnessing these resources will not only provide clean and affordable energy to millions of people but also create job opportunities and drive economic growth. Investing in renewable energy will help mitigate the impacts of climate change and reduce reliance on fossil fuels, contributing to global efforts to combat greenhouse gas emissions. To unlock Africa's energy potential, governments and international organizations must prioritize funding for infrastructure development, technology transfer, and capacity building. This will enable African countries to build a resilient and inclusive energy system that benefits both urban centers and rural communities. By embracing renewable energy solutions and leveraging innovative technologies, Africa can leapfrog traditional energy systems and pave the way for a sustainable future.

Increased financial flows to Africa, the promotion of investment in SSA, and the strategic use of concessional financing to leverage private capital must be given priority by multilateral development banks. This can be achieved by implementing innovative financing mechanisms such as blended finance, which combines public and private funds to address Africa's development challenges. Multilateral development banks should focus on enhancing technical assistance and capacity-building programs in Africa to strengthen local institutions and promote sustainable economic growth. These banks should actively engage with African governments and stakeholders to identify key sectors that require investment, such as infrastructure development, renewable energy, and agriculture. By aligning their priorities with Africa's needs and effectively utilizing concessional finance, multilateral development banks can play a crucial role in unlocking the continent's potential and fostering inclusive and resilient economic development. This approach will not only attract private capital but also contribute to poverty reduction, job creation, and the achievement of the sustainable development goals in Africa.

Issues like cross-cutting risks and high debt loads continue to exist despite the fact that there are more international financial flows as a result of new climate finance and carbon credits. To address these challenges, it is crucial to implement effective risk management strategies. One approach is to diversify funding sources by exploring alternative financing mechanisms. For instance, leveraging public–private partnerships can help mitigate the burden of high debt and attract additional capital. Promoting sustainable investment practices and encouraging responsible lending can minimize financial risks associated with climate finance. Fostering transparency and accountability in financial transactions can enhance investor confidence and attract more international flows.

1.11 AFRICA REQUIRES MORE ROBUST GLOBAL ENERGY ASSISTANCE

Africa requires more robust global energy assistance to meet its growing energy needs and drive sustainable development. With its vast natural resources, including abundant sunlight and wind, Africa has the potential to become a major player in renewable energy production. Lack of access to financing and outdated infrastructure pose significant challenges. To address this, international collaboration is crucial. Developed nations should increase financial support and technology transfer to help African countries build modern and efficient energy systems. This assistance should prioritize the development of renewable energy sources, such as solar and wind power, which can provide clean and affordable electricity to remote communities. Capacity-building programs should be implemented to train local technicians and engineers in the maintenance and operation of these systems. By investing in Africa's energy sector, the global community can not only alleviate poverty but also contribute to mitigating climate change.

At the conference, African leaders can collaborate with international partners to secure funding and support for renewable energy projects in remote communities. This would involve leveraging financial mechanisms such as the Green Climate Fund and engaging with private sector investors to attract sustainable investments. Knowledge-sharing and technology transfer initiatives can be established to enable African countries to adopt innovative solutions for clean energy generation. By fostering partnerships and promoting regional cooperation, African nations can collectively address the energy challenges they face and accelerate their transition toward a low-carbon future. This would not only improve access to electricity but also create employment opportunities, enhance economic growth, and protect the environment for future generations. The COP27 conference presents a crucial opportunity for Africa to showcase its potential as a leader in sustainable development and inspire global action toward a more equitable and climate-resilient world.

1.12 AFRICA'S ENERGY PROSPECTS

Africa's energy prospects in the face of climate change are a complex and pressing issue. As the continent continues to experience rapid population growth and urbanization, the demand for energy is skyrocketing. Africa's reliance on fossil fuels has detrimental effects on the environment and exacerbates climate change. In response, many African countries are exploring renewable energy sources such as solar, wind, and hydroelectric power. These initiatives not only reduce carbon emissions but also provide opportunities for economic growth and job creation. Regional collaborations and investments in clean energy infrastructure are crucial for Africa's sustainable development. By diversifying its energy mix and embracing innovative technologies, Africa can mitigate the impacts of climate change while ensuring access to affordable and reliable energy for its people. Empowering local communities through education and training programs will foster a culture of sustainability and encourage grassroots participation in renewable energy projects. Africa's energy transition is not only an environmental imperative but also a chance to leapfrog outdated technologies and lead the way toward a greener future.

Africa's energy prospects in 2023 are promising, with significant advancements in renewable energy sources. The continent has made great strides in harnessing its abundant solar and wind resources, leading to a substantial increase in clean energy production. Governments across Africa have implemented favorable policies and incentives to attract investment in the renewable sector, resulting in the establishment of numerous solar and wind farms. These projects have not only improved access to electricity for millions of people but also reduced reliance on fossil fuels, contributing to a greener and more sustainable future. Africa has embraced innovative technologies such as mini-grids and off-grid solutions, enabling remote communities to access reliable power for the first time. As a result, economic growth has been stimulated, creating new opportunities for businesses and improving living standards. The development of energy storage systems has addressed the intermittent nature of renewable sources, ensuring a stable and consistent power supply throughout the day. This breakthrough has revolutionized industries such as agriculture and manufacturing, enabling them to operate efficiently and sustainably.

Looking ahead, Africa's energy landscape is poised for even greater transformation. With the increasing adoption of renewable energy sources, such as solar and wind, Africa is gradually reducing its dependence on fossil fuels and embracing a more sustainable future. This transition has not only brought about environmental benefits but has also unlocked immense economic potential. The continent's vast renewable energy resources, combined with advancements in technology and infrastructure, have attracted significant investments from both domestic and international players. These investments have not only bolstered Africa's energy sector but have also spurred job creation and economic diversification. As a result, Africa is now emerging as a global leader in renewable energy deployment, setting an inspiring example for other regions to follow. The development of smart grids and innovative financing models has facilitated the integration of renewable energy into existing power systems, ensuring reliable electricity access for all. This inclusive approach has empowered communities across Africa, particularly those in remote areas that were previously underserved by traditional grid infrastructure.

Despite efforts to improve access, Africa continues to face challenges in developing its energy system. The demand for clean, secure, and affordable modern energy services has outpaced the continent's ability to meet these needs. This disparity has hindered Africa's overall development and economic growth. While progress has been made in certain areas, such as the expansion of renewable energy sources and increased access to electricity in urban areas, there is still a long way to go. The lack of infrastructure, limited investment, and political instability have all contributed to the slow pace of development. The high costs associated with modern energy technologies have made it difficult for many African countries to afford these solutions. To address these challenges, it is crucial for governments, international organizations, and private sector entities to collaborate and invest in sustainable energy projects that can provide reliable and affordable power to all Africans.

The stark contrast between Africa's population and its energy consumption highlights a significant disparity. Despite being home to almost one-fifth of the global population, Africa's share of energy consumption remains disproportionately low. This discrepancy becomes even more apparent when considering that South Africa, with its relatively advanced economy, accounts for a mere 16% of the continent's energy usage (Figure 1.1) (IEA, 2022). This disparity underscores the urgent need for greater investment and development in the energy sector across Africa to meet the growing demands of its population and foster sustainable economic growth.

Despite its relatively low energy consumption, Africa has immense potential for growth and development in the energy sector. With its abundant natural resources, including vast reserves of oil, gas, and renewable energy sources such as solar and wind, the continent holds the key to unlocking a sustainable and prosperous future. By harnessing these resources and investing in modern infrastructure, Africa can not only meet its own energy needs but also become a major player in the global energy market. This would not only fuel economic growth but also improve living standards for millions of people across the continent. By adopting clean and renewable energy

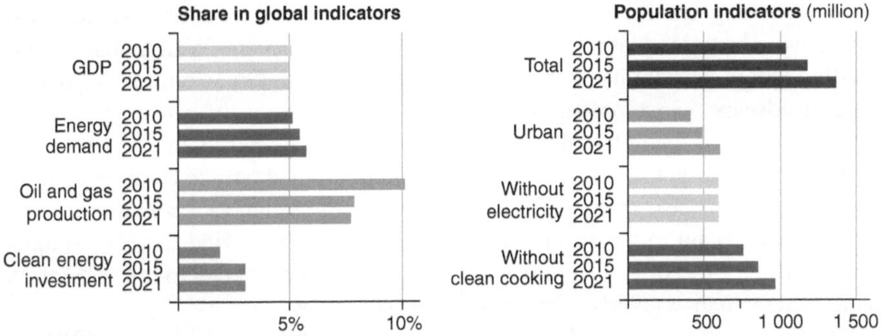

FIGURE 1.1 International Energy Agency-African Energy Outlook, 2022 Africa's share in selected global energy and economic indicators and key population indicators, 2010–2021.

Source: Africa Energy Outlook 2022: International Energy Agency, certified under CC BY-NC-SA 2.0.

technologies, Africa can mitigate the adverse effects of climate change and contribute to global efforts toward a greener future. It is crucial for African nations to prioritize investments in energy infrastructure, promote regional cooperation, and attract foreign investments to fully capitalize on this untapped potential.

While electricity consumption in Africa is rising only 2.3% annually, it is growing more slowly than in other developing regions, where it is rising by an average of 2.4% annually (IEA, 2023). This slower growth rate in electricity consumption in Africa can be attributed to several factors. Firstly, the continent's limited access to reliable and affordable energy sources hinders its ability to meet the increasing demand. Many African countries heavily rely on traditional biomass fuels, such as wood and charcoal, for cooking and heating purposes, which are not only environmentally unsustainable but also inefficient. Secondly, inadequate infrastructure and outdated power grids pose significant challenges to expanding electricity access across the continent. Insufficient transmission and distribution networks result in frequent power outages and limit the potential for industrial development. The high costs associated with building new power plants and extending electricity infrastructure further impede progress in meeting the rising energy needs of Africa's growing population. To address these issues, there is a pressing need for increased investment in renewable energy sources like solar and wind power, as well as advancements in energy storage technologies. Fostering regional cooperation and partnerships can facilitate the sharing of resources and expertise to accelerate progress toward universal access to electricity in Africa.

Sixty-four percent of Africans use wood and waste to cook their food. This heavy dependence on biomass not only contributes to deforestation and environmental degradation but also poses serious health risks due to indoor air pollution. To address this issue, various clean cooking technologies are being introduced, such as improved cookstoves and biogas digesters. These innovations not only reduce emissions but also improve the overall well-being of communities by providing a safer and more efficient cooking environment. The promotion of renewable energy sources like solar power has gained significant momentum in recent years. With abundant sunlight throughout the continent, solar energy has the potential to revolutionize Africa's energy landscape. Governments and organizations are actively investing in large-scale solar projects, including the establishment of solar farms and the distribution of solar panels to households. This shift toward renewables not only reduces greenhouse gas emissions but also creates job opportunities and stimulates economic growth. Africa is tapping into its vast wind resources by developing wind farms in suitable regions. Wind power offers a reliable and sustainable source of electricity generation, contributing to a diversified energy mix for the continent.

Traditional fuels accounted for over 40% of final energy use between 2010 and 2019, potentially contributing to deforestation in some African countries (IEA, 2023), similar to Amazon levels. This alarming statistic highlights the significant reliance on traditional fuels over the past decade. The extensive use of these fuels not only poses a threat to the environment but also has adverse effects on African countries, comparable to the deforestation witnessed in the Amazon rainforest. It is crucial to address this issue promptly and explore sustainable alternatives to mitigate further damage. By transitioning toward renewable energy sources, such as solar and wind power, African nations can reduce their dependence on traditional fuels and alleviate the pressure on their forests. Promoting energy efficiency measures and raising awareness about the environmental consequences of traditional fuel consumption can play a vital role in curbing deforestation rates.

This glaring disparity in access to electricity highlights the urgent need for further efforts to bridge the gap. The progress made in the past decade is commendable, with 160 million Africans finally gaining access to this essential service over 2010–2019 (IEA, 2022). It is disheartening to note that more than 40% of Africans continue to be deprived of electricity (IEA, 2022). This stark reality underscores the importance of continued investment and innovative solutions to ensure that every African has access to reliable and affordable electricity. It is crucial not only for improving living standards but also for fostering economic growth, education, healthcare, and overall development. Efforts must be intensified to address the underlying challenges, such as inadequate infrastructure, high costs, and geographical barriers that hinder electrification efforts in certain regions. By prioritizing this issue and working collaboratively across governments, organizations, and communities, we can strive toward a future where all Africans have equal opportunities and access to electricity.

Africa faces challenges in meeting modern energy demand, accounting for 10% of global energy from 2010 to 2019 and 8% of oil demand growth, relying on inefficient imports (IEA, 2022). To address this issue, Africa is taking steps to transition toward cleaner and more sustainable energy sources. The continent has been investing in renewable energy projects, such as solar and wind power, to reduce its reliance on fossil fuels. African countries are implementing energy efficiency measures to optimize the use of resources and minimize wastage. This shift toward modern energy not only benefits Africa but also has significant implications for global trade. By embracing cleaner technologies and reducing energy consumption, Africa can become a key player in the global green economy. As the demand for renewable energy continues to rise worldwide, Africa's rich natural resources and potential for clean energy production position it as an attractive investment destination. This not only fosters economic growth within the continent but also opens up opportunities for international collaboration and trade partnerships.

This instability further exacerbates the challenges faced by industries and hampers their growth potential. The lack of reliable electricity hinders access to essential services such as healthcare, education, and communication. Without a consistent power supply, hospitals struggle to provide adequate medical care, schools are unable to offer quality education, and businesses find it difficult to connect with global markets. The continent's vast renewable energy potential remains largely untapped, leaving a significant opportunity for sustainable development unrealized. To address these issues, governments and stakeholders must prioritize investments in renewable energy infrastructure and promote policies that encourage its adoption (Proutsos, Tigkas et al., 2023). By harnessing the power of solar, wind, hydroelectricity, and other renewable sources, Africa can overcome its energy challenges while simultaneously reducing its carbon footprint. Fostering regional cooperation and partnerships can facilitate the sharing of resources and expertise in developing efficient transmission networks that span across borders.

Instability has an impact on Africa's significant global energy supply, which accounts for 8% of total oil and gas production and 4% of total demand (IEA, 2022). This disparity highlights Africa's significance as a major energy exporter. The current instability in the region raises concerns about the future of this role. With political and economic uncertainties, there is a risk of disruptions in oil and gas production, which could have far-reaching consequences for the global energy supply. The volatility in Africa's energy sector not only affects its own economy but also impacts the stability of international markets. It underscores the need for diversification and investment in renewable energy sources to reduce reliance on fossil fuels.

Due to insufficient infrastructure investment, Africa's natural resources—which account for 13% of the world's gas and 7% of its oil (IEA, 2023)—are having difficulty increasing production and meeting global demand. This has led to missed opportunities for economic growth and development in the region. The lack of adequate infrastructure, such as pipelines and refineries, hampers the extraction and distribution of these valuable resources. As a result, Africa struggles to fully capitalize on its vast reserves and falls short of meeting the rising global demand for energy. This situation not only limits the continent's potential to generate revenue but also hinders its ability to alleviate poverty and improve living standards for its people. Insufficient infrastructure investment perpetuates a cycle of dependency on foreign markets for energy needs, further impeding Africa's path toward self-sufficiency. To overcome these challenges, it is crucial for African governments and international stakeholders to prioritize infrastructure development in the energy sector.

Africa's renewable energy and mineral resources hold immense potential for the development of clean energy technologies, yet their full utilization remains hindered by inadequate infrastructure. The vast untapped reserves of solar, wind, and hydroelectric power could revolutionize the continent's energy landscape and contribute significantly to global sustainability efforts (Etuk, Agbasi et al., 2016). Africa boasts abundant deposits of critical minerals such as lithium, cobalt, and rare earth elements that are essential for the production of batteries, electric vehicles, and other clean technologies. Without proper transportation networks, reliable electricity grids, and efficient logistics systems, these resources remain largely unexplored. Investment in infrastructure development is crucial to unlocking Africa's renewable energy and mineral wealth, enabling the continent to become a major player in the global clean energy transition while fostering economic growth and job creation.

1.13 INTERNATIONAL COLLABORATION DRIVES GLOBAL EFFORTS TO ACHIEVE ZERO EMISSIONS ACROSS CONTINENTS

Global efforts to achieve zero emissions across various continents have been gaining momentum in recent years. From Europe's commitment to the Green Deal and its ambitious targets to Asia's push for renewable energy and electric transportation, countries across continents are recognizing the urgent need to transition toward a zero-emission future. These efforts are not only crucial for mitigating climate change but also for creating sustainable and resilient economies that prioritize the well-being of both people and the planet. The transition toward a zero-emission future is being driven by a combination of factors, including increasing public awareness and concern about the impacts of climate change, advancements in renewable energy technologies, and the potential for economic growth and job creation in the clean energy sector. Governments, businesses, and individuals are now working together to accelerate this transition through policy measures, investments in clean energy infrastructure, and widespread adoption of sustainable practices. By taking decisive action now, we can pave the way for a greener and more prosperous future for generations to come. Transitioning to clean energy sources not only helps mitigate the effects of climate change but also reduces dependence on fossil fuels, which are finite resources. Investing in renewable energy can lead to long-term cost savings and increased energy efficiency, benefiting both the environment and the economy.

Countries around the world are joining forces to tackle the urgent issue of climate change and work toward achieving zero emissions. Through international collaborations, such as the Paris Agreement, nations are committed to implementing sustainable practices and transitioning to renewable energy sources. These efforts span across various continents, with governments, businesses, and communities actively working together to reduce greenhouse gas emissions and mitigate the impacts of climate change. By sharing knowledge, resources, and technology, countries can accelerate the adoption of clean energy solutions and develop innovative strategies to combat climate change. These collaborations foster a sense of global responsibility and solidarity as nations recognize that addressing climate change requires collective action and cooperation on a global scale (Etuk, Nwokolo et al., 2016). This global cooperation also promotes the transfer of green technologies and expertise, enabling developing countries to leapfrog to cleaner and more sustainable energy systems. These collaborations create opportunities for joint research and development, leading to breakthrough innovations in renewable energy and climate adaptation measures. The IEA (Laura et al., 2023) reported that global collaboration and technology transfer can significantly reduce global greenhouse gas emissions by 40% by 2040. This reduction in emissions would not only mitigate the impacts of climate change but also lead to numerous economic benefits, such as job creation and increased energy security. The IEA emphasized that international cooperation is crucial for developing and implementing effective policies and regulations that can accelerate the transition to a low-carbon economy.

IEA statistics reveal that different countries have varying levels of greenhouse gas emissions, with some being major contributors while others have lower emissions. It is important for countries to work together and share best practices in order to achieve global emission reduction targets. The IEA (2023) highlights the importance of investing in renewable energy sources and promoting energy efficiency measures to further support the transition toward a low-carbon economy. They also revealed that the continent with the highest contributions to greenhouse gas emissions is Asia, followed by North America and Europe. Asia contributes about 52% of global greenhouse gas emissions, while North America and Europe contribute 18% and 15%, respectively. Other continents contributed about 15% of global greenhouse gas emissions, with the remaining 2% distributed among various regions. The country with the highest greenhouse gas emissions is China, accounting for approximately 28% of global emissions. Followed by the United States, which contributes about 15% of global emissions. These two countries alone account for over 40% of the world's greenhouse gas emissions. The European Union is also a significant contributor to global greenhouse gas emissions, accounting for about 10% of the total. Emerging economies such as India and Brazil are becoming increasingly important contributors, with each accounting for around 7% of global emissions. India also plays a significant role, accounting for approximately 7% of global emissions. Africa contributes about 3% of global greenhouse gas emissions, with countries like South Africa and Nigeria being major contributors. It is important to note that these numbers are constantly changing as economies develop and policies evolve. Efforts to reduce emissions and promote sustainable practices are crucial in order to mitigate the impact of climate change on a global scale.

It is worth mentioning that emissions from developing countries are expected to increase as their economies grow. This highlights the need for international cooperation and support to ensure that these countries can adopt cleaner technologies and transition toward low-carbon development pathways. It is essential for developed nations to take responsibility for their historical emissions and provide financial assistance to help developing countries in their efforts to combat climate change. This financial assistance can be used to invest in renewable energy infrastructure, promote sustainable agriculture practices, and enhance climate resilience in vulnerable communities. International cooperation should also focus on knowledge sharing and capacity building to enable developing countries to effectively implement climate change mitigation and adaptation measures. By working together, we can address the global challenge of climate change and create a more sustainable future for all. It is crucial to prioritize the transfer of clean and efficient technologies to developing nations, enabling them to leapfrog traditional, carbon-intensive development pathways. This technology transfer can help accelerate their transition toward low-carbon economies and reduce their greenhouse gas emissions. Supporting research and development efforts in renewable energy and sustainable practices can foster innovation and drive the adoption of more environmentally friendly solutions worldwide.

1.14 HARNESSING INSPIRATION: DENMARK AND BHUTAN PAVE THE WAY FOR AFRICA'S NET ZERO JOURNEY

Denmark and Bhutan have emerged as exemplary models for Africa's net zero journey, providing valuable insights and inspiration. Denmark, known for its strong commitment to renewable energy, has successfully transitioned to a sustainable energy system by investing in wind power and implementing innovative policies. Similarly, Bhutan's focus on environmental conservation and carbon neutrality has made it a leader in sustainable development. By studying their approaches and adapting them to the unique challenges of the African continent, countries can accelerate their progress toward achieving net zero emissions and mitigating the effects of climate change. This could involve harnessing Africa's vast renewable energy potential, such as solar and hydroelectric power, and implementing policies that prioritize sustainable practices in sectors like agriculture and transportation. Fostering international collaborations and sharing best practices with Denmark and Bhutan can further support African countries in their journey toward a greener future. By investing in renewable energy sources like solar and hydroelectric power, African countries can not only reduce their carbon footprint but also create new job opportunities and drive economic growth. Implementing sustainable practices in agriculture and transportation can help minimize greenhouse gas emissions and protect natural resources, ensuring a more resilient and sustainable future for Africa. In addition, promoting sustainable waste management practices can also contribute to a greener future for African countries.

By encouraging recycling and reducing waste generation, African nations can reduce pollution and conserve valuable resources. Investing in green infrastructure projects such as eco-friendly buildings and efficient public transportation systems can further enhance the sustainability efforts of African countries, leading to a more environmentally friendly and prosperous continent. Implementing policies that promote circular economy principles, such as extended producer responsibility and product stewardship, can help African countries shift toward a more sustainable waste management system. These measures would encourage manufacturers to take responsibility for the entire lifecycle of their products, from production to disposal, ultimately reducing the environmental impact of waste. Educating and raising awareness among citizens about the importance of sustainable waste management practices can foster a culture of environmental responsibility and encourage individual actions toward a greener future in Africa. Implementing policies and regulations that incentivize the use of eco-friendly materials and packaging can also contribute to a more sustainable waste management system. By promoting the use of recyclable and biodegradable materials, we can reduce the amount of waste that ends up in landfills or pollutes our oceans.

Investing in research and development of innovative waste management technologies can help find new solutions for recycling and reusing materials, further minimizing the environmental impact of waste in Africa. Implementing effective waste segregation and recycling programs can play a crucial role in diverting waste from landfills and promoting a circular economy. By educating communities about the importance of proper waste disposal and providing accessible recycling facilities, we can encourage individuals to actively participate in sustainable waste management practices. Collaboration with local businesses and industries to adopt sustainable production processes can significantly reduce the overall waste generation in Africa. Implementing effective waste management and recycling programs not only helps protect the environment but also creates economic opportunities. By encouraging the development of recycling industries and supporting local businesses that utilize recycled materials, we can stimulate job growth and contribute to a more sustainable economy in Africa. Investing in research and innovation for waste reduction technologies can further enhance our efforts to minimize waste generation and maximize resource efficiency. These technologies can help us find innovative solutions to tackle the growing waste problem and reduce our reliance on finite resources. Promoting awareness and education about the importance of recycling and waste management can empower individuals and communities to actively participate in sustainable practices, fostering a collective effort toward a greener future in Africa.

1.15 AFRICA'S PIVOTAL ROLE IN ACCELERATING GLOBAL NET ZERO TRANSITION

Africa's abundant natural resources, such as solar and wind energy, are playing a crucial role in driving the global net zero transition. With vast solar potential, Africa has the opportunity to harness clean energy on a large scale, reducing its reliance on fossil fuels and contributing to global efforts to combat climate change. The continent's rich reserves of minerals like lithium and cobalt are essential for the production of batteries used in renewable energy storage systems, further fueling the clean technology revolution worldwide. By investing in renewable energy infrastructure, African countries can not only improve their energy security but also create new job opportunities and stimulate economic growth. The adoption of clean energy technologies can help alleviate the negative impacts of air pollution and improve public health outcomes across the continent. Africa's youthful demographic dividend presents a unique opportunity to drive the global net zero transition through renewable energy.

With a large population of young people, Africa has the potential to become a powerhouse in clean energy production and innovation. By investing in education and training programs that focus on renewable energy technologies, African countries can empower their youth to lead the way toward a sustainable future while also creating jobs and economic growth. This investment in education and training can also help address the high unemployment rates among African youth, providing them with valuable skills and opportunities for employment in the growing renewable energy sector. By harnessing Africa's abundant natural resources, such as solar and wind power, the continent can reduce its reliance on fossil fuels and contribute significantly to global efforts to combat climate change. Africa's geographical advantage, with its abundant sunshine and vast open spaces, positions it as a leader in harnessing solar power for the global net zero transition. The continent's potential to generate clean and renewable energy is immense, offering a sustainable solution to meet increasing energy demands while reducing carbon emissions. Africa's role in driving the global net zero transition extends beyond solar power, as it also possesses significant potential for wind and hydroelectric power generation, further diversifying its renewable energy portfolio. Africa's abundant wind resources make it a prime location for wind power projects, which can contribute to the continent's goal of achieving net zero emissions. Africa's vast network of rivers and water bodies presents an opportunity for harnessing hydroelectric power, further enhancing its renewable energy capabilities, and solidifying its position as a leader in the global net zero transition.

Africa's vast ecosystems hold immense potential for sustainable carbon sequestration and greenhouse gas mitigation, making it a key player in driving the global net zero transition. With its diverse range of forests, wetlands, and grasslands, the continent can contribute significantly to offsetting carbon emissions through natural processes such as photosynthesis and soil carbon storage. Leveraging Africa's rich biodiversity and unique ecosystems can also foster nature-based solutions that promote resilience and adaptation to climate change impacts, further strengthening its role in the global net zero transition. By protecting and restoring its natural habitats, Africa can enhance its capacity to sequester carbon and mitigate the effects of climate change. Investing in renewable energy sources like solar and wind power can help Africa reduce its reliance on fossil fuels and accelerate the transition toward a sustainable, low-carbon future.

1.16 THE TENDENCY OF AFRICA TO SWITCH TO NET ZERO ENERGY

Africa's propensity for a net zero energy transition is a subject of growing importance in the global context. With its vast renewable energy potential, the continent has a unique opportunity to leapfrog traditional fossil fuel–based development and embrace sustainable alternatives. Africa's abundant solar resources, wind patterns, and hydroelectric potential make it an ideal candidate for a clean energy revolution. By harnessing these resources effectively, African nations can not only address their pressing energy needs but also significantly contribute to mitigating climate change. Transitioning to net zero energy would enhance energy security, reduce dependence on volatile fossil fuel markets, and create new economic opportunities through job creation and technology transfer. It is crucial to acknowledge the challenges that lie ahead. Limited access to finance, inadequate infrastructure, and institutional barriers need to be overcome for a successful transition. Collaboration between governments, private sector entities, and international organizations will be essential in driving this transformation forward.

By embracing a net zero energy future, Africa can position itself as a global candidate in sustainable development and contribute significantly to the fight against climate change. This transition will not only benefit the continent's economy but also improve the quality of life for its people. To achieve this, investments in renewable energy sources such as solar, wind, and hydropower need to be prioritized. Efforts should be made to enhance energy efficiency in industries, buildings, and transportation systems. This will not only reduce greenhouse gas emissions but also create new employment opportunities and reduce reliance on fossil fuels. It is important to invest in research and development to drive innovation in clean technologies that can be tailored to Africa's specific needs and conditions. This will enable the continent to leapfrog traditional energy systems and embrace sustainable solutions from the outset. Education and awareness campaigns are also crucial in promoting energy conservation practices among individuals and communities. By empowering its people with knowledge and skills, Africa can build a resilient and sustainable energy sector that serves as a catalyst for overall development.

The transition to a net zero energy future requires long-term planning and collaboration among various stakeholders. Governments, businesses, and civil society organizations must work together to develop comprehensive strategies and policies that prioritize renewable energy sources and energy efficiency measures. Investing in research and development is essential to driving innovation and finding new ways to harness clean energy. Financial support and incentives should be provided to encourage the adoption of sustainable energy solutions, particularly in rural areas where access to electricity is limited. By embracing a holistic approach to the energy transition, Africa can not only mitigate the impacts of climate change but also create new opportunities for economic growth and social development. With a concerted effort toward a net zero energy future, Africa can lead the way in demonstrating the power of sustainable practices on a global scale.

Africa's journey toward a net zero energy transition has been challenging, but it is not impossible. To close this chapter, it is essential to focus on three key areas: policy, technology, and finance. Firstly, African governments must prioritize the development of policies that promote renewable energy sources and discourage the use of fossil fuels. This can be achieved through incentives such as tax breaks and subsidies for renewable energy projects. Secondly, technology plays a crucial role in achieving net zero emissions. African countries must invest in the research and development of clean energy technologies such as solar, wind, and hydroelectric power. Finally, financing is critical to the success of Africa's net zero energy transition. Governments must work with international organizations to secure funding for renewable energy projects while also encouraging private sector investment in the sector. By focusing on these three areas simultaneously, Africa can close this chapter on its proclivity for a net zero energy transition and become a leader in sustainable development.

REFERENCES

Africa Energy Outlook (2022). https://reglobal.org/time-to-fast-track-africas-clean-equitable-energy-future/. https://doi.org/10.1787/2abd9ce2-en

Agyarko, K. A., Opoku, R., & Van Buskirk, R. (2020). Removing barriers and promoting demand-side energy efficiency in households in sub-Saharan Africa: A case study in Ghana. *Energy Policy*. https://doi.org/10.1016/j.enpol.2019.111149

Armah, M., Bossman, A., & Amewu, G. (2023). Information flow between global financial market stress and African equity markets: An EEMD-based transfer entropy analysis. *Heliyon*. https://doi.org/10.1016/j.heliyon.2023.e13899

Babayomi, O. O., Dahoro, D. A., & Zhang, Z. (2022). Affordable clean energy transition in developing countries: Pathways and technologies. *iScience*. https://doi.org/10.1016/j.isci.2022.104178

Bester, V., & Uys, T. (2023). Artisanal mining and its drivers in the South African context. *Extractive Industries and Society*. https://doi.org/10.1016/j.exis.2023.101278

Casati, P., Moner-Girona, M., Khaleel, S. I., Szabo, S., & Nhamo, G. (2023). Clean energy access as an enabler for social development: A multidimensional analysis for sub-Saharan Africa. *Energy for Sustainable Development*. https://doi.org/10.1016/j.esd.2022.12.003

Chapungu, L., Nhamo, G., Chikodzi, D., & Maoela, M. A. (2022). BRICS and the race to net-zero emissions by 2050: Is COVID-19 a barrier or an opportunity? *Journal of Open Innovation: Technology, Market, and Complexity*. https://doi.org/10.3390/joitmc8040172

Chisale, S. W., & Lee, H. S. (2023). Evaluation of barriers and solutions to renewable energy acceleration in Malawi, Africa, using AHP and fuzzy TOPSIS approach. *Energy for Sustainable Development*. https://doi.org/10.1016/j.esd.2023.101272

Cooper, A. C. G. (2018). Evaluating energy efficiency policy: Understanding the 'energy policy epistemology' may explain the lack of demand for randomised controlled trials. *Energy Efficiency*. https://doi.org/10.1007/s12053-018-9618-8

Copiello, S. (2015). Achieving affordable housing through energy efficiency strategy. *Energy Policy*. https://doi.org/10.1016/j.enpol.2015.06.017

Delbeke, J., Runge-Metzger, A., Slingenberg, Y., & Werksman, J. (2019). The paris agreement. In *Towards a Climate-Neutral Europe: Curbing the Trend*. https://doi.org/10.4324/9789276082569-2

Elabbas, M. A. E., de Vries, L., & Correljé, A. (2023). African power pools and regional electricity market design: Taking stock of regional integration in energy sectors. *Energy Research and Social Science*. https://doi.org/10.1016/j.erss.2023.103291

Etuk, S., Agbasi, O., & Nwokolo, S. C. (2016). Modelling and estimating photosynthetically active radiation from measured global solar radiation at Calabar, Nigeria. *Physical Science International Journal*. https://doi.org/10.9734/psij/2016/28446

Etuk, S., Nwokolo, S. C., Agbasi, O., & John-Jaja, S. A. (2016). Analysis of photosynthetically active radiation over six tropical ecological zones in Nigeria. *Journal of Geography, Environment and Earth Science International*. https://doi.org/10.9734/jgeesi/2016/27945

Foster, J. A., & Poston, A. (2023). Domestic energy consumption: Temporal unregulated electrical energy consumption in kitchens in Scottish affordable and social housing. *Energy Efficiency*. https://doi.org/10.1007/s12053-023-10143-3

Guo, Q., Abbas, S., AbdulKareem, H. K. K., Shuaibu, M. S., Khudoykulov, K., & Saha, T. (2023). Devising strategies for sustainable development in sub-Saharan Africa: The roles of renewable, non-renewable energy, and natural resources. *Energy*. https://doi.org/10.1016/j.energy.2023.128713

Gwagwa, A., Kazim, E., & Hilliard, A. (2022). The role of the African value of Ubuntu in global AI inclusion discourse: A normative ethics perspective. *Patterns*. https://doi.org/10.1016/j.patter.2022.100462

International Renewable Energy Agency (IRENA). (2019). *Future of solar photovoltaic: Deployment, investment, grid integration and socio-economic aspects*. International Renewable Energy Agency (IRENA).

International Renewable Energy Agency (IRENA). (2020). *Renewable energy and climate pledges: Five years after the Paris Agreement*. International Renewable Energy Agency (IRENA).

International Renewable Energy Agency (IRENA). (2022). *Renewable power generation costs in 2021*. ISBN 978-92-9260-452-3. International Renewable Energy Agency.

Ituen, E. E., Esen, N. U., Nwokolo, S. C., & Udo, E. G. (2012). Prediction of global solar radiation using relative humidity, maximum temperature and sunshine hours in Uyo, in the Niger Delta Region, Nigeria. *Advances in Applied Science Research, 3*(4), 1923–1937.

Kelly, A. M., & Ngo Nguéda Radler, R. D. (2024). Does energy consumption matter for climate change in Africa? New insights from panel data analysis. *Innovation and Green Development, 3*(3). https://doi.org/10.1016/j.igd.2024.100132

Kichere, C. E. (2020). Performance audit report on the management of accessibility and reliability of electricity supply services. *National Audit Office*.

Laura, C. et al. (2023). World Energy Outlook. In *World Energy Outlook*. International Energy Agency. www.iea.org/weo

Liu, Y. (2016). Seasonal relationship of peak demand and energy impacts of energy efficiency measures—A review of evidence in the electric energy efficiency programmes. *Energy Efficiency*. https://doi.org/10.1007/s12053-015-9407-6

Lo, K., & Kibalya, B. (2023). Electric cooperatives and the political economy of rural electrification in Africa: Insights from Uganda. *Electricity Journal*. https://doi.org/10.1016/j.tej.2023.107238

Mirzania, P., Gordon, J. A., Balta-Ozkan, N., Sayan, R. C., & Marais, L. (2023). Barriers to powering past coal: Implications for a just energy transition in South Africa. *Energy Research and Social Science*. https://doi.org/10.1016/j.erss.2023.103122

Mkhize, D. N. (2023). Reconceptualising the notion of cross-linguistic transfer in multilingual spaces: A Global South perspective from South Africa. *Language Sciences*. https://doi.org/10.1016/j.langsci.2023.101573

Mohammed, B. U., Wiysahnyuy, Y. S., Ashraf, N., Mempouo, B., & Mengata, G. M. (2023). Pathways for efficient transition into net zero energy buildings (nZEB) in sub-Sahara Africa. Case study: Cameroon, Senegal, and Côte d'Ivoire. *Energy and Buildings*. https://doi.org/10.1016/j.enbuild.2023.113422

Nwokolo, S. C. (2017). A comprehensive review of empirical models for estimating global solar radiation in Africa. *Renewable and Sustainable Energy Reviews, 78*(October), 955–995. https://doi.org/10.1016/j.rser.2017.04.101

Nwokolo, S. C., Amadi, S. O., Obiwulu, A. U., Ogbulezie, J. C., & Eyibio, E. E. (2022). Prediction of global solar radiation potential for sustainable and cleaner energy generation using improved Angstrom-Prescott and Gumbel probabilistic models. *Cleaner Engineering and Technology, 6*(January), 100416. https://doi.org/10.1016/j.clet.2022.100416

Nwokolo, S. C., Echeng, E., Obiwulu, A., Meyer, E. L., Christian, C., Ogbulezie, J. C., & Proutsos, N. (2024). A multi-model approach based on CARIMA-SARIMA-GPM for assessing the impacts of climate change on concentrated photovoltaic (CPV) potential. *Physics and Chemistry of the Earth, 134*(January), 103560. https://doi.org/10.1016/j.pce.2024.103560

Nwokolo, S. C., Meyer, E. L., & Ahia, C. C. (2024). Exploring the interactive influences of climate change and urban development on the fraction of absorbed photosynthetically active radiation. *Atmosphere, 15*(3), 253.

Nwokolo, S. C., Obiwulu, A. U., & Ogbulezie, J. C. (2023). Machine learning and analytical model hybridization to assess the impact of climate. *Physics and Chemistry of the Earth, 130*(February), 103389. https://doi.org/10.1016/j.pce.2023.103389

Nwokolo, S. C., & Ogbulezie, J. C. (2018a). A qualitative review of empirical models for estimating diffuse solar radiation from experimental data in Africa. *Renewable and Sustainable Energy Reviews, 92*(18), 353–393. https://doi.org/10.1016/j.rser.2018.04.118

Nwokolo, S. C., & Ogbulezie, J. C. (2018b). A quantitative review and classification of empirical models for predicting global solar radiation in West Africa. *Beni-Suef University*

Journal of Basic and Applied Sciences, *7*(4), 367–396. https://doi.org/10.1016/j.bjbas.2017.05.001

Nwokolo, S. C., Ogbulezie, J. C., & Obiwulu, A. U. (2022). Impacts of climate change and meteo-solar parameters on photosynthetically active radiation prediction using hybrid machine learning with Physics-based models. *Advances in Space Research*, *70*(11), 3614–3637. https://doi.org/10.1016/j.asr.2022.08.010

Nwokolo, S. C., Ogbulezie, J. C., & Ushie, O. J. (2023). A multi-model ensemble-based CMIP6 assessment of future solar radiation and PV potential under various climate warming scenarios. *Optik*, *285*(May), 170956. https://doi.org/10.1016/j.ijleo.2023.170956

Nwokolo, S. C., & Otse, C. (2019). Impact of sunshine duration and clearness index on diffuse solar radiation estimation in mountainous climate. *Trends in Renewable Energy*, *5*(3), 307–332. https://doi.org/10.17737/tre.2019.5.3.00107

Nwokolo, S. C., Proutsos, N., & Meyer, E. L. (2023). Machine learning and Physics-based hybridization models for evaluation of the effects of climate change and urban expansion on photosynthetically active radiation. *Atmosphere*, *14*(687), 1–36. https://doi.org/doi.org/10.3390/atmos14040687

Obiwulu, A. U., Chendo, M. A. C., Erusiafe, N., & Nwokolo, S. C. (2020). Implicit meteorological parameter-based empirical models for estimating back temperature solar modules under varying tilt-angles in Lagos, Nigeria. *Renewable Energy*, *145*. https://doi.org/10.1016/j.renene.2019.05.136

Obiwulu, A. U., Erusiafe, N., Olopade, M. A., & Nwokolo, S. C. (2020). Modeling and optimization of back temperature models of mono-crystalline silicon modules with special focus on the effect of meteorological and geographical parameters on PV performance. *Renewable Energy*. https://doi.org/10.1016/j.renene.2020.02.103

Obiwulu, A. U., Erusiafe, N., Olopade, M. A., & Nwokolo, S. C. (2022). Modeling and estimation of the optimal tilt angle, maximum incident solar radiation, and global radiation index of the photovoltaic system. *Heliyon*, *8*(6), e09598. https://doi.org/10.1016/j.heliyon.2022.e09598

O'Connor, N., & Macur, R. (2018). Engaging residents in affordable housing—Resident engagement pilot at Denver Housing Authority Westridge apartments. *Energy Efficiency*, *11*, 1871–1886. https://doi.org/10.1007/s12053-018-9641-9

Okpanachi, E., Ambe-Uva, T., & Fassih, A. (2022). Energy regime reconfiguration and just transitions in the Global South: Lessons for West Africa from Morocco's comparative experience. *Futures*, *139*(August 2021). https://doi.org/10.1016/j.futures.2022.102934

Proutsos, N., Liakatas, A., Alexandris, S., Nwokolo, S. C., Solomou, A. D., & Amadi, S. O. (2023). Assessing the impact of atmospheric attributes on the effectiveness of solar irradiance for photosynthesis of urban vegetation in Attica, Greece. *Theoretical and Applied Climatology*, *155*(2), 1–13 https://doi.org/10.1007/s00704-023-04700-0

Proutsos, N., Tigkas, D., Tsevreni, I., Alexandris, S. G., Solomou, A. D., Bourletsikas, A., Stefanidis, S., & Nwokolo, S. C. (2023). A thorough evaluation of 127 potential evapotranspiration models in two Mediterranean Urban Green Sites. *Remote Sensing*, *15*(14), 1–41. https://doi.org/10.3390/rs15143680

Steynor, A., & Pasquini, L. (2023). Exploring east African climate change risk perceptions to inform effective climate services. *Climate Services*, *32*(January 2022), 100419. https://doi.org/10.1016/j.cliser.2023.100419

Tesega, M. (2022). Does financial globalization contribute to financial development in developing countries? Evidence from Africa. *Heliyon*. https://doi.org/10.1016/j.heliyon.2022.e10974

Wang, J., Shan, Y., Cui, C., Zhao, C., Meng, J., & Wang, S. (2024). Investigating the fast energy-related carbon emissions growth in African countries and its drivers. *Applied Energy*, *357*(September 2023), 122494. https://doi.org/10.1016/j.apenergy.2023.122494

Wang, M., Wei, Y., Azumah, G., & Wang, C. L. (2023). African returnees in international knowledge transfer: A social capital perspective. *Journal of International Management*, *30*(1), 101118. https://doi.org/10.1016/j.intman.2023.101118

2 Africa's Current Economic Issues

2.1 INTRODUCTION

Africa is currently facing several economic challenges that are hindering its growth and development. One of the key issues is the high levels of poverty and inequality prevalent across the continent. This has resulted in limited access to basic services such as healthcare, education, and clean water, further exacerbating the cycle of poverty. Africa's heavy reliance on commodity exports makes it vulnerable to global market fluctuations, leading to economic instability. Inadequate infrastructure and a lack of diversification in the economy contribute to the continent's economic challenges. Insufficient roads (Sumberg et al., 2021), electricity (Lopes and te Velde, 2021), and telecommunications networks hinder trade (Assan, 2023) and investment opportunities (Baarsch et al., 2020). The overreliance on commodities leaves African countries susceptible to price volatility (Adwuyi and Awodumi, 2021) and limits their ability to generate sustainable economic growth (Trotter and Abdullah, 2018) through other sectors such as manufacturing (John, 2021) or services.

This chapter discusses the continent of Africa's current economic issues. Africa's current economic woes and potential solutions highlight the need for infrastructure development investments to boost connectivity and ease trade both within and outside of the continent. It emphasizes the importance of diversifying African economies by promoting innovation, entrepreneurship, and skill development to create new opportunities and reduce reliance on commodities. The chapter emphasizes the significance of addressing governance and corruption issues to foster a conducive business environment and attract foreign investments. It also underscores the importance of fostering regional cooperation and integration to enhance economic growth and stability across Africa. By promoting innovation, entrepreneurship, and skill development, African economies can tap into untapped potential and drive sustainable economic growth. This approach not only creates new opportunities but also reduces the vulnerability of countries solely reliant on commodities. Fostering regional cooperation and integration can lead to economies of scale, increased market access, and improved infrastructure, further boosting economic growth and stability in Africa. Investing in education and healthcare systems is crucial for the long-term development of African economies. By improving access to quality education and healthcare services, countries can empower their citizens and create a skilled workforce that can contribute to

DOI: 10.1201/9781003483175-3

economic growth. Strengthening governance and combating corruption are essential for creating a conducive business environment that attracts both domestic and foreign investments, ultimately driving economic prosperity in Africa.

2.2 HISTORICAL CONTEXT OF ECONOMIC DEVELOPMENT IN AFRICA

2.2.1 Colonialism and Its Impact on Africa's Economy

Colonialism refers to the historical period when European powers established political and economic control over African territories. This had a profound impact on Africa's economy, as it was primarily geared toward serving the interests of the colonizers rather than the indigenous population. The extraction of valuable resources (John et al., 2023), such as minerals (Akyeampong and Bates, 2014) and agricultural products (Nyoka, 2023), was prioritized, leading to the exploitation (Zeleza, 2011) and depletion of Africa's natural wealth (Magu, 2021). Colonial powers implemented exploitative labor systems (Karodia and Soni, 2014) and disrupted local industries (Kohnert, 2009), hindering the development of self-sustaining economies in Africa (Smith, 2013). This economic exploitation also resulted in the displacement (Baker, 2018) and impoverishment of many indigenous communities (Ouma, 2012), further exacerbating the inequalities and injustices imposed by colonial rule (Meintjes, 2004). The lack of investment in infrastructure and education during this period left African nations at a significant disadvantage when they eventually gained independence, perpetuating a cycle of economic dependency on former colonial powers.

This cycle of economic dependency continues to hinder the progress and growth of African nations today.

Despite efforts to promote sustainable development and attract foreign investment, many countries still struggle to achieve self-sufficiency and overcome the legacies of colonialism. Addressing these historical injustices and investing in infrastructure, education, and local industries are crucial steps toward building strong and independent economies in Africa. It is essential for African nations to diversify their economies and reduce their reliance on a few key exports. This will help them become more resilient to global economic fluctuations and promote long-term stability. Promoting regional integration and cooperation can foster economic growth by creating larger markets and facilitating the movement of goods and services across borders. By promoting regional integration and cooperation, African nations can also benefit from economies of scale and increased competitiveness in the global market. This can lead to improved productivity, job creation, and overall economic development. Investing in education and skills development is crucial for building a skilled workforce that can drive innovation and contribute to the growth of diverse sectors within the economy.

Investing in infrastructure development is essential to facilitate the movement of goods and services as well as attract foreign direct investment (FDI). Adequate transportation networks, such as roads, railways, and ports, are necessary for efficient trade and economic growth. Fostering a business-friendly environment through policies that promote entrepreneurship and reduce bureaucratic barriers can encourage domestic and foreign investment, leading to further economic development. Investing in modern and reliable communication networks, such as high-speed internet and mobile connectivity, is crucial for businesses to connect with global markets and access information quickly. Prioritizing sustainable infrastructure development that minimizes environmental impact can not only support economic growth but also contribute to long-term sustainability and resilience. Fostering innovation and supporting research and development initiatives can drive technological advancements and enhance competitiveness in the global market. Promoting a skilled workforce through education and training programs can ensure that businesses have access to qualified individuals who can contribute to economic growth and productivity.

Implementing policies and regulations that encourage sustainable practices can lead to cost savings for businesses in the long run. For instance, adopting renewable energy sources can reduce reliance on fossil fuels and lower energy expenses. Promoting the circular economy, such as recycling and waste reduction, can not only minimize environmental impact but also create new business opportunities and job prospects in the green sector. Investing in research and development of green technologies can spur innovation and create a competitive advantage for businesses. This can lead to the development of new products and services that meet the growing demand for sustainable solutions, opening up new markets and increasing profitability. Adopting sustainable practices can enhance a company's reputation and brand image, attracting environmentally conscious consumers and investors who prioritize companies with strong environmental commitments.

Integrating green technologies into business operations can also result in cost savings through reduced energy consumption and waste management. By implementing sustainable practices, companies can minimize their environmental footprint and contribute to the overall well-being of the planet, aligning with global efforts to combat climate change and preserve natural resources. Adopting green technologies can also enhance a company's reputation and attract top talent, as more and more individuals are seeking to work for environmentally responsible organizations. By embracing sustainability, companies can stay ahead of changing regulations and market trends, ensuring long-term viability and competitiveness in the evolving business landscape.

2.2.1.1 The Case of South Africa

In the case of South Africa, colonialism had a profound impact on its economy. The arrival of European powers in the 17th century led to the establishment of settler colonies, which resulted in the displacement and marginalization of indigenous populations. This, coupled with the extraction of valuable resources such as gold and diamonds, led to significant wealth accumulation for colonizers while leaving the local population impoverished and dispossessed. The exploitation of labor through systems like apartheid further exacerbated economic inequality, as indigenous people were forced into low-paying jobs and denied access to education and opportunities for upward mobility. These historical legacies continue to shape South Africa's economy today, with persistent disparities in wealth distribution and limited economic opportunities for marginalized communities.

The legacy of apartheid also resulted in the concentration of wealth and resources in the hands of a few privileged individuals and corporations, perpetuating a cycle of economic inequality. The lack of access to quality education and limited opportunities for marginalized communities have hindered their ability to participate fully in the economy, further exacerbating the disparities in wealth distribution. Systemic discrimination and the unequal distribution of land ownership have further contributed to the economic challenges faced by marginalized communities. This has led to limited access to productive resources and hindered their ability to generate sustainable income and wealth.

The lack of access to financial services and credit has also played a significant role in perpetuating economic challenges for marginalized communities. Without access to affordable loans or banking services, it becomes difficult for individuals and businesses within these communities to invest in education, start or expand businesses, or save for the future. This further widens the gap between marginalized communities and more privileged ones, reinforcing the cycle of limited opportunities and economic inequality. The lack of financial literacy and education in marginalized communities exacerbates the problem. Many individuals are not equipped with the knowledge and skills to make informed financial decisions, leading to a higher likelihood of falling into predatory lending practices or accumulating high levels of debt. This perpetuates a cycle of financial instability and hinders their ability to break free from economic challenges. In addition, the absence of accessible banking services and affordable credit options in these communities further deepens the cycle of limited opportunities. Without access to traditional financial institutions, individuals may resort to using alternative financial services that often come with exorbitant fees and interest rates, trapping them in a never-ending cycle of debt. This systemic issue calls for comprehensive efforts to address both the lack of financial education and the need for inclusive financial services in marginalized communities.

One possible solution is to establish community-based financial literacy programs that provide individuals with the knowledge and skills necessary to make informed financial decisions. Policymakers should work toward creating policies that promote the development of affordable and accessible financial services tailored to the specific needs of marginalized communities. By addressing these underlying issues, we can begin to break the cycle of limited opportunities and empower individuals in these communities to achieve financial stability. It is crucial to collaborate with community organizations and financial institutions to ensure that these literacy programs and policies are effectively implemented and reach the target population. This can be done through partnerships that offer resources, support, and guidance to individuals seeking financial education and services. By prioritizing financial inclusion and empowerment, we can foster economic growth and reduce inequality in marginalized communities. These partnerships can also help in identifying the specific needs and challenges faced by the target population, allowing for tailored approaches and solutions to be developed. Engaging community organizations and financial institutions can help build trust and credibility within the community, making individuals more likely to participate in these literacy programs and take advantage of the offered services.

2.2.1.2 The Case of Nigeria

Colonialism played a significant role in shaping Africa's economic development, including in the case of Nigeria. The arrival of European powers in the late 19th century marked the beginning of a period where African resources were exploited for the benefit of colonial powers. In Nigeria, colonial powers focused on exporting raw materials such as palm oil, rubber, and cocoa, which resulted in a lopsided economy heavily dependent on these commodities. Nigeria, rich in natural resources such as oil, became a prime target for extraction and trade. This led to the establishment of extractive industries that prioritized exporting raw materials, hindering the growth of local industries and stifling economic diversification. This exploitation not only disrupted local economies but also hindered the growth of indigenous industries, leaving a lasting legacy that continues to shape Nigeria's economy today. Despite being blessed with abundant natural resources, Nigeria's overreliance on the export of raw materials has made it vulnerable to fluctuations in global commodity prices. This has created a cycle of boom and bust, where the country experiences periods of economic growth when commodity prices are high but faces significant challenges during times of price decline. The lack of investment in local industries has limited job creation and technological advancement, further perpetuating Nigeria's dependence on extractive industries.

The lack of diversification in the economy has hindered Nigeria's ability to withstand external shocks and adapt to changing global market dynamics. Without a strong manufacturing base and value-added industries, the country struggles to generate sustainable economic growth and reduce poverty levels. To break this cycle, Nigeria needs to prioritize investment in sectors that promote innovation, skill development, and job creation, fostering a more resilient and inclusive economy. By investing in sectors such as technology, renewable energy, and agriculture, Nigeria can tap into new sources of economic growth and create opportunities for its growing population. This will not only help diversify the economy but also reduce its reliance on oil exports, making it less vulnerable to fluctuations in global oil prices. Promoting entrepreneurship and supporting small and medium-sized enterprises (SMEs) can contribute to job creation and empower local communities to participate actively in the country's economic development. Investing in infrastructure development, such as improving transportation networks and expanding access to electricity, can further stimulate economic growth and attract foreign investment. Implementing policies that prioritize education and skills training can enhance the country's human capital, enabling Nigerians to compete in the global market and drive innovation in various sectors. Fostering entrepreneurship and supporting SMEs can also contribute to job creation and economic development. By providing financial assistance, training, and mentorship programs, the government can encourage the growth of local businesses and create a more vibrant entrepreneurial ecosystem. Promoting sustainable practices and investing in renewable energy sources can not only boost economic growth but also address environmental challenges and contribute to a more sustainable future for Nigeria.

2.2.2 Post-independence Economic Strategies and Their Outcomes in Africa

Post-independence economic strategies and their outcomes in Africa have played a crucial role in shaping the continent's development trajectory. Following the wave of decolonization in the mid-20th century, African nations embarked on various economic approaches to foster growth and reduce dependency on former colonial powers. These strategies encompassed various approaches, ranging from import substitution industrialization (ISI) to digital transformation efforts. While some countries achieved significant economic growth and diversification, others faced challenges such as high levels of debt, dependency on primary commodities, and limited technological advancements. Understanding the diverse outcomes of these strategies is essential in analyzing Africa's economic landscape and identifying potential avenues for sustainable development.

In this section, post-independence economic strategies and their outcomes in Africa encompassed diverse models, including (1) ISI, which focused on developing domestic industries to replace imported goods; (2) export-led growth, which aimed to increase exports and attract foreign investment; (3) import substitution with export promotion, which combined elements of both strategies; (4) structural adjustment programs, which involved implementing market-oriented reforms and reducing government intervention in the economy; (5) regional integration, which seeks to promote economic cooperation and integration among neighboring countries through trade agreements and common policies. These efforts aim to address the challenges posed by rapid technological advancements and the need for digital skills in the workforce. Digital transformation initiatives also aim to bridge the digital divide and ensure equal access to technology and information for all segments of society.

2.2.2.1 Import Substitution Industrialization

ISI was a popular economic strategy adopted by many African countries after gaining independence. The main objective of ISI was to reduce dependency on foreign imports by promoting domestic industries and manufacturing. This approach aimed to stimulate economic growth, create employment opportunities, and achieve self-sufficiency in essential goods. The outcomes of ISI varied across African nations, with some experiencing limited success due to challenges such as inadequate infrastructure, a lack of skilled labor, and limited access to capital. Despite these challenges, some African nations were able to successfully implement ISI and achieve significant economic growth. For example, countries like South Africa and Egypt were able to develop strong manufacturing sectors and reduce their reliance on foreign imports. The implementation of ISI also led to the emergence of new industries and the diversification of economies in some African countries. These countries in Africa include Kenya, Ethiopia, and Ghana. These nations were able to leverage their abundant natural resources and invest in infrastructure development to support the growth of domestic industries. As a result, they experienced increased employment opportunities and improved standards of living for their citizens. The implementation of ISI allowed these countries to gain more control over their economies and reduce their vulnerability to external economic shocks.

Other nations faced difficulties in implementing ISI due to political instability and corruption, which hindered the growth of domestic industries. These countries in Africa include Nigeria, Kenya, Angola, and Zimbabwe. Despite their potential for economic growth, political instability and corruption have hindered their ability to effectively implement ISI policies. This has resulted in limited industrial development and a lack of employment opportunities for their citizens. These countries have struggled to attract foreign investment and have faced challenges in diversifying their economies beyond traditional sectors such as agriculture. The lack of infrastructure and a skilled labor force has further impeded their efforts to attract foreign investors. The reliance on natural resources as the main source of revenue has made these countries vulnerable to fluctuations in global commodity prices, exacerbating their economic challenges. The limited access to financial services and the high cost of doing business have also hindered these countries' ability to attract foreign investment. Political instability and corruption have created an unfavorable business environment, further deterring potential investors. These countries must prioritize investing in education and infrastructure development to enhance their competitiveness and attract diversified foreign investments.

Despite these obstacles, some African countries, such as Egypt and South Africa, were able to successfully implement ISI, leading to significant economic growth and development in their respective industries. One key factor that contributed to the success of ISI in Egypt and South Africa was their ability to effectively leverage their domestic resources and establish strong domestic industries. Both countries had a well-developed infrastructure and skilled workforce, which allowed them to effectively compete in the global market. They implemented policies that encouraged local entrepreneurship and innovation, further driving economic growth. As a result, these countries were able to attract foreign investment and diversify their economies, creating more employment opportunities for their citizens. Both countries prioritized education and invested heavily in research and development, which helped them stay at the forefront of technological advancements. This enabled them to produce high-quality goods and services that were in demand globally, contributing to their economic success. They fostered a favorable business environment by implementing transparent regulations and reducing bureaucratic hurdles, attracting even more foreign investors and boosting their economies further.

Overall, while there were challenges and varying outcomes, the pursuit of ISI in Africa played a crucial role in promoting economic development and reducing dependency on foreign imports. The global economic changes in the 1980s and 1990s, such as liberalization policies and globalization, also had an impact on the effectiveness of ISI in Africa. These changes opened up new opportunities for foreign investments and trade, making it harder for domestic industries to compete. The structural adjustment programs imposed by international financial institutions forced many African countries to abandon their ISI strategies and focus on export-oriented economies instead. This shift further weakened the implementation of ISI and hindered its potential benefits for economic development in Africa. As a result, African countries faced challenges in developing their own industries and became heavily dependent on imported goods. The lack of infrastructure and technological advancements in many African countries makes it difficult for them to effectively implement ISI policies and compete with more developed nations.

2.2.2.2 Export-Led Growth

Export-led growth was one of the key economic strategies adopted by many African countries after gaining independence. This strategy aimed to boost economic development by focusing on increasing exports of goods and services. It involved diversifying the export base, improving trade relations with other countries, and attracting FDI. Export-led growth was seen as a way to reduce dependence on external aid and promote self-sufficiency in African economies. These countries focused on increasing their exports, particularly in sectors such as agriculture, mining, and manufacturing, to drive economic growth and development. The outcomes of this strategy varied across different African countries. Some nations experienced significant success, with their exports contributing to a rapid increase in GDP and foreign exchange earnings. This led to improvements in infrastructure, education, and healthcare, ultimately raising living standards for their citizens.

Countries that succeed in Africa include Nigeria, which saw a boost in its oil exports, leading to increased revenue and investments in sectors such as telecommunications and banking. South Africa's diversified export base, including minerals, machinery, and vehicles, played a key role in its economic growth and job creation. Another success story is Ethiopia, where the government's focus on exporting textiles and agricultural products resulted in job creation and poverty reduction. These examples demonstrate the potential of exports to drive economic transformation and improve the lives of people in African nations. Exports have also been instrumental in promoting industrialization and attracting FDI in African countries. For instance, the manufacturing sector in countries like Kenya and Nigeria has experienced significant growth due to increased exports of manufactured goods. This has not only created employment opportunities but also contributed to the overall economic development of these nations. The expansion of export markets has allowed African countries to diversify their economies and reduce their dependence on a single industry, thereby increasing their resilience to external shocks. The growth of the manufacturing sector has also attracted FDI to these countries. This influx of FDI has not only brought in capital and technology but has also facilitated knowledge transfer and skill development among the local workforce. As a result, African countries have been able to enhance their industrial capabilities and competitiveness on the global stage.

Other countries faced challenges in diversifying their export base and expanding their market access. These countries in Africa include countries such as Malawi, Lesotho, and Burundi. Limited infrastructure, political instability, and inadequate access to finance have hindered their ability to attract FDI and develop their industrial sectors. Consequently, these countries have struggled to compete in the global market and have faced difficulties in achieving sustainable economic growth. These challenges have also led to high levels of unemployment and poverty in these countries. In order to overcome these obstacles, they have been working toward implementing policies that promote investment in infrastructure, improve political stability, and enhance access to finance. By addressing these issues, they aim to create a more conducive environment for attracting FDI and fostering the growth of their industrial sectors, ultimately leading to sustainable economic development. These countries have been focusing on diversifying their economies by investing in sectors such as technology, renewable energy, and tourism. This approach aims to reduce their dependence on a single industry and create new opportunities for employment and economic growth. They are also prioritizing education and skill development programs to equip their workforce with the necessary knowledge and expertise to thrive in a rapidly changing global economy.

2.2.2.3 Import Substitution with Export Promotion

Import substitution with export promotion was a post-independence economic strategy implemented in Africa that aimed to reduce dependence on foreign imports while simultaneously promoting exports. This strategy involved the development of domestic industries to produce goods that were previously imported, thereby boosting local production and reducing reliance on foreign markets. Export promotion initiatives were introduced to encourage the growth of African exports and generate foreign exchange earnings. These combined efforts were expected to enhance economic self-sufficiency and stimulate overall economic development in Africa. By focusing on domestic industries, African countries aimed to create job opportunities and improve income distribution among their populations. The export promotion initiatives aimed to diversify the African economies by encouraging the production and export of value-added goods rather than solely relying on raw material exports. This approach sought to foster long-term economic growth and stability in Africa.

These initiatives also aimed to reduce Africa's dependency on foreign aid and increase self-sufficiency. By developing domestic industries and promoting exports, African countries sought to strengthen their economies and become more economically independent. This approach also aimed to attract foreign investment and enhance international trade partnerships, further contributing to economic development in Africa. These initiatives focused on improving infrastructure and creating a favorable business environment to attract both local and foreign investors. This included investing in transportation networks, telecommunications systems, and energy resources to support the growth of industries and facilitate trade within and outside of Africa. Efforts were made to enhance education and skills training programs to ensure a skilled workforce capable of driving economic growth in various sectors. The results of this strategy differed across African countries.

Some countries experienced significant improvements in their business environment, attracting a large influx of both local and foreign investors. This led to a surge in economic growth and job creation. African countries that have seen substantial improvements in their business environment include Rwanda, which implemented a series of reforms to streamline business registration and reduce bureaucracy. As a result, the country has witnessed a significant increase in FDI and entrepreneurship. Another example is Ethiopia, which has made strides in improving its infrastructure and promoting industrialization. These efforts have attracted multinational companies and boosted exports, contributing to the country's economic expansion. Ghana has made notable progress in creating a more business-friendly environment by implementing policies that promote transparency and ease of doing business. This has resulted in increased foreign investment and a thriving entrepreneurial ecosystem.

Rwanda has emerged as a shining example of economic transformation through its focus on innovation and technology. The government's investment in digital infrastructure and the promotion of tech startups have attracted international attention and bolstered the country's economic growth. Similarly, Kenya has become a regional hub for entrepreneurship and innovation, with its vibrant startup scene attracting significant venture capital funding and fostering a culture of creativity and innovation. These advancements have not only propelled economic growth but also created job opportunities for the youth population. The government's focus on digital literacy and training programs has further enhanced the skills of the workforce, making Kenya an attractive destination for multinational companies looking to outsource their operations. The success stories of local tech entrepreneurs have inspired a new generation of aspiring innovators, leading to a surge in entrepreneurial activity and a thriving ecosystem for startups in the country.

Other countries faced challenges in implementing these reforms effectively, resulting in limited progress in attracting investments and fostering economic development. Some African countries that faced challenges in implementing these reforms effectively include Nigeria, Ethiopia, and Zambia. Despite their potential, these countries struggled with bureaucratic hurdles, corruption, and inadequate infrastructure, hindering their ability to create a conducive environment for startups and attract foreign investments. In Nigeria, for example, bureaucratic hurdles and corruption have been major obstacles to implementing effective reforms. The complex and time-consuming processes for obtaining permits and licenses have discouraged many potential investors, while corruption has undermined the trust and confidence of both domestic and foreign businesses. Inadequate infrastructure, such as unreliable power supplies and poor transportation networks, has further hampered economic development in these countries. Limited access to quality education and healthcare services has hindered the development of a skilled workforce and overall human capital. This lack of investment in human resources has contributed to high levels of unemployment and underemployment, exacerbating poverty and inequality in these nations. The absence of transparent and fair legal systems has deterred FDI, as investors are concerned about their ability to enforce contracts and protect their assets.

2.2.2.4 Structural Adjustment Programs

Structural adjustment programs were implemented in many African countries as a response to the economic challenges faced after gaining independence. These programs aimed to promote economic stability and growth by implementing market-oriented policies, reducing government intervention, and encouraging foreign investment. The outcomes of these programs varied across the continent. Positive benefits, such as increased GDP growth and improved macroeconomic indicators, were recorded in several nations. Ghana was one of the African countries that achieved positive results, as it implemented structural adjustment programs in the 1980s and enjoyed significant economic progress. Another example is Rwanda, which implemented a series of economic reforms and witnessed rapid economic growth and poverty reduction. Countries like Botswana and Mauritius have successfully diversified their economies and attracted foreign investment, leading to sustained economic development. These success stories demonstrate that implementing structural adjustment programs and economic reforms can lead to positive outcomes for African nations. It is crucial for other countries in the region to learn from these examples and prioritize economic diversification and attracting foreign investment to achieve sustained development.

Other countries in Africa faced challenges like rising unemployment rates and widening income inequalities. These challenges highlight the need for comprehensive strategies that address social and economic disparities, such as investing in education and skills training programs to create more job opportunities and implementing progressive taxation policies to reduce income inequalities. Fostering regional cooperation and partnerships can also help address common challenges and promote inclusive growth across the continent. African countries that have experienced difficulties as a result of the implementation of structural adjustment initiatives include Zimbabwe, which faced severe economic decline and hyperinflation after implementing structural adjustment programs in the 1990s. The country's reliance on exports and lack of diversification contributed to its vulnerability to external shocks. Similarly, Nigeria also struggled with the effects of structural adjustment, as the reduction in government spending and removal of subsidies led to social unrest and economic instability. These examples highlight the importance of carefully considering the potential consequences and implementing measures to mitigate any negative impacts when pursuing economic reforms.

Critics argue that structural adjustment programs often lead to social unrest and political instability, as they can exacerbate existing inequalities and create winners and losers within societies. This can further deepen divisions and fuel discontent among marginalized groups, potentially leading to protests and even violent conflicts. Therefore, it is crucial for policymakers to carefully consider the social implications of structural adjustment programs and ensure that measures are in place to mitigate any negative consequences for vulnerable populations. One way to achieve this is by implementing targeted social safety nets that provide support and assistance to those most affected by the reforms. Policymakers should actively engage with marginalized communities and involve them in the decision-making process to ensure their voices are heard and their needs are addressed. By taking these steps, policymakers can work toward creating a more inclusive and equitable society, reducing the risk of social unrest and conflict. It is crucial for policymakers to invest in education and skill development programs that empower individuals to adapt to the changing economic landscape. This will not only enhance their employability but also promote social mobility and reduce inequality. By prioritizing these measures, policymakers can foster a sense of hope and opportunity among marginalized communities, fostering a more cohesive and harmonious society.

2.2.2.5 Regional Integration

Regional integration has been a key focus of post-independence economic strategies in Africa. Countries have sought to foster closer economic ties and cooperation with neighboring nations through initiatives such as regional trade agreements and the establishment of regional economic communities. The outcomes of these efforts have varied, with some regions experiencing significant economic growth and increased trade, while others have faced challenges due to political instability, limited infrastructure, and disparities in development levels among member countries. Countries in Africa that experienced significant economic growth and increased trade include Rwanda, which has been praised for its efforts to improve the business environment and attract foreign investments. Countries like Kenya and Ethiopia have also witnessed notable economic growth, driven by sectors such as agriculture, manufacturing, and services. African nations such as Somalia, South Sudan, Zimbabwe, and the Democratic Republic of the Congo have faced difficulties as a result of political unrest, inadequate infrastructure, and differences in the levels of development among their constituents. These countries have struggled to attract foreign investments and develop their economies due to ongoing conflicts and weak governance. Efforts are being made by regional and international organizations to support these countries in overcoming their challenges and promoting sustainable economic development.

It is important to note that not all African nations have benefited equally from regional trade agreements. For instance, countries like Somalia and South Sudan continue to face political instability and conflicts that hinder their economic progress. Disparities in development levels among member countries can create challenges. These challenges can include unequal access to resources and markets, as well as differing regulatory frameworks that can impede trade. The lack of infrastructure and connectivity in some regions can limit the potential benefits of regional trade agreements for certain African nations. To address these disparities, it is crucial for regional trade agreements to prioritize inclusive growth and provide targeted support to the most vulnerable economies. This can involve implementing special provisions or assistance programs that aim to uplift disadvantaged countries and ensure that the benefits of regional trade are distributed more equitably.

Nonetheless, regional integration remains an important avenue for African nations to enhance their economic development and overcome the challenges of limited market access and fragmented economies. By pooling resources, sharing expertise, and harmonizing trade policies, African nations can create a more conducive environment for trade and investment, leading to increased economic growth and job creation. Regional integration can also promote regional stability and cooperation, fostering peace and security on the continent. Regional integration can help African nations address common challenges such as infrastructure development, energy security, and climate change. By working together, countries can leverage their collective resources to build robust transportation networks, ensure reliable access to energy sources, and implement sustainable practices that protect the environment. This collaborative approach not only strengthens economic development but also contributes to the overall well-being and resilience of the region.

2.2.3 Shift toward Market-Oriented Reforms and Globalization

The shift toward market-oriented reforms and globalization has brought both opportunities and challenges to Africa's economy. While these reforms have opened up new avenues for trade and investment, they have also exposed the continent to increased competition and vulnerability to global economic fluctuations. The rapid pace of globalization has highlighted the need for African countries to enhance their competitiveness, improve infrastructure, and invest in human capital to fully benefit from the evolving global economy. Globalization has led to a shift in economic power dynamics, with African countries needing to adapt and navigate the complexities of global markets. This requires strategic planning and the development of robust policies that promote sustainable growth and inclusive development. African nations must also address social and environmental concerns to ensure that the benefits of globalization are shared equitably and do not come at the expense of local communities or natural resources.

African countries should prioritize investments in education and skills development to enhance their competitiveness in the global economy. By equipping their workforce with the necessary knowledge and expertise, they can attract FDI and foster innovation, ultimately driving economic diversification and reducing reliance on traditional sectors. Fostering regional integration and cooperation among African nations can strengthen their collective bargaining power in global trade negotiations and enable them to better navigate the challenges of globalization. In addition, promoting entrepreneurship and supporting SMEs can also contribute to enhancing competitiveness in the global economy. By providing resources, training, and access to markets, African nations can empower local businesses to thrive and compete on an international scale. Investing in infrastructure development, such as transportation networks and digital connectivity, can further enhance Africa's integration into the global economy and facilitate trade across borders.

Fostering innovation and promoting technological advancements can drive economic growth and create new opportunities for African entrepreneurs. By encouraging research and development, as well as nurturing a culture of creativity and risk-taking, African nations can position themselves as hubs for innovation and attract foreign investment. Fostering collaboration between governments, private sector entities, and educational institutions can ensure a steady supply of skilled workers equipped with the necessary knowledge and expertise to drive entrepreneurship forward. This collaboration can also lead to the establishment of incubators and accelerators that provide support and resources to aspiring entrepreneurs, helping them turn their ideas into successful businesses. Implementing policies that promote ease of doing business and reduce bureaucratic hurdles can further enhance the entrepreneurial ecosystem in African nations, making them more attractive to both local and international investors.

2.2.4 Past Economic Challenges Faced by Africa

Some of the past economic challenges faced by African countries include a lack of diversification in their economies, which has made them vulnerable to fluctuations in commodity prices. Corruption and political instability have hindered economic growth and foreign investment. These challenges have further exacerbated the issues of poverty, unemployment, and income inequality in the region. Inadequate infrastructure and limited access to education and healthcare have also contributed to the economic challenges faced by African countries. These factors have hindered productivity and hindered the development of a skilled workforce, making it difficult for these countries to compete globally and attract investment. Political instability and corruption have posed significant obstacles to economic growth in Africa. The lack of transparent governance and accountability has deterred both domestic and foreign investors, further perpetuating the cycle of poverty and inequality. The continent's heavy reliance on commodity exports has made its economies vulnerable to global market fluctuations, making it crucial for African countries to diversify their industries and promote sustainable development. Inadequate infrastructure and limited access to basic services such as healthcare and education have hindered human capital development in Africa. This has resulted in a lack of skilled workers, further impeding economic progress and hindering the continent's ability to compete globally.

Therefore, it is imperative for African governments to prioritize investments in infrastructure and human development to create a conducive environment for sustainable economic growth. By investing in infrastructure, such as roads, bridges, and telecommunications networks, African governments can improve connectivity and facilitate trade both within the continent and with the rest of the world. By focusing on human development through initiatives like improving access to quality education and healthcare, African countries can nurture a skilled workforce that can drive innovation and productivity, ultimately leading to economic prosperity. Promoting entrepreneurship and creating a favorable business environment can attract FDI and stimulate job creation. This can further contribute to economic growth by diversifying the economy and reducing dependency on a single sector. Fostering regional integration and cooperation can enhance market access and promote the flow of goods, services, and investments across African nations, fostering economic development on a larger scale.

2.2.5 Impact of Colonization and Its Long-Lasting Effects

The impact of colonization on the current African economy cannot be understated. Colonization resulted in the exploitation of Africa's natural resources, leading to the depletion of valuable commodities such as minerals and timber. Colonial powers implemented economic systems that were designed to benefit their own countries, often at the expense of African nations. These systems hindered the development of local industries and infrastructure, leaving many African economies heavily reliant on exporting raw materials rather than diversifying their economies. As a result, Africa continues to face persistent economic challenges, including high levels of poverty, unemployment, and income inequality. The exploitation of natural resources has led to environmental degradation and the loss of biodiversity in many African countries. These issues have hindered sustainable development and hindered the ability of African nations to fully harness their potential for economic growth and prosperity. In addition, the overreliance on natural resources has made African economies vulnerable to fluctuations in global commodity prices, further exacerbating their economic challenges. The lack of investment in education and infrastructure has limited the capacity of African nations to develop competitive industries and attract foreign investment, perpetuating their dependence on resource extraction.

These systemic issues require comprehensive strategies that prioritize economic diversification, sustainable development, and inclusive growth to uplift Africa from its current economic struggles. Addressing the systemic issues in Africa's economy requires not only diversifying industries and promoting sustainable development but also implementing effective governance and reducing corruption. By improving transparency and accountability, African nations can create an environment that encourages foreign investment and fosters economic growth. Investing in human capital through quality education and healthcare will empower the African workforce to participate in higher value industries, ultimately driving long-term economic prosperity. African nations should prioritize infrastructure development to enhance connectivity and facilitate trade both within the continent and with global markets. This will not only attract more foreign investment but also enable local businesses to expand their reach and competitiveness.

Fostering innovation and technological advancements can help African countries leapfrog traditional development stages and capitalize on emerging opportunities in sectors such as renewable energy, the digital economy, and e-commerce. Investing in education and skills development is crucial to ensuring that the African workforce is equipped with the necessary knowledge and expertise to thrive in these emerging sectors. By prioritizing education, African countries can empower their citizens to participate fully in the digital revolution and contribute to economic growth. Promoting regional integration and cooperation can create economies of scale, attract larger investments, and foster a more harmonized business environment across the continent. Investing in research and development can lead to the creation of innovative solutions tailored to the specific needs and challenges of African industries. This can drive technological advancements and enhance competitiveness in the global market. Fostering partnerships between governments, private sector entities, and educational institutions can facilitate knowledge transfer and skills development, further boosting the potential for economic growth in Africa. These partnerships can also attract FDI and promote the exchange of expertise and best practices.

2.2.6 Difficulties in Achieving Economic Growth and Development

One of the main challenges Africa faces in achieving economic growth and development is the lack of infrastructure. Insufficient transportation networks, inadequate power supply, and limited access to clean water hinder productivity and impede trade. Political instability and corruption pose significant obstacles to attracting foreign investment and fostering a conducive business environment. The lack of infrastructure also affects the delivery of essential services such as healthcare and education, exacerbating social inequalities and hindering human development. Without proper roads, hospitals, and schools, it becomes difficult for African countries to provide basic necessities to their citizens and improve their overall quality of life. Addressing these infrastructure gaps is crucial for Africa to unlock its full economic potential and create a sustainable future for its people. The lack of infrastructure hinders economic growth and foreign investment, as businesses require reliable transportation networks and access to utilities in order to thrive. This perpetuates a cycle of poverty and limits opportunities for job creation and income generation.

Therefore, investing in infrastructure development is not only essential for improving living conditions but also for fostering economic prosperity and reducing dependency on aid in African nations. By investing in infrastructure development, African nations can improve the quality of life for their citizens. Access to reliable transportation networks and utilities can enhance access to education, healthcare, and other essential services. A well-developed infrastructure can attract foreign investment and promote economic growth, leading to job creation and increased income opportunities for the population. Infrastructure development can also facilitate regional integration and trade among African nations. Improved roads, ports, and airports can enhance connectivity and enable the movement of goods and services across borders, fostering economic cooperation and development. A well-connected infrastructure network can encourage tourism, opening up new avenues for revenue generation and cultural exchange within the continent.

Infrastructure development can attract FDI by creating a conducive environment for businesses to operate in. This can lead to job creation and technology transfer, contributing to economic growth and reducing unemployment rates. A well-developed infrastructure network can improve access to basic services such as healthcare and education, ultimately improving the quality of life for the population. Infrastructure development can enhance connectivity and facilitate trade between different regions, promoting economic integration and regional cooperation. It can also help address transportation challenges and reduce logistical costs, making it easier for goods and services to reach markets efficiently. A well-developed infrastructure network can attract tourists and promote tourism industry growth, generating revenue and creating employment opportunities in the hospitality sector.

2.3 CURRENT ECONOMIC CHALLENGES IN AFRICA

2.3.1 Demographic Challenges Currently Facing Africa

Demographic challenges currently facing Africa include rapid population growth (Kundu and Pandey, 2020), urbanization (Nwokolo, 2017), and youth unemployment (Enock, 2018). These challenges pose significant pressure on limited resources, infrastructure, and job opportunities. They also present a need for quality education, healthcare, and social services to support the growing population. The rapid population growth and urbanization in Africa also contribute to environmental concerns such as deforestation, pollution, and increased demand for energy. Addressing these challenges requires sustainable development strategies that promote economic growth while ensuring the preservation of natural resources and a healthy environment for future generations. The increasing urbanization in Africa has led to a strain on infrastructure, including transportation and housing. This calls for the implementation of efficient urban planning and development strategies to accommodate the growing population and ensure habitable cities. It is crucial to prioritize investments in renewable energy sources and promote eco-friendly practices to mitigate the negative impact of increased energy demand on the environment. Investing in education and skills development is essential to equip future generations with the knowledge and abilities needed to address the challenges of urbanization and sustainability. Fostering collaboration between governments, communities, and businesses is crucial to finding innovative solutions and creating a shared responsibility for the well-being of both present and future populations. In this section, we will discuss how rapid population growth, urbanization, and youth unemployment pose significant challenges to Africa today.

2.3.1.1 Rapid Population Growth and Its Implications for Economic Development

Rapid population growth in Africa has both positive and negative implications for economic development. On one hand, a larger population can lead to a larger labor force, which can stimulate economic growth and innovation. A growing consumer base can drive domestic demand and attract foreign investment. African countries with rapid population growth and data include Nigeria, Ethiopia, and the Democratic Republic of the Congo. Statistically, the percentage growth for these countries between 2020 and 2023 includes an estimated 2.6% for Nigeria, 2.9% for Ethiopia, and 3.1% for the Democratic Republic of the Congo (Carbone, 2018). This growth presents opportunities for job creation and increased productivity, potentially reducing poverty levels and improving living standards. Rapid population growth also poses challenges such as strain on infrastructure, healthcare systems, and natural resources, which need to be addressed to ensure sustainable development in these countries. Is rapid population growth in Nigeria, Ethiopia, and the Democratic Republic of the Congo having a significant impact on economic development? Rapid population growth in Nigeria, Ethiopia, and the Democratic Republic of the Congo is indeed having a significant impact on economic development. While it can stimulate economic growth through increased consumer demand and a larger labor force, it also puts pressure on limited resources and infrastructure. As a result, these countries need to implement effective policies and investments to harness the potential benefits of population growth while mitigating its challenges.

Rapid population growth also puts pressure on limited resources such as food, water, and infrastructure, which can hinder economic progress. It is crucial for African governments to implement effective policies that promote sustainable development and address the challenges posed by population growth. These policies should focus on improving access to education and healthcare, as well as promoting family planning and reproductive health services. By investing in these areas, African governments can ensure that their growing population is equipped with the necessary skills and resources to contribute to economic development while also protecting the environment and preserving natural resources for future generations. It is crucial for African governments to prioritize sustainable infrastructure development and promote renewable energy sources. By investing in clean and efficient technologies, they can mitigate the negative impact of population growth on the environment and reduce carbon emissions. Fostering partnerships with international organizations and neighboring countries can facilitate knowledge exchange and resource sharing, leading to more effective strategies for addressing the challenges of population growth in Africa. Implementing policies that promote family planning and reproductive health services can help manage population growth in a sustainable manner. These initiatives can empower individuals to make informed decisions about their family size and access necessary healthcare, ultimately contributing to the overall well-being of communities and the environment. Integrating education on sustainable practices into school curricula can raise awareness and instill a sense of responsibility toward the environment among future generations, ensuring long-term sustainability in Africa.

2.3.1.2 Youth Unemployment and the Need for Job Creation

Youth unemployment and the need for job creation in Africa have become pressing issues in recent years. With a rapidly growing population and limited employment opportunities, many young Africans are struggling to find meaningful work. This not only hampers their personal growth and economic independence but also poses a significant challenge to the overall development of the continent. Countries in Africa with the highest statistics of youth unemployment include Nigeria, South Africa, and

Egypt. In Egypt, South Africa, and Nigeria, the unemployment rates from 2020 to 2023 are as follows: in Egypt, the unemployment rate is projected to increase from 9.6% in 2020 to 11.2% in 2023 (World Bank, 2022). Similarly, South Africa is expected to see a rise in unemployment from 29.1% in 2020 to 31.7% in 2023 (World Bank, 2022). Nigeria, on the other hand, faces a more severe situation with an estimated unemployment rate of 27.1% in 2020, which is projected to reach 33 by 2023 (World Bank, 2022). These high unemployment rates pose significant challenges for these countries' economies and governments. The increasing unemployment rates could lead to social unrest and economic instability, making it crucial for these nations to implement effective policies and strategies to address this issue.

These countries face unique challenges such as inadequate infrastructure, a mismatch between skills and job opportunities, and limited access to quality education and training programs. These high unemployment rates pose significant challenges for these countries, including social unrest and economic instability. The governments of Egypt, South Africa, and Nigeria will need to implement effective policies and strategies to address this issue and create job opportunities for their growing populations. One potential strategy could be to invest in infrastructure development, such as improving transportation networks and expanding access to electricity. This would not only create immediate job opportunities in construction and maintenance but also attract foreign investment and stimulate economic growth in other sectors. Governments could prioritize investments in education and vocational training programs to bridge the skills gap and equip their populations with the necessary tools for employment in emerging industries. By investing in education and vocational training programs, governments can ensure that their populations are prepared for the demands of emerging industries. This would not only address the skills gap but also lead to a more competitive workforce, attract further investment, and drive economic growth. Governments could consider implementing policies that support research and development initiatives, foster innovation, and drive technological advancements that can further propel economic growth in various sectors.

2.3.1.3 Urbanization and Its Impacts on Infrastructure and Service Provision

Urbanization in Africa has been rapidly increasing over the past few decades, leading to significant challenges in infrastructure and service provision. As more people migrate to cities, there is a growing strain on existing systems such as transportation, housing, water supply, and sanitation. This has resulted in overcrowded cities, inadequate access to basic services, and increased pressure on limited resources in Africa. The rapid pace of urbanization often outpaces the ability of governments to plan and provide necessary infrastructure, exacerbating the problem further in Africa. The lack of affordable housing options in urban areas has led to the proliferation of informal settlements and slums, further exacerbating the strain on existing infrastructure. The unequal distribution of resources and services within cities has created stark disparities in living conditions, with marginalized communities often bearing the brunt of inadequate infrastructure and service provision. The African nations with the highest rates of urbanization include Nigeria, South Africa, and Kenya. Rapid urbanization in these countries has resulted in overcrowding, strained public transportation systems, and increased pressure on already limited resources. In Nigeria, South Africa, and Kenya, the urbanization rates from 2020 to 2023 are as follows: Nigeria is projected to have an urbanization rate of 4.2% from 2020 to 2023, indicating a significant increase in the urban population (Akinsola et al., 2022). Similarly, South Africa and Kenya are expected to experience urbanization rates of 2.8 and 3.5% (Akinsola et al., 2022), respectively, during the same period, highlighting the ongoing trend of rapid urban growth in these nations. This rapid urban growth poses numerous challenges for these countries, including the need for adequate infrastructure to support the growing urban population. It puts a strain on social services such as healthcare and education, as well as increasing competition for jobs and housing in urban areas.

This rapid urbanization poses challenges for governments and policymakers to address the increasing demand for housing, transportation, and basic services in order to ensure sustainable development and improve the quality of life for urban residents. It also requires careful planning and investment in infrastructure to accommodate the growing population and avoid issues such as overcrowding and inadequate resources. Effective urban governance and policies are needed to manage the social, economic, and environmental impacts of urbanization, including addressing issues of inequality, unemployment, and environmental degradation. Promoting inclusive and equitable urban development is crucial to ensuring that all residents, regardless of their socio-economic background, have access to essential services and opportunities. This can be achieved through the implementation of affordable housing initiatives, improved public transportation systems, and the creation of green spaces for recreational activities. Engaging local communities in decision-making processes can help foster a sense of ownership and responsibility toward the development of their cities, leading to more sustainable and habitable urban environments. Investing in education and healthcare facilities is crucial to ensuring equal access to quality services for all residents. By providing affordable and accessible education, individuals from diverse socio-economic backgrounds can have equal opportunities for personal and professional growth. Similarly, accessible healthcare services can contribute to the overall well-being of the community, reducing health disparities and promoting a healthier urban population.

2.3.2 Resource Dependency and Natural Resource Management

Resource dependency refers to the reliance of a country or region on the extraction and export of natural resources for economic development. In Africa, many countries heavily depend on the exploitation of their abundant natural resources, such as oil, minerals, and timber. This heavy reliance often leads to challenges in natural resource management, including issues of sustainability, environmental degradation, and economic vulnerability. Resource dependency can also contribute to political instability and conflict as competing interests vie for control over these valuable resources. It is essential for African countries to diversify their economies and reduce their reliance on natural resources. This can be achieved through investing in sectors such as agriculture, manufacturing, and tourism, which have the potential to create more sustainable and inclusive growth. Promoting good governance, transparency, and accountability in resource management is crucial to ensuring that the benefits of these resources are equitably distributed among the population and contribute to long-term development.

According to the World Bank, countries that successfully diversify their economies tend to experience higher levels of economic stability and resilience. Investing in human capital through education and skills development is essential for countries to effectively transition to more diversified economies. By equipping their citizens with the necessary knowledge and skills, countries can attract investment, foster innovation, and create opportunities for sustainable economic growth. Unfortunately, Africa has been severely reliant on the export of natural resources such as oil, minerals, and agricultural goods for many decades. This has made African economies vulnerable to fluctuations in global commodity prices and hindered their ability to diversify and develop other sectors. African countries mostly affected by this include Nigeria, Angola, and South Africa. These countries have experienced economic volatility and struggled to break free from the cycle of resource dependence. There is a growing recognition among African leaders of the need to shift toward knowledge-based economies and invest in sectors such as technology, manufacturing, and services.

In line with this, the World Bank predicts that Africa's economic growth will continue to be hindered unless there is a shift toward developing a knowledge-based economy. This would involve investing in education, research and development, and technology infrastructure to promote innovation and attract industries that rely on intellectual capital rather than natural resources. Fostering entrepreneurship and supporting SMEs can also contribute to diversifying African economies and reducing their dependence on commodity exports. By promoting entrepreneurship and supporting SMEs, African countries can create a more dynamic and competitive business environment. This would not only lead to job creation but also encourage the growth of innovative industries that can drive economic diversification. Investing in education, research, and development would equip the workforce with the necessary skills and knowledge to participate in a knowledge-based economy, enabling Africa to compete globally in sectors such as technology, healthcare, and renewable energy.

2.3.2.1 Challenges Related to Overreliance on Commodity Exports

One of the major challenges related to overreliance on commodity exports in Africa is the vulnerability to price fluctuations in global markets. As African countries heavily depend on the export of commodities such as oil, minerals, and agricultural products, any drop in global prices can have a significant impact on their economies. This can lead to economic instability, budget deficits, and difficulties in funding essential social services such as education and healthcare. Overreliance on commodity exports often hinders diversification efforts in African economies. This lack of diversification leaves African countries vulnerable to shocks in global markets, as they have limited sources of income to rely on. Without a diverse range of industries and sectors, these countries struggle to adapt and recover from economic downturns, further exacerbating the impact of price fluctuations. The reliance on commodity exports can also lead to environmental degradation and unsustainable resource extraction practices, which can have long-term negative effects on the environment and local communities.

African countries are mostly affected by challenges related to overreliance on commodity exports, including Nigeria, Angola, and South Africa. These countries heavily rely on the export of oil, minerals, and other commodities, making them vulnerable to global market fluctuations. The lack of economic diversification leaves them exposed to the risks associated with price volatility and limited revenue streams. The overexploitation of natural resources often leads to deforestation, pollution, and displacement of indigenous populations, exacerbating social and environmental challenges in these regions. The dependence on extractive industries hinders the development of other sectors such as manufacturing and services, which could provide more sustainable and inclusive economic growth. Without diversifying their economies, these countries may struggle to adapt to future changes in global demand for their primary exports, potentially leading to economic instability and social unrest. According to World Bank statistics, the problems associated with Nigeria, Angola, and South Africa's excessive reliance on commodity exports are clear from the data on the contributions of the various economic sectors between 2020 and 2023 in each of these countries (World Bank, 2022). In Nigeria, for example, the oil sector accounted for over 90% of total exports in 2020, leaving the country vulnerable to fluctuations in global oil prices (World Bank, 2022). Similarly, Angola heavily relies on oil exports, which made up around 70% of its total exports in the same year (World Bank, 2022). South Africa's economy, on the other hand, is heavily dependent on mineral resources such as gold and platinum, with mining contributing significantly to its export earnings. As global demand for these commodities fluctuates and alternative energy sources become more popular, countries like Angola and South Africa may face challenges in sustaining their economies. Diversification of their export sectors becomes crucial to reduce reliance on a single commodity and mitigate the impact of price fluctuations. Investing in other industries, such as manufacturing or tourism, can provide alternative sources of revenue and promote long-term economic stability.

Diversifying their export sectors can also help these countries create more job opportunities and reduce unemployment rates. By investing in industries like manufacturing or tourism, Angola and South Africa can attract foreign investments and stimulate economic growth. Investing in manufacturing or tourism can also lead to technology transfer and knowledge exchange, enhancing the overall competitiveness of these countries. This can further strengthen their position in the global market and attract more FDI. By diversifying their export sectors, Angola and South Africa can reduce their reliance on a single industry, making them less vulnerable to economic shocks and ensuring sustained growth in the long run. By investing in manufacturing or tourism, Angola and South Africa can create more job opportunities for their citizens, reducing unemployment rates and improving living standards. Diversifying their export sectors can also foster innovation and creativity within these countries, as different industries bring new ideas and approaches to the table. This can ultimately drive economic growth and foster a more dynamic and resilient economy for both Angola and South Africa.

2.3.2.2 Weak Diversification of Economies and Vulnerability to Commodity Price Fluctuations

Weak diversification of economies and vulnerability to commodity price fluctuations can hinder sustainable development efforts in Africa. The heavy reliance on a limited number of resources leaves countries exposed to the volatility of global markets, making it difficult to plan and implement long-term development strategies. The lack of diversification limits job creation opportunities and hampers efforts to reduce poverty and inequality in the region. In addition, overreliance on a few key resources can also lead to environmental degradation and the depletion of natural resources. This further exacerbates the challenges faced by African countries in achieving sustainable development goals, as it undermines their ability to sustainably manage and utilize their resources for future generations. The dependence on a narrow range of resources makes African economies vulnerable to external shocks and fluctuations in global commodity prices. This volatility can disrupt economic stability and hinder long-term growth prospects, making it even more crucial for African countries to diversify their economies and explore alternative sources of income. The lack of diversification can also hinder technological advancements and innovation, as countries may become reliant on outdated industries and fail to invest in new sectors with higher potential for growth. A lack of economic diversification can also lead to increased inequality within African countries. When an economy relies heavily on a single industry or sector, the benefits and opportunities generated by that industry may not be evenly distributed among the population. This can result in a widening wealth gap and limited social mobility, ultimately hindering overall development and prosperity. Without diversification, African countries may miss out on opportunities for sustainable development and green technologies, which are becoming increasingly important in the global economy.

African countries with the weakest diversification of economies and vulnerability to commodity price fluctuations include countries such as Nigeria, Angola, and Zambia. According to the World Bank, Nigeria, Angola, and Zambia are statistically projected to experience weak diversification of economies and vulnerability to commodity price fluctuations at varying rates, as follows: Nigeria, being heavily dependent on oil exports, is expected to face significant challenges in diversifying its economy and reducing its vulnerability to commodity price fluctuations. Angola, similarly reliant on oil, will also need to prioritize diversification efforts to mitigate the impact of fluctuating oil prices. Zambia, with its dependence on copper exports, will need to explore alternative sectors and strategies to reduce its exposure to commodity price volatility. These countries must focus on sustainable development and green technologies as part of their long-term economic growth plans. By embracing sustainable development and green technologies, these countries can not only reduce their reliance on volatile commodity prices but also contribute to global efforts to combat climate change. Diversifying their economies will not only create new job opportunities but also foster innovation and resilience in the face of future economic challenges.

2.3.2.3 Sustainable Management of Natural Resources for Long-Term Development

Sustainable management of natural resources for long-term development in Africa is crucial for the continent's sustainable development. Africa is rich in natural resources, including minerals, oil, and gas, which have the potential to drive economic growth and alleviate poverty. Resource dependency can also lead to challenges such as environmental degradation, social conflicts, and economic instability. Therefore, effective natural resource management strategies are needed to ensure that these resources are utilized in a sustainable and equitable manner, benefiting both present and future generations. Promoting the diversification of the economy and investing in renewable energy can help reduce the reliance on finite resources and mitigate the negative impacts of resource dependency. By encouraging innovation and supporting industries that are less resource-intensive, countries can create new opportunities for job creation and economic development.

Investing in renewable energy sources such as solar and wind power can not only reduce greenhouse gas emissions but also provide a stable and sustainable source of energy in the long run. The adoption of renewable energy can also enhance energy security by diversifying the energy mix and reducing vulnerability to price fluctuations in fossil fuels. Transitioning to renewable energy can contribute to improving public health by reducing air pollution and its associated health risks, leading to cost savings in healthcare expenses. Investing in renewable energy technologies can stimulate economic growth and create new job opportunities in the clean energy sector. This can help boost local economies and reduce dependence on foreign energy sources. The development of renewable energy infrastructure can also enhance resilience to natural disasters and other disruptions to the traditional energy grid, ensuring a more reliable and resilient energy system for communities.

The best performing African nation for long-term development in terms of sustainable resource management is Rwanda. With its commitment to sustainable resource management, Rwanda has made significant progress in developing renewable energy sources such as hydroelectric power and solar energy. The country's investment in clean energy has not only contributed to economic growth but also improved access to electricity for its citizens, promoting social development and reducing poverty levels. Rwanda's focus on sustainable resource management has positioned it as a leader in environmental conservation and climate change mitigation efforts, setting an example for other African nations to follow. Rwanda's commitment to renewable energy has also resulted in a significant reduction in greenhouse gas emissions, contributing to global efforts to combat climate change. The country's emphasis on sustainable resource management has helped preserve its natural beauty and biodiversity, attracting eco-tourism and boosting the tourism industry. Rwanda's government has implemented policies and initiatives to promote sustainable agriculture practices, such as agroforestry and organic farming. These efforts not only support local farmers but also contribute to food security and resilience against climate change impacts. Overall, Rwanda's holistic approach to environmental conservation has not only benefited its own citizens but has also positioned the country as a role model for sustainable development in Africa.

In terms of long-term development and sustainable resource management, South Sudan and Nigeria are the two African nations that perform the worst. South Sudan has faced significant challenges in establishing effective governance systems and infrastructure, hindering its ability to prioritize sustainable resource management. Similarly, Nigeria has struggled with issues such as corruption and inadequate investment in renewable energy, impeding its progress toward long-term development and sustainability. These challenges have resulted in both countries experiencing high levels of poverty, political instability, and environmental degradation. South Sudan's ongoing conflict has further exacerbated these issues, diverting resources away from sustainable development efforts. In Nigeria, the reliance on oil as the main source of revenue has led to a neglect of other sectors and a failure to diversify the economy for long-term stability and resource management. As a result, both countries face significant challenges in achieving sustainable development goals and addressing the needs of their populations. The lack of economic diversification and overreliance on specific industries has hindered the ability to create stable and inclusive economies, leaving many vulnerable to poverty and political unrest. The environmental degradation caused by unsustainable practices further threatens the long-term viability of these nations. The inadequate investment in education and healthcare systems exacerbates the social and economic disparities within these countries. This lack of access to quality education and healthcare perpetuates cycles of poverty and limits opportunities for individuals to improve their livelihoods. The absence of effective governance and corruption in both nations undermines efforts to implement sustainable development strategies and hampers progress toward achieving long-term stability and prosperity.

2.3.3 Infrastructure and Connectivity

2.3.3.1 Inadequate Transportation Networks and Limited Access to Markets

Inadequate transportation networks and limited access to markets in Africa have hindered economic growth and development in the region. This lack of infrastructure has made it difficult for businesses to transport goods efficiently and has increased transportation costs, making African products less competitive in global markets. Limited connectivity, especially in rural areas, has restricted access to information and technology, further exacerbating the development gap between Africa and other regions of the world. The lack of access to markets has also hindered foreign investment in Africa. Without efficient transportation and connectivity, multinational companies are less likely to establish operations in the region, depriving African economies of potential job opportunities and technological advancements. As a result, Africa continues to face significant challenges in achieving sustainable economic growth and reducing poverty levels. The limited infrastructure in Africa, such as inadequate road networks and an unreliable power supply, further exacerbates the gap between the region and other parts of the world. These infrastructure challenges make it difficult for businesses to operate efficiently and for goods to be transported effectively, hindering economic development.

The lack of access to reliable internet connectivity also hampers technological advancements and innovation in Africa, limiting its ability to compete globally. The inadequate road networks and unreliable power supply hinder not only economic development but also access to basic services such as healthcare and education. This perpetuates a cycle of poverty and inequality in the region. The lack of reliable internet connectivity also limits access to online education and digital platforms, further widening the knowledge gap between Africa and other parts of the world. The lack of investment in technological infrastructure and digital literacy programs exacerbates the digital divide in Africa. Without proper access to information and communication technologies, individuals and businesses struggle to keep up with the rapidly evolving global economy. This further hamper Africa's ability to compete on a global scale and achieve sustainable development.

As the world becomes increasingly interconnected, the digital divide in Africa becomes a pressing issue that needs to be addressed. In addition to hindering economic growth, the lack of access to technology also limits educational opportunities for African youth. Bridging this gap requires not only investment in infrastructure but also the development of comprehensive digital literacy programs to empower individuals and communities with the skills needed to thrive in the digital age. By closing the digital divide in Africa, not only can the continent unlock its full economic potential, but it can also foster innovation and entrepreneurship among its population. Providing access to technology and digital literacy programs can enable African youth to connect with the global community, exchange knowledge, and participate in online learning platforms, ultimately contributing to their personal and professional development. Bridging the digital divide in Africa can lead to increased job opportunities, as digital skills are increasingly in demand in today's global economy. By equipping African youth with the necessary digital skills, they can become competitive in the job market and contribute to the growth of various industries. Access to technology and digital literacy programs can empower individuals to become agents of change within their communities, using technology to address social and economic challenges and drive sustainable development.

2.3.3.2 Lack of Reliable Energy Supply and Its Impact on Industrialization

One of the major challenges faced in terms of infrastructure and connectivity is the lack of a reliable energy supply. This not only hampers the growth of industries but also affects the overall industrialization process. Without a consistent and stable energy source, businesses struggle to operate efficiently, leading to decreased productivity and limited expansion opportunities. An unreliable energy supply can deter potential investors who seek a stable environment for their ventures, further hindering industrial development. The lack of a reliable energy supply can also result in higher production costs for industries. Businesses may have to rely on expensive alternative energy sources or invest in backup generators, increasing their operational expenses. This additional financial burden can make it difficult for industries to remain competitive in the global market and hinder their ability to invest in research and development, further impeding industrialization.

The unreliability of energy supply can lead to disruptions in production schedules and delays in meeting customer demands. These interruptions can damage a company's reputation and customer relationships, potentially leading to a loss of market share and decreased profitability. The lack of a reliable energy supply can discourage FDI, as potential investors may be hesitant to establish operations in an area with uncertain energy availability. This can further hinder industrial development by limiting access to capital and technological advancements from foreign sources. An unreliable energy supply can also have adverse effects on the environment. Inadequate infrastructure and outdated technology may result in higher emissions and pollution levels, contributing to climate change and compromising the overall sustainability of the region. Consequently, investing in a stable and sustainable energy system is crucial for both economic growth and environmental preservation. A lack of access to reliable and affordable energy can hinder industrial development and job creation in the region. This can further exacerbate poverty and inequality, making it even more important to prioritize investments in a stable and sustainable energy system.

One African country that has experienced a significant lack of reliable energy supply is Nigeria. This has had a detrimental impact on its industrialization efforts, as frequent power outages and inadequate infrastructure have hindered the growth of industries and discouraged foreign investment. As a result, Nigeria has struggled to fully harness its economic potential and create sustainable job opportunities for its growing population. The unreliable energy supply in Nigeria has also led to increased reliance on expensive and environmentally harmful alternatives such as diesel generators, further exacerbating the country's economic and environmental challenges. The lack of a consistent power supply has limited access to modern technology and hindered the development of innovative industries, preventing Nigeria from fully participating in the global economy. This lack of consistent power supply has also hindered the growth of small businesses and entrepreneurship in Nigeria, as they struggle to operate efficiently and compete on a global scale. The unreliable power supply has resulted in increased production costs for businesses, as they are forced to rely on expensive alternative sources of energy. This has further hindered Nigeria's ability to attract foreign investment and create job opportunities, perpetuating the cycle of economic challenges in the country.

One African country that has a relatively reliable supply is South Africa. This consistent power supply has played a crucial role in driving industrialization and economic growth in the country. It has attracted foreign investments, allowed for the establishment of large-scale manufacturing industries, and facilitated the development of innovative technologies. As a result, South Africa has been able to participate more actively in the global economy and create employment opportunities for its citizens. The reliable power supply in South Africa has also improved the quality of life for its citizens by ensuring access to electricity for households, schools, and healthcare facilities. This has contributed to advancements in education, healthcare, and overall infrastructure development within the country. The consistent power supply has helped reduce reliance on fossil fuels and promote the use of renewable energy sources, making South Africa a leader in sustainable development initiatives. The improved power supply has attracted foreign investments and stimulated economic growth in various sectors, such as manufacturing and technology. This has created job opportunities and increased income levels for the population, ultimately reducing poverty and improving living standards. The reliable electricity supply has enabled the expansion of industries, leading to increased productivity and competitiveness on a global scale.

2.3.3.3 Utilizing Technology and Digital Connectivity for Economic Advancement

Utilizing technology and digital connectivity for economic advancement in Africa has become a key priority for governments and organizations across the continent. By investing in infrastructure development, such as building reliable internet networks and expanding access to mobile technology, African countries can foster innovation, attract foreign investment, and create new opportunities for their citizens. Improving connectivity can also enhance education and healthcare services, allowing for better access to information and resources that can improve the overall quality of life. Increased connectivity can facilitate e-commerce and digital entrepreneurship, enabling African businesses to reach a wider customer base and participate in the global economy. It can also promote collaboration and knowledge-sharing among different sectors, fostering cross-border partnerships and accelerating economic growth in the region. Increased connectivity can enhance the efficiency of government services, such as online tax filing and digital governance, leading to improved public administration and transparency. It can empower individuals by providing them with opportunities for online learning and remote work, ultimately reducing unemployment rates and driving economic empowerment.

Increased connectivity can also promote innovation and entrepreneurship by providing a platform for startups and small businesses to reach a wider audience and access global markets. This can lead to the creation of new jobs and the development of a thriving digital economy in the region. Enhanced connectivity can facilitate better communication and collaboration between different stakeholders, such as businesses, governments, and civil society organizations, fostering greater cooperation and collective problem-solving for societal challenges. Improved connectivity can also enable the delivery of essential services, such as healthcare and education, to remote areas, bridging the gap between urban and rural communities. It can empower individuals by providing them with access to information and resources that were previously inaccessible, promoting knowledge sharing, and empowering marginalized groups. Improved connectivity can enhance economic opportunities by enabling remote work and e-commerce, opening up new markets, and reducing geographical barriers. This can lead to increased job creation and economic growth, particularly in underdeveloped regions. It can strengthen democratic processes by facilitating access to information and promoting citizen engagement, allowing for more transparent and inclusive governance.

Kenya is one of the African countries with the highest utilization of technology and digital connectivity for economic advancement. Kenya has made significant strides in utilizing technology and digital connectivity to drive economic growth. The country is known for its innovative mobile money platform, M-Pesa, which has revolutionized financial services and enabled millions of Kenyans to access banking services through their mobile phones. Kenya has a thriving tech startup ecosystem, with companies like Safaricom and BRCK gaining international recognition for their innovative solutions. These advancements have not only boosted the economy but also improved access to education, healthcare, and other essential services for many Kenyans. The use of mobile technology has allowed for the development of e-learning platforms, telemedicine services, and mobile banking, making education, healthcare, and financial services more accessible to remote and underserved areas of the country. These advancements have had a significant impact on improving the overall quality of life for many Kenyans and have contributed to the country's continued economic growth.

The World Intellectual Property Organization (WIPO) lists Kenya as one of the African nations that use digital connections and technology the most to boost its economy (Bannerman, 2020). With a booming digital sector that has drawn investment from both domestic and foreign sources, Kenya has also emerged as a center for innovation and entrepreneurship. This has created job opportunities and driven economic development, further solidifying Kenya's position as a leader in technological advancement in Africa. Kenya has implemented various initiatives to promote digital literacy and inclusion, ensuring that more citizens have access to technology and can benefit from its advantages. The government has also established innovation hubs and incubation centers, fostering a supportive ecosystem for startups and encouraging collaboration among entrepreneurs. These efforts have contributed to the country's reputation as a breeding ground for innovative ideas and solutions.

2.3.4 Governance and Institutions

One of the key challenges that Africa faces in its economic development is the issue of governance and institutions. Weak governance structures and ineffective institutions hinder the continent's ability to attract investment, promote business growth, and ensure equitable distribution of resources. Corruption and a lack of transparency further exacerbate these issues, leading to a lack of trust in the government and hindering economic progress. The absence of strong regulatory frameworks and rules of law discourages both domestic and foreign investors from engaging in long-term economic activities. Without a conducive environment for business, Africa struggles to create sustainable jobs and improve living standards for its growing population. Addressing these governance and institutional challenges is crucial for unlocking Africa's full economic potential and fostering inclusive growth. In addition, corruption and a lack of transparency in government institutions further exacerbate the challenges faced by Africa's economy. This not only deters foreign investments but also hampers domestic entrepreneurship and innovation. Therefore, implementing effective anti-corruption measures and promoting accountability within the public sector are essential steps toward creating a favorable business environment in Africa. By doing so, the continent can attract more investments, stimulate economic growth, and uplift the lives of its people. Addressing corruption in government institutions will help to build trust and confidence among both local and international investors. This will create a stable and predictable business environment, encouraging long-term investments and sustainable economic development in Africa.

According to the World Bank, one of the African countries that has mitigated the issue of governance and institutions to a reasonable extent is Rwanda. The country has implemented various measures to combat corruption and improve governance, such as establishing an independent anti-corruption authority and implementing strict accountability measures for public officials. These efforts have resulted in increased investor confidence and a steady inflow of FDI into Rwanda's economy. Rwanda has made significant progress in strengthening its institutions by investing in education and training programs for public servants. This has helped build a skilled and competent workforce that is capable of effectively implementing policies and delivering public services. As a result, the country has seen improvements in areas such as healthcare, education, and infrastructure, which have further contributed to its economic growth and development. Rwanda has also prioritized innovation and technology as key drivers of economic growth. The government has implemented various initiatives to promote entrepreneurship and attract foreign investment in sectors such as information technology and telecommunications. This focus on innovation has not only created new job opportunities but has also enhanced the country's competitiveness in the global market. As a result, Rwanda has become a hub for tech startups and has seen a significant increase in digital infrastructure, further fuelling its economic progress. The government has also prioritized investment in education and skills development to ensure a skilled workforce that can contribute to the growth of the technology sector. This has led to the establishment of numerous training programs and partnerships with international organizations, fostering a culture of innovation and knowledge-sharing in Rwanda.

One of the African countries most affected by the issue of governance and institutions is Nigeria. The government should prioritize implementing transparency and accountability measures to combat corruption and ensure that public funds are used for the benefit of the people. By doing so, Nigeria can attract more FDI, create job opportunities, and improve the overall well-being of its citizens. Strengthening the rule of law and promoting good governance will help to build a conducive business environment, fostering innovation and entrepreneurship in the country. These efforts will also enhance investor confidence and encourage domestic businesses to thrive, leading to economic growth and development. By promoting transparency and accountability, Nigeria can set a positive example for other countries in the region, inspiring them to adopt similar measures and collectively combat corruption on a larger scale.

Accordingly, the World Bank estimates that Nigeria will experience significant economic growth if it successfully addresses corruption and improves governance (World Bank, 2022). By implementing transparent and accountable practices, Nigeria can attract more FDI, stimulate job creation, and diversify its economy beyond oil dependency. This would ultimately lead to improved living standards and a brighter future for its citizens. Tackling corruption and improving governance would also enhance the country's reputation on the global stage, increasing its credibility and trustworthiness among international partners. This could open up opportunities for increased collaboration, trade, and partnerships with other countries, further boosting Nigeria's economic growth and development. Investing in education and technology would also play a crucial role in diversifying Nigeria's economy. By equipping its citizens with the necessary skills and knowledge, Nigeria can foster innovation and entrepreneurship, leading to the creation of new industries and job opportunities.

Improving infrastructure, such as transportation and energy systems, would attract foreign investment and facilitate the movement of goods and services within the country, further stimulating economic growth. Investing in research and development would enable Nigeria to stay at the forefront of technological advancements, making it a competitive player in the global market. This would not only attract foreign investors but also encourage domestic businesses to innovate and expand. Promoting a culture of lifelong learning and continuous skill development would ensure that Nigeria's workforce remains adaptable and capable of meeting the demands of a rapidly evolving economy. Investing in research and development would also foster collaboration between academia, industry, and government, leading to the creation of new knowledge and cutting-edge technologies. This would not only enhance

Nigeria's overall economic growth but also create a conducive environment for entrepreneurship and job creation.

2.3.5 Youth Unemployment

2.3.5.1 High Rate of Youth Unemployment in Africa

The high rate of youth unemployment in Africa can be attributed to several factors. Firstly, there is a lack of job opportunities and limited economic growth in many African countries. This leads to a scarcity of formal employment options for young people. The education system often fails to equip youth with the necessary skills and qualifications demanded by the job market, exacerbating the unemployment issue. The rapid population growth in Africa adds to the challenge of youth unemployment. With a large number of young people entering the job market each year, the competition for limited job opportunities becomes even more intense. The lack of access to financial resources and support systems further hinders young Africans from starting their own businesses or pursuing entrepreneurial ventures as an alternative to formal employment. This lack of access to resources and support systems creates a cycle of dependency on traditional employment, making it difficult for young Africans to break free from the constraints of the job market.

The limited availability of quality education and skills training exacerbates the problem, as many young people lack the necessary qualifications and knowledge to compete in today's global economy. The stigma surrounding failure and risk-taking in African societies often discourages young people from taking the leap into entrepreneurship. This fear of failure hinders innovation and prevents young Africans from exploring their full potential outside of traditional employment structures. The lack of investment in research and development further limits opportunities for young Africans to develop innovative solutions and contribute to economic growth. The limited access to capital and financial resources adds another barrier for young Africans aspiring to become entrepreneurs. Without adequate funding, it becomes challenging for them to start or expand their businesses, hindering their ability to create jobs and stimulate economic growth. The absence of supportive policies and regulatory frameworks in many African countries creates an unfavorable environment for entrepreneurship, making it difficult for young people to navigate the complexities of starting and running a business.

The African country with the lowest rate of youth unemployment from 2020 to 2023, according to World Bank data, is Ghana, which, with a youth unemployment rate of 12.5% in 2020, is projected to have the lowest rate of youth unemployment in Africa from 2020 to 2023 (World Bank, 2022). This can be attributed to the country's efforts to implement policies that promote entrepreneurship and provide support for young entrepreneurs, such as access to finance and business development programs. These initiatives have helped create a conducive environment for youth employment and economic growth in Ghana. The government has also focused on improving the quality of education and vocational training programs to equip young people with the necessary skills for employment and entrepreneurship. By investing in education and vocational training, the government aims to bridge the skills gap and ensure that young people have the knowledge and expertise needed to succeed in the job market.

The government has also encouraged collaboration between educational institutions and industry partners to align curriculum with industry needs, fostering a seamless transition from education to employment for young entrepreneurs.

According to World Bank statistics, the African country with the second-lowest rate of youth unemployment is Rwanda, with a youth unemployment rate of 13.1% as of 2020 (World Bank, 2022). This can be attributed to the government's efforts to promote entrepreneurship and implement policies that support youth employment. Despite this progress, there is still a need for more comprehensive strategies and investments to address the underlying challenges and create more opportunities for young entrepreneurs across the continent. Rwanda's success in reducing youth unemployment can also be attributed to its focus on skills development and vocational training programs, which equip young people with the necessary tools to succeed in the job market. The government has actively encouraged foreign investment and created a favorable business environment, further stimulating job creation and economic growth. Ongoing efforts are required to ensure sustainable and inclusive growth that benefits all segments of the population and reduces the overall unemployment rate in Africa. These efforts can include implementing policies that promote entrepreneurship and small business development, as well as investing in education and infrastructure to attract more industries and create a diverse range of job opportunities. Addressing issues such as gender inequality and regional disparities can also contribute to reducing unemployment and fostering inclusive growth in Africa.

2.3.5.2 Impact of Youth Unemployment on Economic Growth and Social Stability

Youth unemployment in Africa has a significant impact on both economic growth and social stability. Firstly, high levels of youth unemployment hinder economic development as it limits the potential workforce and reduces productivity. With a large portion of the population unable to find employment, there is a decrease in consumer spending, which directly affects businesses and overall economic growth. Youth unemployment can lead to social instability in Africa. When young people are unable to secure jobs and earn a living, they may become disillusioned and frustrated. This can lead to increased crime rates, social unrest, and political instability. The lack of opportunities for young people can also result in a "brain drain," as talented individuals may choose to migrate to countries with better employment prospects, further exacerbating the issue of youth unemployment in Africa. The economic burden of unemployment can strain government resources and hinder investments in crucial sectors such as education and healthcare. This perpetuates a cycle of poverty and limited opportunities, making it even more challenging for young people to break free from the cycle of unemployment and contribute to their communities' development. The lack of job opportunities for young people can lead to social unrest and political instability. When a large portion of the population is unable to find work, frustration and disillusionment can build up, potentially leading to protests or even violence. This not only hampers economic growth but also creates an unsafe environment for both individuals and businesses, further deterring potential investments and exacerbating the issue of youth unemployment.

The consequences of youth unemployment extend beyond immediate economic setbacks. The long-term effects can include a loss of skills and experience, as young people are unable to gain valuable work-related knowledge. This can perpetuate a cycle of unemployment and hinder future career prospects, ultimately impacting the overall productivity and competitiveness of a nation. The burden of supporting unemployed youth often falls on the government, straining public resources and diverting funds that could be allocated toward other pressing social issues. The lack of employment opportunities for young people can also lead to increased social and economic inequalities within a society. As a result, marginalized groups may face even greater barriers to accessing decent jobs and improving their living conditions. The negative impact on the mental health and well-being of unemployed youth should not be overlooked, as prolonged unemployment can lead to feelings of frustration, hopelessness, and low self-esteem.

South Africa has the highest percentage of youth unemployment in Africa, with over 60% of young people aged 15–24 unemployed in 2020, according to World Bank estimates (World Bank, 2022). The rate reached a startling 63.30% in the first quarter of 2021, exacerbating concerns about social stability and economic growth in the country (World Bank, 2022). This alarming trend highlights the urgent need for targeted interventions and policies to address youth unemployment and provide opportunities for this vulnerable population. The consequences of high youth unemployment extend beyond economic implications, as they can lead to increased social unrest and crime rates. Without adequate support and opportunities, young people may face long-term challenges in accessing education, acquiring skills, and securing stable employment. It is crucial for the government and stakeholders to collaborate in implementing effective strategies that prioritize job creation and skill development for the youth. By investing in youth employment initiatives, governments can not only address the immediate economic impact but also foster a more inclusive and equitable society. Providing mentorship programs, vocational training, and entrepreneurship opportunities can empower young individuals to contribute to their communities and build a brighter future. Promoting partnerships between educational institutions and industries can ensure that young people are equipped with the relevant skills needed in the evolving job market. By investing in these initiatives, governments can create a workforce that is adaptable and prepared for the future. By prioritizing inclusivity and equity in these programs, governments can bridge the gap between different socioeconomic backgrounds and promote equal opportunities for all individuals.

2.3.5.3 Strategies for Addressing Youth Unemployment

One potential strategy for addressing youth unemployment in Africa is to focus on improving access to quality education and vocational training. By investing in educational infrastructure and curriculum development, young people can acquire the skills and knowledge needed to enter the job market. It is crucial to address the gender gap in education and ensure equal opportunities for girls and women. By implementing policies that promote gender equality in education, such as providing scholarships and creating safe learning environments, Africa can empower its female population and contribute to overall economic growth. Collaboration with international organizations and partnerships with private sector entities can bring in additional resources and expertise to further enhance the quality of education and vocational training in Africa. The World Bank reports that between 2020 and 2023, Africa's literacy rates range from 62 to 79% (World Bank, 2022), highlighting the need for continued efforts to improve education. By investing in teacher training programs and modernizing educational infrastructure, Africa can ensure that its youth have access to quality education and are equipped with the skills needed for a competitive job market. Promoting digital literacy and incorporating technology into classrooms can help bridge the educational gap and prepare students for the digital age. By addressing the root causes of low literacy rates, such as poverty and gender inequality, Africa can further enhance its education system. Fostering partnerships with international organizations and implementing inclusive policies can contribute to sustainable development and long-term progress in education across the continent. These efforts can also lead to increased access to quality education for marginalized communities, including children with disabilities and those living in remote areas. Investing in teacher training programs and providing them with the necessary resources can empower educators to effectively utilize technology and innovative teaching methods, ultimately improving learning outcomes for students.

According to the World Bank, the African country with the highest literacy rates between 2020 and 2023 is Mauritius, with a projected literacy rate of 90% (World Bank, 2022). This achievement can be attributed to the government's commitment to investing in education and prioritizing inclusive policies. By focusing on improving access to education and supporting teachers, Mauritius has been able to make significant progress in ensuring that all children have the opportunity to develop essential literacy skills. In addition to government efforts, community involvement and partnerships with non-governmental organizations have also played a crucial role in promoting literacy in Mauritius. These collaborations have helped create innovative programs and initiatives that target marginalized groups, such as children from low-income families or those living in remote areas, ensuring that no one is left behind in the country's pursuit of high literacy rates. These programs and initiatives often include mobile libraries, after-school tutoring programs, and community reading events. By reaching out to these marginalized groups and providing them with accessible resources and support, Mauritius is working toward a more inclusive and equitable society where everyone has the opportunity to develop their literacy skills.

According to the World Bank, the African country with the lowest literacy rates between 2020 and 2023 is South Sudan, with an estimated literacy rate of only 27% (World Bank, 2022). This highlights the urgent need for similar collaborative efforts and investments in education to address the literacy challenges faced by marginalized communities in various African countries. Improving literacy rates is crucial for the overall development and empowerment of individuals, as it not only enhances their ability to access information and opportunities but also promotes social and economic progress at the national level. It is imperative for governments and organizations to prioritize literacy programs and initiatives that target underserved populations, ensuring that no one is left behind in the pursuit of education and knowledge. Investing in literacy programs can have a ripple effect on other areas of development, such as healthcare and poverty reduction. By equipping individuals with the ability to read and write, they are better equipped to understand health information and make informed decisions about their well-being. Literacy can also serve as a tool for economic empowerment, as it opens up doors for employment opportunities and entrepreneurship. By prioritizing literacy, African countries can pave the way for sustainable development and an improved quality of life.

Promoting entrepreneurship and creating an enabling environment for small businesses can also be effective strategies. This can include providing financial support, mentorship programs, and reducing bureaucratic barriers to starting a business for individuals and communities. By equipping people with the necessary skills and resources, they can not only secure employment but also create their own businesses, contributing to economic growth and poverty reduction. Fostering a culture of entrepreneurship can lead to innovation and the diversification of industries, ultimately driving economic development in African countries. In addition, promoting entrepreneurship can also address the issue of youth unemployment, which is a significant challenge in many African countries. By encouraging young people to pursue entrepreneurial ventures, they can become self-employed and create job opportunities for themselves and others. This can help alleviate the strain on the formal job market and provide alternative avenues for income generation. Fostering a culture of entrepreneurship can also attract foreign investment and stimulate economic growth through increased business activities and market competitiveness.

The African country with the highest rates of promoting entrepreneurship and creating an enabling environment for small and large-scale businesses between 2020 and 2023, according to the World Bank, is Rwanda (World Bank, 2022). The government of Rwanda has implemented various policies and initiatives to support entrepreneurship, such as streamlining business registration processes, providing access to finance for startups, and promoting innovation through technology hubs. These efforts have resulted in a thriving entrepreneurial ecosystem and have attracted both local and foreign investors, further boosting economic growth in the country. In addition to these measures, the government has also focused on improving infrastructure and investing in education and skills development to create a conducive environment for businesses. As a result, Rwanda has seen an increase in the number of successful startups and a rise in job opportunities, contributing to poverty reduction and overall socio-economic development. The government has implemented policies to promote innovation and entrepreneurship, providing support and resources for aspiring entrepreneurs. This has led to the emergence of vibrant startup incubators and accelerators, fostering a culture of innovation and technological advancement in Rwanda. These initiatives have not only created new business opportunities but have also attracted international attention, positioning Rwanda as a hub for innovation in Africa.

2.3.6 Climate Change and Environmental Sustainability

Climate change and environmental sustainability are two of the most pressing economic issues confronting Africa. These issues have significant implications for the continent's development and future. The impacts of climate change, such as extreme weather events and rising sea levels, pose a threat to Africa's agriculture, water resources, and overall ecosystem. Climate change exacerbates existing challenges such as poverty and food insecurity, as it disproportionately affects vulnerable populations in Africa. Without effective measures to mitigate and adapt to these changes, the continent's economic growth and social stability could be severely compromised. It is crucial for African nations to prioritize environmental sustainability and implement sustainable development strategies that address both climate change and economic development. By investing in renewable energy sources and promoting sustainable agriculture practices, African nations can reduce their greenhouse gas emissions and build resilience against the impacts of climate change. International cooperation and financial support are essential to assist African countries in their efforts to combat climate change and achieve sustainable development goals. These efforts can be further enhanced by implementing policies that encourage energy efficiency and conservation, as well as promoting the use of clean technologies. Capacity-building programs and knowledge-sharing initiatives can empower African nations to effectively implement and monitor their sustainable development strategies, ensuring long-term success in mitigating climate change and fostering economic growth.

According to the Intergovernmental Panel on Climate Change (IPCC), the region in Africa most affected by climate change is the Sahel region (IPCC, 2021), which is experiencing increased droughts and desertification. These impacts have severe consequences for agriculture, food security, and the livelihoods of millions of people in the region. This region includes countries such as Niger, Chad, Mali, Mauritania, and Sudan. The Sahel region is already facing significant challenges in terms of poverty and political instability, and climate change exacerbates these issues. It is crucial for international cooperation and support to be provided to these countries in order to build resilience and adapt to the changing climate conditions. Investing in sustainable agricultural practices and water management systems can help mitigate the impacts of climate change in the Sahel region and ensure a more sustainable future for its inhabitants. By addressing poverty and political instability, international cooperation can create a foundation for long-term development in the Sahel region. This can involve initiatives such as promoting economic growth, improving governance, and fostering social inclusion. Empowering local communities and implementing education programs can enhance their capacity to respond to climate change and build resilience from within. In addition, investing in renewable energy sources and promoting sustainable agricultural practices can contribute to reducing the region's vulnerability to climate change. By diversifying the economy and creating job opportunities, these measures can also help alleviate poverty and reduce dependence on natural resources. Ultimately, a comprehensive approach that combines social, economic, and environmental strategies is crucial for ensuring a more sustainable future for the Sahel region and its inhabitants.

One example of a successful African development program to alleviate the effects of climate change in the Sahel area is the Great Green Wall initiative. This program aims to combat desertification and land degradation by planting a belt of trees across the Sahel region, stretching from Senegal to Djibouti. By restoring degraded land and promoting sustainable land management practices, the Great Green Wall not only helps to mitigate the impacts of climate change but also provides economic opportunities for local communities through agroforestry and eco-tourism initiatives. The

Great Green Wall initiative plays a crucial role in improving food security in the Sahel region. The trees planted along the belt help to retain moisture in the soil, making it more fertile for agriculture and increasing crop yields. This not only ensures a stable food supply for local communities but also reduces their dependence on external sources of food aid. The Great Green Wall initiative promotes sustainable land management practices, such as terracing and water conservation techniques, which help to mitigate the effects of drought and desertification. These practices not only contribute to food security but also enhance the resilience of local communities against climate change impacts. By diversifying income sources and reducing vulnerability to external shocks, the initiative fosters long-term economic stability and self-sufficiency in the Sahel region. The Wall initiative encourages the participation of local communities in decision-making processes, ensuring that their traditional knowledge and expertise are valued and incorporated into sustainable land management practices. This approach not only empowers the communities but also strengthens their sense of ownership and responsibility toward the land, leading to more effective and lasting results in combating desertification and promoting environmental sustainability.

Promoting environmental sustainability is crucial for ensuring the long-term well-being of African communities and preserving their natural resources for future generations. By investing in renewable energy sources and adopting green technologies, African nations can reduce their dependence on fossil fuels and mitigate the negative impacts of climate change. Sustainable development practices can also create new job opportunities and stimulate economic growth in sectors such as renewable energy, agriculture, and eco-tourism. Implementing sustainable development practices can improve access to clean water and sanitation, which are essential for the health and well-being of African communities. By promoting sustainable agriculture techniques and supporting local farmers, African nations can enhance food security and reduce the vulnerability of their populations to hunger and malnutrition. Embracing sustainable development can also lead to the creation of green jobs and opportunities for entrepreneurship, empowering individuals and communities to become self-sufficient and economically stable. By conserving natural resources and preserving biodiversity, African nations can protect their unique ecosystems and contribute to global efforts to mitigate climate change.

Agencies promoting environmental sustainability in Africa include the African Union Commission, which has established the African Ministerial Conference on the Environment (AMCEN) to coordinate and promote sustainable development initiatives across the continent. Another important agency is the United Nations Environment Programme (UNEP), which provides technical assistance and capacity-building support to African countries in their efforts to achieve environmental sustainability (IEA and UNEP, 2018). These agencies work together with local governments, NGOs, and other stakeholders to implement policies and programs that address environmental challenges while promoting economic growth and social well-being in Africa. In addition to these agencies, the African Development Bank (AfDB) plays a crucial role in supporting sustainable development in Africa. The AfDB provides financial resources and expertise to fund projects that focus on renewable energy, climate change adaptation, and natural resource management. By collaborating with regional governments and stakeholders, the AfDB aims to create a more resilient and environmentally conscious Africa. Together, these agencies form a strong network of support for the continent's sustainable development goals.

One example of a successful development initiative in Africa is the AfDB's Desert to Power program, which aims to provide clean energy access to 250 million people in the Sahel region (African Development Bank, 2022). This program not only contributes to environmental sustainability but also helps alleviate poverty and improve living conditions in the region. These agencies also prioritize the conservation of natural resources and biodiversity through initiatives such as protected area management and sustainable agriculture practices. These initiatives not only protect the unique ecosystems and wildlife of Africa but also promote sustainable economic development by ensuring the long-term availability of resources for local communities. By integrating conservation efforts with poverty alleviation strategies, these agencies are working toward a more balanced and prosperous future for the region. These agencies also collaborate with local communities to raise awareness about the importance of conservation and provide training and support for sustainable livelihoods. Through these partnerships, they empower communities to actively participate in the preservation of natural resources and biodiversity, fostering a sense of ownership and responsibility toward their environment. Ultimately, these efforts contribute to the overall well-being and resilience of both the ecosystems and the people who depend on them.

2.3.7 Rising Indebtedness in Africa

Rising indebtedness in Africa has become a growing concern in recent years. This increase in debt levels has been driven by a combination of factors, including infrastructure development projects, economic downturns, and borrowing from international financial institutions. As a result, many African countries are now facing the challenge of servicing their debts while also meeting the needs of their populations for essential services such as healthcare and education. Rising indebtedness can also limit the ability of these countries to attract foreign investment and achieve sustainable economic growth. The burden of debt servicing can lead to reduced government spending on social welfare programs, exacerbating poverty and inequality within these countries. The high levels of debt can also increase vulnerability to external shocks and economic instability, making it harder for African nations to recover from crises and achieve long-term development goals. The heavy reliance on borrowing can crowd out domestic investment as governments prioritize debt repayment over funding critical infrastructure projects and productive sectors of the economy. This can hinder job creation and hinder the diversification of African economies, further perpetuating their dependence on a few key industries or commodities. The inability to effectively manage and service debt can damage a country's credit rating, making it more expensive and difficult to access international capital markets in the future.

According to World Bank data on the continent's debt load, the average debt-to-GDP ratio in Africa is predicted to climb from 63 to 70% between 2020 and 2023 (World Bank, 2022). Although the total amount of foreign debt is expected to climb by 7.9% each year to an incredible $700 billion by 2023 (World Bank, 2022), this rising debt load not only significantly strains government budgets but also limits cash available for important public services such as healthcare and education. This increasing debt burden also poses a risk to economic stability and hampers the ability of African countries to invest in infrastructure development and poverty reduction programs. It may make it more challenging for these countries to attract foreign investment and access international capital markets in the future, potentially hindering their economic growth prospects. The high levels of debt can lead to a decrease in government spending on social welfare programs, further exacerbating inequality and poverty within these countries. The reliance on external borrowing may result in a loss of sovereignty as African nations become more dependent on foreign lenders and are subject to their terms and conditions. The burden of servicing these debts can divert funds away from crucial sectors such as education and healthcare, impeding the development of human capital and overall societal progress. The volatility in global financial markets can expose African nations to sudden shifts in interest rates or currency fluctuations, making it even more challenging for them to manage their debt obligations effectively.

Statistics from the World Bank show that one of the African nations most impacted by debt loads is Zambia (World Bank, 2022), where debt levels have surged in recent years to an unbelievable amount of over 100% of its GDP. This has severely constrained the government's ability to invest in essential sectors such as education and infrastructure, hindering the country's long-term economic growth prospects. The high debt burden also puts Zambia at a higher risk of defaulting on its loans, further damaging its credibility in international markets and making it harder to attract foreign investment. The country's high debt-to-GDP ratio has led to a decrease in public investment and an increase in borrowing costs, hindering economic growth and exacerbating poverty levels. The burden of debt servicing has limited the government's ability to allocate resources toward crucial social programs, further deepening inequality and hampering development efforts. The country's reliance on external financing has left it vulnerable to volatile global market conditions, increasing the risk of financial instability. This has resulted in a lack of confidence from investors and a decrease in FDI, further constraining the country's economic prospects and potential for long-term sustainable development.

One of the African countries least affected by debt loads, according to World Bank statistics, is Rwanda. Despite being one of the poorest countries in the world (World Bank, 2022), Rwanda has managed to maintain a relatively low debt-to-GDP ratio through prudent fiscal management and strategic investments. This has allowed the government to prioritize social programs and infrastructure development, leading to significant improvements in healthcare, education, and overall living standards for its citizens. Rwanda has actively sought international aid and partnerships to support its development goals. By fostering strong relationships with donor countries and organizations, the government has been able to secure funding for key projects and initiatives. This approach has not only helped Rwanda reduce its reliance on debt but has also fostered sustainable growth and economic stability in the long run. Rwanda has implemented innovative policies and programs to address specific challenges in healthcare and education. For instance, the government has invested heavily in building a robust healthcare infrastructure, including the construction of modern hospitals and clinics across the country. This has significantly improved access to quality healthcare services for its citizens. Similarly, Rwanda has prioritized education by implementing initiatives such as free primary education and vocational training programs, ensuring that its citizens have the necessary skills for economic growth and development. The government has also focused on improving healthcare outcomes by investing in medical research and development, leading to advancements in treatments and technologies. This has further enhanced the overall quality of healthcare services provided to citizens. Similarly, Rwanda's commitment to education is evident through its efforts to promote gender equality in schools and provide scholarships for higher education, enabling more individuals to pursue advanced degrees and contribute to the country's progress.

2.3.8 African Effects of the COVID-19 Pandemic and Russia's Invasion of Ukraine

African effects of the COVID-19 pandemic and Russia's invasion of Ukraine have had significant global implications. The COVID-19 pandemic has hit African countries particularly hard, exacerbating existing health and economic challenges. The lack of adequate healthcare infrastructure and limited access to vaccines have further hindered efforts to control the spread of the virus in the region. The pandemic has disproportionately affected vulnerable populations in Africa, such as those living in poverty or with underlying health conditions. This has led to a higher mortality rate and increased strain on already stretched healthcare systems. The economic impact of the pandemic has been severe, with many African countries experiencing a decline in GDP and a rise in unemployment rates. These challenges have highlighted the urgent need for international support and collaboration to address the health and economic crises in Africa. The lack of access to healthcare resources and infrastructure in many African countries has further exacerbated the impact of the pandemic. The disruption of global supply chains has made it difficult for African nations to procure essential medical supplies and equipment, hindering their ability to effectively respond to the crisis.

The World Bank reports that the nation in Africa most impacted by the COVID-19 pandemic is South Africa (World Bank, 2022). The country has experienced a significant increase in COVID-19 cases and has struggled to contain the spread of the virus due to various factors, such as high population density and limited healthcare capacity. The economic impact has also been severe, with many businesses closing down and unemployment rates skyrocketing. South Africa's pre-existing socio-economic inequalities have exacerbated the effects of the pandemic, as marginalized communities face greater challenges in accessing healthcare and adhering to preventive measures. The government has implemented various measures to mitigate the impact, including financial relief packages and restrictions on movement, but the road to recovery remains challenging as the country grapples with both health and economic crises simultaneously. The pandemic has also exposed the underlying structural issues in South Africa's healthcare system, such as a shortage of healthcare workers and inadequate infrastructure. These challenges have further strained the country's ability to effectively respond to the crisis and provide adequate care to those in need. Despite these obstacles, the government is working toward long-term solutions to address these systemic issues and ensure a more resilient healthcare system for future emergencies.

As per the World Bank, the nation in Africa that has experienced the least impact from the COVID-19 pandemic is Rwanda (World Bank, 2022). The country's proactive approach to implementing strict lockdown measures and widespread testing has helped to contain the spread of the virus. Rwanda's strong healthcare system and effective communication strategies have played a crucial role in minimizing the impact on its population. Despite these efforts, the long-term effects on the economy and vulnerable communities cannot be overlooked, highlighting the need for continued support and resilience in the face of ongoing challenges. Rwanda's proactive approach to contact tracing and quarantine measures has also contributed to the containment of the virus. The government's swift response and ability to adapt to changing circumstances have been instrumental in preventing a surge in cases and protecting its citizens. It is important to recognize that the road to recovery will require sustained efforts to rebuild the economy and address the needs of marginalized groups who have been disproportionately affected by the pandemic.

Russia's invasion of Ukraine has sparked international tensions and raised concerns about territorial integrity and sovereignty. The conflict has led to a humanitarian crisis, with thousands of people displaced and an escalation in violence that threatens regional stability. The international community has called for diplomatic negotiations and sanctions to de-escalate the situation and restore peace in the region. The ongoing conflict also underscores the importance of upholding international laws and norms to prevent further aggression and protect the sovereignty of nations. This has impacted the African economy in a number of ways, including a decrease in foreign investment and trade, as investors and businesses are hesitant to engage with a region experiencing such instability. The influx of displaced people has put a strain on local resources and infrastructure, further hindering economic development. The humanitarian crisis has also led to increased spending on relief efforts, diverting funds that could have been allocated toward development projects. The conflict has had detrimental effects on the African economy, hindering its growth and potential for prosperity.

Given that Russia and Ukraine provide food supplies to Africa, including grains, vegetables, and meat products, the ongoing conflict between the two countries has disrupted these crucial supply chains. This has resulted in food shortages and price hikes in African markets, exacerbating the already dire situation. The instability caused by the conflict has deterred foreign investors from engaging in trade with African nations, further impeding economic growth and prosperity in the region. The disruption of food supplies from Russia and Ukraine has forced African countries to rely on more expensive imports from other regions, putting a strain on their already limited resources. The conflict has also hindered the development of agricultural infrastructure in Africa, making it even more difficult for the region to achieve food security and self-sufficiency in the long term. This has led to increased vulnerability to food shortages and malnutrition among African populations. The ongoing conflict has created a hostile environment for foreign investment in the region, deterring potential investors from contributing to the development of African agriculture and exacerbating the economic challenges faced by these nations.

The World Bank predicts that as a result of the disruption of food supplies from Russia and Ukraine to African countries (World Bank, 2022), food prices in Africa will continue to rise, exacerbating the already high levels of food insecurity and poverty in the region. This could lead to increased malnutrition and health issues among vulnerable populations, further perpetuating the cycle of poverty and dependence on foreign aid. In order to address this challenge, it is crucial for African countries to invest in agricultural development and infrastructure, promote sustainable farming practices, and enhance local food production. Efforts should be made to diversify food sources by exploring alternative crops and improving irrigation systems to mitigate the impact of climate change on agricultural productivity. By investing in agricultural development and infrastructure, African countries can improve their food security and reduce their reliance on foreign aid. This can be achieved by providing farmers with access to modern farming techniques, technologies, and resources. Promoting sustainable farming practices will not only protect the environment but also ensure long-term productivity and resilience in the face of changing climate conditions.

According to World Bank data, the African country most affected by food supply disruptions caused by the Russia–Ukraine conflict is Uganda (World Bank, 2022). The conflict has led to a decrease in agricultural exports, which has had a significant impact on the country's food security. In order to address this issue, Uganda should focus on increasing domestic food production and improving storage and distribution systems to ensure a stable food supply for its population. Uganda could also explore diversifying its agricultural sector by promoting the cultivation of alternative crops that are less affected by global conflicts. This would not only help mitigate the impact of future disruptions but also provide opportunities for economic growth and resilience in the face of external shocks. Investing in agricultural research and development can lead to the adoption of innovative technologies and practices that can enhance productivity and improve the overall efficiency of the sector. These advancements can also contribute to sustainable farming practices, such as reducing water usage and minimizing the reliance on chemical inputs. Investing in agricultural education and training programs can empower farmers with the knowledge and skills needed to adapt to changing market demands and navigate potential challenges in the sector.

In terms of food supply disruptions brought on by the conflict between Russia and Ukraine, the African nation least affected, according to World Bank data, is Ethiopia. This is due to its diverse agricultural practices and self-sufficiency in food production. The country has made significant investments in irrigation systems and crop diversification, allowing it to withstand external shocks and maintain a stable food supply. Ethiopia has implemented policies to promote sustainable farming practices and improve agricultural productivity. The government has also prioritized the development of rural infrastructure, such as roads and storage facilities, to enhance the efficiency of food distribution within the country. These measures have contributed to Ethiopia's resilience in the face of potential food supply disruptions caused by conflicts in other regions. Ethiopia has also invested in research and development to enhance crop varieties and introduce new technologies that can withstand climate change impacts. This proactive approach has helped the country adapt to changing environmental conditions and ensure a consistent food supply for its population.

Ethiopia has actively engaged in regional partnerships and international collaborations to strengthen its food security efforts, including sharing best practices and knowledge exchange with other countries facing similar challenges.

2.4 GLOBAL ECONOMIC CHALLENGES AND AFRICA

2.4.1 Impact of Globalization on African Economies

Globalization has had both positive and negative impacts on African economies. On one hand, it has opened up new opportunities for trade and investment, leading to increased economic growth and development in some countries. On the other hand, African economies have also faced challenges such as increased competition from foreign companies, unequal distribution of benefits, and vulnerability to global economic shocks. It is important for African countries to effectively navigate these challenges and leverage the opportunities presented by globalization to ensure sustainable and inclusive economic growth. One way African countries can navigate these challenges is by implementing policies that promote domestic industries and protect local businesses from unfair competition. This can help create a more level playing field and ensure that the benefits of globalization are shared more equitably. Investing in education and skills development can help African economies adapt to the changing demands of the global market and increase their competitiveness on a global scale. Fostering innovation and entrepreneurship can also play a crucial role in driving economic growth in African countries. By encouraging the development of new ideas and supporting small businesses, governments can stimulate job creation and attract foreign investment. Strengthening regional cooperation and trade agreements can enhance market access for African countries, allowing them to tap into larger consumer bases and diversify their export markets. Investing in infrastructure and improving connectivity within and between African countries can facilitate trade and boost economic development. This includes developing transportation networks, such as roads, railways, and ports, as well as expanding access to reliable electricity and internet services. These efforts can help overcome logistical challenges and enable African countries to fully participate in global value chains, attracting more FDI and driving economic growth.

African nations that took advantage of the new trade and investment opportunities created by globalization to experience greater economic growth and development include Rwanda, which implemented various reforms to improve its business environment and attract foreign investment. The country has seen significant improvements in infrastructure, such as the construction of new roads and the expansion of its airport, which have facilitated trade and connectivity. Rwanda has also focused on developing its human capital by investing in education and skills training programs. This has resulted in a more skilled workforce, which has attracted multinational companies to set up operations in the country. As a result, Rwanda's economy has experienced sustained growth and diversification, leading to improved living standards for its citizens. Rwanda has implemented policies to promote entrepreneurship and innovation, fostering the growth of SMEs. This has created more job opportunities and contributed to the overall economic development of the country. The government has prioritized sustainable development initiatives, such as renewable energy projects and conservation efforts, which have not only benefited the environment but also attracted foreign investment in these sectors.

According to World Bank statistics, globalization is expected to increase new trade and investment opportunities as well as economic growth and development in Rwanda between 2020 and 2023 by a percentage of 10 to 15% (World Bank, 2022). This is due to the government's efforts to improve infrastructure, promote business-friendly policies, and invest in human capital development. These measures have attracted foreign investors and stimulated domestic entrepreneurship, leading to increased job creation and a more diversified economy. Rwanda's strategic location within the East African Community provides access to a larger market and enhances its competitiveness as a regional hub for trade and investment. Rwanda has also implemented various initiatives to promote innovation and technology adoption, further boosting its economic growth. The government has established innovation hubs and incubators, encouraging the development of startups and fostering a culture of entrepreneurship. These efforts have not only attracted FDI in the technology sector but have also helped local businesses adopt new technologies and improve their productivity. As a result, Rwanda has witnessed significant advancements in sectors such as fintech, e-commerce, and digital services, contributing to its overall economic success. The government's support for innovation has led to the creation of job opportunities for the country's youth and reduced unemployment rates. The advancements in technology have enhanced access to financial services for previously underserved populations, promoting financial inclusion and reducing poverty levels in Rwanda.

Countries like Ethiopia have also made strides in attracting foreign investment by investing in industrial parks and promoting export-oriented industries. These efforts have resulted in increased job opportunities and economic diversification. The government has implemented policies to improve the ease of doing business, such as streamlining bureaucratic processes and reducing red tape. This has created a more favorable environment for foreign investors, leading to a surge in FDI in recent years. As a result, Ethiopia has experienced sustained economic growth and is becoming an increasingly attractive destination for foreign businesses looking to expand their operations. In addition to attracting foreign investment, Ethiopia has also focused on developing its domestic industries and promoting entrepreneurship. The government has launched initiatives to support SMEs, providing them with access to funding and training programs. This has led to the growth of a vibrant private sector, contributing to job creation and overall economic development. With its favorable business environment and diverse economy, Ethiopia is poised to continue its upward trajectory in the global market. In addition to supporting SMEs, the Ethiopian government has implemented policies to attract FDI. These policies include tax incentives and streamlined bureaucratic processes, making it easier for foreign companies to set up operations in the country. As a result, Ethiopia has seen an increase in foreign investment, further boosting its economic growth and international competitiveness.

According to World Bank statistics, globalization is expected to increase new trade and investment opportunities as well as economic growth and development in Ethiopia between 2020 and 2023 by a percentage of 10% (World Bank, 2022). This is due to the country's efforts to diversify its economy and attract FDI. Ethiopia's strategic location and access to regional markets through trade agreements, such as the African Continental Free Trade Area, contribute to its potential for increased globalization benefits. Ethiopia's investment in infrastructure development, such as transportation and telecommunications, enhances its connectivity and ability to participate in global trade networks. This, coupled with the government's commitment to implementing business-friendly policies and reforms, creates a favorable environment for foreign investors looking to tap into Ethiopia's growing market. Ethiopia's strategic location in the Horn of Africa also plays a crucial role in its potential for increased globalization benefits. Being a gateway to other African countries, it offers opportunities for trade and investment not only within the continent but also with international markets. Ethiopia's young and rapidly growing population provides a large consumer base, further attracting foreign investors seeking to capitalize on the country's expanding market potential.

2.4.2 Challenges of Trade Imbalances and Unequal Partnerships

One of the major challenges faced by Africa in the global economy is trade imbalances. Many African countries struggle to compete with more developed nations, resulting in a significant trade deficit. This imbalance often leads to a heavy reliance on imports and a limited ability to export goods and services. Unequal partnerships also pose a challenge for Africa in the global economy. Many international agreements and trade deals tend to favor developed nations, leaving African countries at a disadvantage. This can hinder their economic growth and perpetuate dependency on foreign aid and assistance. The lack of infrastructure and technological advancements in many African countries further exacerbates the challenges they face in competing globally. Without adequate resources and support, it becomes difficult for these nations to diversify their economies and attract foreign investment. This cycle of dependency can hinder long-term sustainable development and hinder Africa's ability to fully participate in the global economy. The limited access to quality education and healthcare services in Africa contributes to the challenges faced by these nations. The lack of a skilled workforce and a healthy population hinders their ability to innovate and increase productivity, further perpetuating their dependency on foreign aid.

Addressing these underlying issues is crucial for Africa to break free from the cycle of dependency and achieve sustainable economic growth. Investing in education and healthcare infrastructure is essential to equipping the African population with the necessary skills and good health to drive economic development. By improving access to quality education and healthcare services, African nations can nurture a skilled workforce and a healthy population, fostering innovation and productivity. This, in turn, will reduce their reliance on foreign aid and pave the way for self-sufficiency and sustainable economic growth in Africa. Investing in education and healthcare infrastructure can also contribute to reducing poverty and inequality in African countries. By providing equal opportunities for education and healthcare, marginalized communities can have a chance to improve their socio-economic status and participate actively in the development of their nations. Ultimately, this will lead to a more inclusive and prosperous society for all Africans.

According to the World Bank, the African country most affected by trade imbalances and unequal partnerships is Nigeria (World Bank, 2022). Despite being one of the largest economies in Africa, Nigeria faces significant challenges in terms of infrastructure development and the equitable distribution of resources. These imbalances have hindered the country's ability to effectively address poverty and inequality, leading to a widening gap between the rich and the poor. With targeted investments in infrastructure, Nigeria has the potential to unlock its economic potential and create opportunities for all its citizens. Improving infrastructure would not only enhance connectivity and facilitate economic growth but also improve access to basic services such as healthcare and education. By promoting inclusive policies and ensuring equitable resource allocation, Nigeria can work toward reducing poverty and narrowing the wealth gap, ultimately fostering a more prosperous and equitable society. Investing in renewable energy sources and sustainable practices can help Nigeria mitigate the effects of climate change and reduce its reliance on fossil fuels. This would not only benefit the environment but also create new job opportunities in the green sector. By prioritizing these initiatives, Nigeria can position itself as a leader in sustainable development and attract foreign investment, further boosting its economic potential.

2.4.3 Effects of Economic Crises and Fluctuations in Global Markets

The effects of economic crises and fluctuations in global markets on the African continent can be significant and wide-ranging. These events can lead to a decline in foreign investment, reduced trade opportunities, and increased unemployment rates. They can exacerbate existing social and economic inequalities, hinder poverty reduction efforts, and negatively impact government budgets, making it difficult to fund essential services such as healthcare and education. Economic crises and market fluctuations can also strain the agricultural sector, which is a vital source of income for many African countries. This can result in food insecurity and further deepen poverty levels in already vulnerable communities. The lack of economic stability may discourage foreign aid and assistance, making it even more challenging for African nations to recover from these setbacks and achieve sustainable development goals. The lack of access to proper infrastructure and technology in rural areas can hinder agricultural productivity and limit the ability of farmers to adapt to changing environmental conditions. This further exacerbates the challenges faced by African countries in ensuring food security and economic growth. The limited availability of healthcare and education services in these regions can perpetuate a cycle of poverty and hinder human capital development, making it difficult for individuals to escape poverty and contribute to their country's overall development. Inadequate infrastructure, such as roads and transportation systems, can impede the distribution of goods and services, hindering economic growth and development. Political instability and corruption in some African countries can create an uncertain business environment, discouraging foreign investment and further limiting opportunities for economic growth. Limited access to quality education and healthcare in many African countries can perpetuate the cycle of poverty and hinder human capital development. The lack of diversification in economies heavily reliant on a single sector, such as agriculture or natural resources, can make them vulnerable to external shocks and fluctuations in global markets.

According to some international agencies like the World Bank and the International Monetary Fund (IMF), these challenges can be addressed through targeted policies and investments in infrastructure, education, and healthcare (World Bank, 2022). By improving access to quality education and healthcare, African countries can enhance their human capital and productivity, leading to increased economic growth. Promoting diversification in economies can help reduce reliance on a single sector and make them more resilient to external shocks. Fostering innovation and technological advancements can drive economic transformation and create new opportunities for African countries. This can be achieved through investments in research and development, as well as promoting entrepreneurship and creating a conducive environment for innovation. Strengthening regional integration and trade partnerships can enhance market access and promote economic cooperation among African nations, fostering sustainable growth and development. Building strong digital infrastructure and improving access to affordable internet can also play a crucial role in driving innovation and technological advancements in Africa. This will enable more individuals and businesses to connect, collaborate, and leverage digital tools for innovation and economic growth. Investing in education and skills development programs can empower the African workforce with the necessary knowledge and expertise to thrive in a rapidly evolving technological landscape. By equipping individuals with digital literacy skills, they can effectively utilize the available resources and contribute to the overall development of the continent. Fostering partnerships between local businesses and international technology companies can further enhance knowledge transfer and create opportunities for African entrepreneurs to scale their innovations globally.

2.5 INNOVATIVE SOLUTIONS FOR AFRICA'S ECONOMIC CHALLENGES

2.5.1 Importance of Entrepreneurship and Innovation in Driving Economic Growth

Entrepreneurship and innovation play a pivotal role in fueling economic growth in Africa. By fostering a culture of entrepreneurship, individuals are empowered to create their own businesses, generate employment opportunities, and contribute to the overall development of their communities. Innovation drives economic growth by introducing new products, services, and technologies that address pressing challenges and improve the quality of life for African citizens. These factors not only stimulate economic activity but also attract foreign investment and foster sustainable development across the continent. When individuals are empowered to create their own businesses, it leads to increased competition in the market, which ultimately benefits consumers through lower prices and improved product quality. The promotion of entrepreneurship and innovation can also lead to the diversification of industries, reducing reliance on a single sector and making economies more resilient to external shocks. The promotion of entrepreneurship and innovation can also contribute to job creation, as new businesses often require a skilled workforce. This can help reduce unemployment rates and provide individuals with greater economic opportunities. By encouraging entrepreneurship, countries can tap into the potential of their own citizens and harness their creativity and talent, driving overall economic growth and prosperity. Fostering entrepreneurship and innovation can lead to the development of new industries and technologies, which can further diversify and strengthen the economy. This can attract foreign investment and create a competitive advantage in the global market.

Promoting entrepreneurship can also foster a culture of self-reliance and independence, empowering individuals to take control of their own economic destiny. Supporting entrepreneurship can lead to job creation and reduce unemployment rates, as new businesses often require a workforce to support their operations. This can have a positive impact on the overall standard of living and quality of life for individuals within a society. Fostering entrepreneurship can encourage creativity and problem-solving skills, as entrepreneurs are often faced with unique challenges that require innovative solutions. This can contribute to the overall development and advancement of society as a whole. Promoting entrepreneurship can also lead to the development of new industries and technologies, driving economic growth and competitiveness on a global scale. By encouraging individuals to take risks and pursue their innovative ideas, societies can benefit from increased productivity and a more diverse range of products and services.

According to the WIPO, entrepreneurship plays a crucial role in promoting technological innovation and creating jobs (Bannerman, 2020). It enables countries to harness their human capital and resources effectively, leading to increased productivity and economic prosperity. Fostering entrepreneurship can also address social challenges by providing innovative solutions and empowering marginalized communities. In Africa, the WIPO predicts that between 2020 and 2023, the number of entrepreneurs is expected to increase significantly (Bannerman, 2020), which presents a great opportunity for economic growth and poverty reduction. By supporting and investing in entrepreneurship, African countries can unlock their full potential and drive sustainable development in the region. This can be achieved through policies that promote entrepreneurship education, access to finance, and the creation of a conducive business environment. According to statistics from the World Bank, the percentage rise in entrepreneurship in African nations ranges between 10 and 15% annually (World Bank, 2022). This demonstrates the strong entrepreneurial spirit and potential for innovation in the region. It also indicates a growing interest and willingness among Africans to start their own businesses and contribute to the economy. It is important for governments and stakeholders to continue fostering an enabling environment that encourages entrepreneurship, as well as providing the necessary support and resources for aspiring entrepreneurs to thrive. This growth in entrepreneurship is particularly evident in sectors such as technology and e-commerce, where African startups are making significant strides. The rise of mobile technology and internet connectivity has also played a crucial role in enabling entrepreneurship, as it has opened up new opportunities for businesses to reach customers and markets both locally and globally.

2.5.2 Role of Technology and Digitalization in Transforming African Economies

The role of technology and digitalization in transforming African economies cannot be overstated. These advancements have the potential to revolutionize various sectors, such as agriculture, finance, healthcare, and education. By embracing technology and digital solutions, African countries can leapfrog traditional development stages and accelerate economic growth. These innovations can promote inclusivity by bridging the digital divide and creating opportunities for marginalized communities to participate in the digital economy. The adoption of technology in African economies can also enhance efficiency and productivity, leading to improved competitiveness in the global market. Digital solutions can streamline supply chains, facilitate online transactions, and enable remote work, ultimately driving economic progress and attracting foreign investment. By investing in digital infrastructure and skills development, African countries can empower their citizens with the tools and knowledge necessary to thrive in the digital era. These investments can also foster innovation and entrepreneurship, as individuals and businesses are equipped with the resources to develop and scale new technologies and solutions.

The integration of digital technologies can help bridge the digital divide within African societies, ensuring that all citizens have equal access to opportunities and resources in the digital era. By investing in digital infrastructure and providing access to affordable internet, countries can create a level playing field for their citizens, regardless of their geographical location or socio-economic background. This can lead to increased participation in the digital economy, improved education and healthcare services, and overall societal development. The integration of digital technologies can enhance governance and transparency, enabling governments to better serve their citizens and foster trust in public institutions. Affordable internet access can also stimulate entrepreneurship and innovation, as it provides individuals with the tools and resources necessary to start their own businesses and contribute to economic growth. It can also facilitate communication and collaboration among different sectors of society, fostering social cohesion and inclusivity.

According to the IMF, the role of technology and digitalization in transforming African economies is undeniable (IMF, 2020). It has the potential to drive innovation, increase productivity, and create new job opportunities. Digitalization can enhance financial inclusion by providing access to banking and financial services for previously underserved populations. The organization predicts that African economies that embrace digitalization could experience a significant boost in GDP growth, with estimates ranging from 1 to 5% per year. This growth can be attributed to the increased efficiency and effectiveness of digital technologies in various sectors, such as agriculture, healthcare, and education. The IMF emphasizes the importance of investing in digital infrastructure and skills development to fully harness the benefits of digitalization and ensure inclusive growth for all African countries. According to the IMF, the African country that has benefited the most from digitalization and technology is Kenya (IMF, 2020). Kenya has made significant strides in digital innovation, particularly in the mobile money sector, with the introduction of M-Pesa. This has revolutionized financial inclusion and enabled millions of Kenyans to access banking services through their mobile phones. Kenya has also seen advancements in e-commerce and digital entrepreneurship, creating new opportunities for economic growth and job creation.

The Organization for Economic Cooperation and Development (OECD) predicts that African economies that embrace digitalization could experience a significant boost in productivity and economic growth (Woodward, 2022). This is supported by evidence from other regions, such as Asia, where digitalization has played a crucial role in driving innovation and increasing competitiveness. Therefore, it is crucial for African countries to prioritize digital transformation and create an enabling environment for businesses to adopt and leverage digital technologies. By doing so, African countries can unlock their full potential and overcome existing barriers to economic development. Embracing digitalization can also lead to job creation and the development of new industries, further stimulating economic growth in the region. According to the OECD, the African countries that have benefited the least from digitalization and technology include Zimbabwe, Ethiopia, and Chad (Woodward, 2022). These countries face significant challenges in terms of limited access to the internet and technology infrastructure, as well as low digital literacy rates among their populations. With targeted investments in these areas and the implementation of policies that promote digital inclusion, these countries have the potential to leapfrog traditional development pathways and harness the power of digitalization for sustainable economic growth. Fostering partnerships with international organizations and neighboring countries that have made strides in digitalization can provide valuable knowledge-sharing opportunities and support for capacity-building initiatives in these African nations.

For instance, Chad can learn from countries like Kenya and Rwanda, which have successfully implemented digital strategies to drive economic growth and improve public services. By leveraging their experiences and expertise, Chad can accelerate its own digital transformation journey and avoid potential pitfalls. Collaboration with regional partners can lead to the development of cross-border digital infrastructure, enabling seamless connectivity and trade opportunities that further enhance economic development in the region. Ethiopia can learn from the experiences of countries like Kenya and Rwanda, which have successfully implemented digital inclusion strategies and seen significant economic benefits as a result. By studying their best practices and adapting them to suit their own context, Ethiopia can accelerate its digital transformation and bridge the digital divide within its population. Collaborating with international organizations such as the World Bank and the United Nations can provide access to funding and expertise to further support Ethiopia's digital inclusion efforts (World Bank, 2022). Zimbabwe can learn from countries like Kenya and Rwanda, which have successfully implemented digital inclusion strategies and have seen significant improvements in their economies as a result. By studying their best practices and adapting them to the Zimbabwean context, the country can accelerate its own digital transformation and reap the benefits of increased connectivity and access to digital services. Forging partnerships with regional organizations such as the African Union can help Zimbabwe tap into a wider network of expertise and resources, further enhancing its digitalization efforts.

2.6 TECHNOLOGICAL ADVANCEMENTS IN AFRICA

2.6.1 Digital Divide and the Need for Improved Access to Technology and Internet Connectivity

The digital divide in Africa refers to the gap between those who have access to technology and internet connectivity and those who do not. This divide is often influenced by factors such as economic disparities, infrastructure limitations, and educational opportunities. Bridging this gap is crucial, as it can unlock immense potential for economic growth, innovation, and social development on the continent. By investing in improved access to technology and internet connectivity, African countries can empower their citizens with the tools and resources necessary to participate fully in the digital age and leverage the opportunities it presents. This can lead to increased job opportunities, entrepreneurship, and access to global markets. It can enhance education and healthcare services, promote cultural exchange, and facilitate communication and collaboration among individuals and communities across Africa.

By investing in technology and internet infrastructure, African countries can bridge the digital divide and ensure that all citizens have equal access to information and opportunities. This can also foster innovation and creativity, as individuals can tap into a wealth of knowledge and connect with experts from around the world. Embracing digital technologies can help African countries address various societal challenges, such as poverty alleviation, climate change mitigation, and sustainable development. By leveraging digital technologies, African countries can create new economic opportunities and job prospects. This can lead to increased entrepreneurship and economic growth, ultimately improving the overall standard of living for citizens. The adoption of digital technologies can enhance the delivery of essential services such as healthcare and education, making them more accessible and efficient for all. Digital technologies can also facilitate the collection and analysis of data, enabling African countries to make informed decisions and policies regarding climate change adaptation and mitigation strategies. This can help in identifying vulnerable areas and implementing effective measures to reduce the impact of climate change on communities and ecosystems.

The use of digital platforms can promote knowledge sharing and collaboration among African countries, fostering regional cooperation in addressing common challenges related to climate change and sustainable development. The integration of climate change considerations into national development plans can ensure that adaptation and mitigation strategies are mainstreamed across sectors such as agriculture, energy, and infrastructure. This approach can enhance the resilience of African countries to climate change impacts and contribute to their long-term sustainable development goals. International partnerships and financial support can play a crucial role in supporting African countries' efforts to implement effective climate change adaptation and mitigation measures.

According to the WIPO, Africa's economy can be expanded simply by leveraging its intellectual property (IP) assets (Bannerman, 2020). By promoting innovation and protecting IP rights, African countries can attract foreign investment, foster technological advancements, and stimulate economic growth. WIPO suggests that encouraging the development of local IP ecosystems and providing capacity-building programs can help African nations unlock the full potential of their IP, leading to job creation, increased competitiveness, and a more diversified economy. A strong IP framework can also encourage domestic entrepreneurship and the growth of local industries. By providing legal protection for inventions, trademarks, and copyrights, African countries can create a conducive environment for homegrown businesses to thrive and contribute to the overall economic development of the region. Fostering collaboration between academia, research institutions, and industry players can further enhance the utilization of IP assets and drive innovation-led growth in Africa. This collaboration can involve initiatives such as joint research projects, technology transfer programs, and mentorship opportunities for aspiring entrepreneurs. By leveraging the expertise and resources of academia and research institutions, African businesses can gain access to cutting-edge knowledge and technologies, fuelling their growth and competitiveness in the global market. Promoting IP rights awareness and education among entrepreneurs can help them understand the value of their creations and encourage them to protect their innovations, fostering a culture of innovation and entrepreneurship in Africa.

2.6.2 Potential of Technology in Fostering Innovation, Improving Productivity, and Driving Economic Growth

In recent years, Africa has witnessed a remarkable surge in technological advancements that have the potential to transform various sectors of the economy. From mobile banking and e-commerce to renewable energy solutions and agricultural innovations, technology is playing a crucial role in fostering innovation and driving economic growth across the continent. These advancements not only improve productivity but also create new opportunities for entrepreneurship and job creation, ultimately leading to more inclusive and sustainable development in Africa. By leveraging technology, African countries are able to overcome traditional barriers to growth, such as limited access to financial services and infrastructure. For example, mobile banking has allowed individuals in remote areas to access financial services and participate in the formal economy. The use of renewable energy solutions is not only reducing reliance on fossil fuels but also creating new business opportunities in the clean energy sector.

The adoption of digital platforms and e-commerce has opened up new markets for African entrepreneurs, enabling them to reach customers both locally and globally. This has led to an increase in job opportunities and economic empowerment for individuals in previously underserved communities. The integration of technology in healthcare has improved access to quality medical services, leading to better health outcomes and overall well-being for African populations. The digital revolution has also facilitated the growth of the gig economy in Africa, allowing individuals to leverage their skills and talents to earn income through online platforms. This has created a flexible and inclusive work environment, particularly for marginalized groups such as women and youth. Technology has played a crucial role in enhancing education in Africa, with online learning platforms and digital resources providing access to quality education for those who were previously disadvantaged by limited educational opportunities.

In many African nations, including Nigeria, Kenya, and South Africa, technology has the potential to promote innovation, increase productivity, and propel economic growth, according to the World Bank (2022). The use of technology has not only transformed the education sector but has also opened up new avenues for entrepreneurship and job creation. For instance, mobile banking services have allowed individuals in remote areas to access financial services and start their own businesses, empowering them to become self-sufficient and contribute to the economic development of their communities. The integration of technology in various industries has led to the development of advanced manufacturing processes and automation, resulting in increased efficiency and cost-effectiveness. This has not only boosted productivity but has also created new job opportunities in fields such as artificial intelligence and data analysis, driving economic growth and fostering a culture of innovation. The integration of technology has also improved access to education and information, empowering individuals to acquire new skills and knowledge. This has further enhanced their ability to participate in the global economy and adapt to rapidly changing market demands. Overall, the combination of technological advancements and increased job opportunities has played a crucial role in uplifting communities and promoting sustainable economic development.

REFERENCES

Adewuyi, A. O., & Awodumi, O. B. (2021). Environmental pollution, energy import, and economic growth: Evidence of sustainable growth in South Africa and Nigeria. *Environmental Science and Pollution Research*. https://doi.org/10.1007/s11356-020-11446-z

African Development Bank. (2022). African Development Bank – Building today, a better Africa tomorrow. www.afdb.org/en/

Akinsola, F. A., Ologundudu, M. M., Akinsola, M. O., & Odhiambo, N. M. (2022). Industrial development, urbanization and pollution nexus in Africa. *Heliyon*. https://doi.org/10.1016/j.heliyon.2022.e11299

Akyeampong, E., Bates, R. H., Nunn, N., & Robinson, J. A. (2014). Africa's development in historical perspective. In *Africa's Development in Historical Perspective*. https://doi.org/10.1017/CBO9781139644594

Assan, N. (2023). Socio-cultural, economic, and environmental implications for innovation in sustainable food in Africa. *Frontiers in Sustainable Food Systems*. https://doi.org/10.3389/fsufs.2023.1192422

Baarsch, F., Granadillos, J. R., Hare, W., Knaus, M., Krapp, M., Schaeffer, M., & Lotze-Campen, H. (2020). The impact of climate change on incomes and convergence in Africa. *World Development*. https://doi.org/10.1016/j.worlddev.2019.104699

Baker, B. W. (2018). The death of Wakanda: The impact of chattel slavery on the advancement of the descendants of the African enslaved. *Africology: The Journal of Pan African Studies*, *11*(9), 61–63.

Bank, E. I. (2020, February 26). *Banking in Africa: Financing Transformation Amid Uncertainty*. European Investment Bank. http://books.google.ie/books?id=zxXTDwAAQBAJ&printsec=frontcover&dq=A+vision +of+Africa%27s+future:+Mapping+change,+transformations+and+trajectories+towards+ 2030&hl=&cd=8&source=gbs_api

Bannerman, S. (2020). The World Intellectual Property Organization and the sustainable development agenda. *Futures*. https://doi.org/10.1016/j.futures.2020.102586

Carbone, G. (2018, January 1). *A Vision of Africa's Future*. Ispi Publications. http://books.google.ie/books?id=MopFvgEACAAJ&dq=A+vision+of+Africa%27s+future:+Mapping+change,+transformations+and+trajectories+towards+2030&hl=&cd=2&source=gbs_api

Enock, Y. M. (2018). Breeding rice for grain yield, earliness and grain quality in Eastern Africa. *Pakistan Research Journal of Management Sciences*, *7*(5), 1–250. http://erepository.uonbi.ac.ke/handle/11295/104365

Goldstone, B., & Obarrio, J. (2017, January 31). *African Futures*. University of Chicago Press. http://books.google.ie/books?id=nUWkDQAAQBAJ&printsec=frontcover&dq=A+vision+of+Africa%27s+future:+Mapping+change,+transformations+and+trajectories+towards+2030&hl=&cd=4&source=gbs_api

IEA & UNEP. (2018). International Energy Agency and the United Nations Environment Programme – Global status report 2018: Towards a zero-emission, efficient and resilient buildings and construction sector. *Global Status Report*.

IMF. (2020). World Economic Outlook Database. www.imf.org

IPCC. (2021). *IPCC: Summary for Policymakers. Climate Change 2021: The Physical Science Basis. Contribution of Working Group I to the Sixth Assessment Report of the Intergovernmental Panel on Climate Change*.

John, O. I., Messina, G.-M., & Odumegwu, A. C. (2023). The effects of neocolonialism on Africa's development. *Pan African Journal of Governance and Development (PJGD)*. https://doi.org/10.46404/panjogov.v4i2.4846

John, S. F. (2021). Technology governance: Minding and closing the gaps in Africa. *African Journal of Governance and Development (AJGD)*. https://doi.org/10.36369/2616-9045/2021/v10i2a5

Karodia, A. M., & Soni, D. (2014). South African MBA's must focus on development issues not business and finance only and must redefine the dominance of the Eurocentric approach. *Mediterranean Journal of Social Sciences*. https://doi.org/10.5901/mjss.2014.v5n27p108

Kohnert, D. (2009). EU-African economic relations: Continuing dominance, traded for aid? In *A Global Security Triangle: European, African and Asian Interaction*. https://doi.org/10.4324/9780203866870-17

Kundu, D., & Pandey, A. K. (2020). World urbanisation: Trends and patterns. In *Developing National Urban Policies: Ways Forward to Green and Smart Cities*. https://doi.org/10.1007/978-981-15-3738-7_2

Lopes, C., & te Velde, D. W. (2021). *Structural Transformation, Economic Development and Industrialization in Post-Covid-19 Africa*. Institute for New Economic Thinking.

Magu, S. M. (2021). Africa's post-colonial foreign policy: Assessing history, imagining the future. In *Explaining Foreign Policy in Post-Colonial Africa*. https://doi.org/10.1007/978-3-030-62930-4_9

Meintjes, S. (2004). Gender, nationalism and transformation: Difference and commonality in South Africa's past and present. In *Women, Ethnicity and Nationalism: The Politics of Transition*. https://doi.org/10.4324/9780203169582-7

Nwokolo, S. C. (2017). A comprehensive review of empirical models for estimating global solar radiation in Africa. *Renewable and Sustainable Energy Reviews*, *78*(October), 955–995. https://doi.org/10.1016/j.rser.2017.04.101

Nyoka, B. (2023). Bernard Magubane on the political economy of race and class in South Africa. *Ethnic and Racial Studies*. https://doi.org/10.1080/01419870.2023.2195051

Onuoha, F. C., & Agbede, M. O. (2019). Impact of disagregated public expenditure on unemployment rate of selected African countries: A panel dynamic analysis. *Journal of Economics, Management and Trade*. https://doi.org/10.9734/jemt/2019/v24i530175

Ouma, S. O. (2012). The African renaissance and discourse ownership: Challenging debilitating discourses on Africa. In *African intellectuals and decolonization*. **LIT Verlag.**

Skea, J., Shukla, P., Al Khourdajie, A., & McCollum, D. (2021). Intergovernmental panel on climate change: Transparency and integrated assessment modeling. *Wiley Interdisciplinary Reviews: Climate Change*. https://doi.org/10.1002/wcc.727

Smith, K. (2013). Contrived boundaries, kinship and ubuntu: A (South) African view of "the international." In *Thinking International Relations Differently*. https://doi.org/10.4324/9780203129920-24

Sumberg, J., Fox, L., Flynn, J., Mader, P., & Oosterom, M. (2021). Africa's "youth employment" crisis is actually a "missing jobs" crisis. *Development Policy Review*. https://doi.org/10.1111/dpr.12528

The World Bank Group. (2022). *Risk of Global Recession in 2023 Rises Amid Simultaneous Rate Hikes*. The World Bank Group.

Trotter, P. A., & Abdullah, S. (2018). Re-focusing foreign involvement in sub-Saharan Africa's power sector on sustainable development. *Energy for Sustainable Development*. https://doi.org/10.1016/j.esd.2018.03.003

Woodward, R. (2022). *The Organisation for Economic Cooperation and Development (OECD)*. https://doi.org/10.4324/9781351025867

Zeleza, P. T. (2011). Typologies of War in Twentieth-Century Africa. *Tropical Ecology*.

3 Repercussions of the Economic Crisis on the Energy Sector

3.1 INTRODUCTION

The economic crisis has had far-reaching consequences across various sectors globally, and the energy sector in Africa is no exception. The repercussions of this crisis on the energy sector in Africa have been significant, leading to both immediate and long-term effects that have impacted the region's energy production (Schulhof et al., 2022), distribution (Pombo, 2022), and access (Bano et al., 2022). Decreased investment in the energy sector has resulted in a decline in infrastructure development (Pretorious et al., 2015) and maintenance (Guina et al., 2021), hindering the expansion of energy access to remote and underserved areas (Asiri et al., 2022). The reduced funding has also affected research and development efforts for renewable energy sources, slowing down the transition toward a more sustainable and environmentally friendly energy system in Africa. As a result, many African countries continue to heavily rely on fossil fuels, contributing to environmental degradation and exacerbating climate change (Nwokolo, 2017). The lack of investment in the energy sector has also led to frequent power outages (Obiwulu, Chendo et al., 2020) and unreliable electricity supply (Obiwulu, Erusiafe et al., 2020), further impeding economic growth and development in the region (Obiwulu et al., 2022). Furthermore, the reliance on fossil fuels has negative health impacts on African communities (Nwokolo and Ogbulezie, 2018a), as the burning of these fuels releases harmful pollutants into the air (Nwokolo and Ogbulezie, 2018b). This leads to increased respiratory diseases and other health issues among the population (Nwokolo, Ogbulezie, & Obiwulu, 2022). The lack of reliable electricity hinders access to education (Nwokolo, Meyer et al., 2023) and healthcare services (Obafemi et al., 2022), limiting opportunities for social (Watts, 2018) and economic progress in Africa (Buckland et al., 2023).

Impaired infrastructure development and maintenance has not only hindered the expansion of energy access to remote and underserved areas but has also led to frequent power outages (Nwokolo, Obiwulu et al., 2023) and unreliable electricity supply in many parts of Africa (Nwokolo, Ogbulezie, & Ushie, 2023). This lack of reliable energy infrastructure further hampers economic growth and development, as industries and businesses struggle to operate efficiently without a consistent and stable source of power (Nwokolo, Amadi et al., 2022). Furthermore, the unreliable electricity supply also affects the daily lives of individuals (Nwokolo and Otse, 2019),

DOI: 10.1201/9781003483175-4

making it difficult to access basic services such as healthcare and education (Pradella and Cillo, 2021). The lack of reliable energy infrastructure hinders the adoption of renewable energy sources, preventing Africa from fully harnessing its vast renewable energy potential (Pendame, 2021) and transitioning toward a more sustainable and environmentally friendly energy system (Sonebi et al. 2022). This reliance on unreliable electricity also hampers economic growth and development in Africa. Industries and businesses struggle to operate efficiently and expand without a consistent power supply, leading to limited job opportunities (Khan et al., 2014) and reduced productivity (Danso-Wiredu et al., 2016). The lack of reliable energy infrastructure makes it challenging for Africa to attract foreign investments (Nen and Nagy, 2021) and participate fully in global markets (Tadesse, 2016), further hindering its overall economic progress.

Limited access to electricity for the population also has significant implications for education and healthcare. Without reliable electricity, schools struggle to provide quality education, as students are unable to access computers and other technological resources. Healthcare facilities are hindered in their ability to provide essential services, such as refrigeration for vaccines and medical equipment (Howard et al., 2014). This lack of electricity access perpetuates a cycle of poverty and inequality (Hakan et al., 2023), as individuals and communities are unable to fully participate in economic and social opportunities (Ballentyne et al., 2011). Furthermore, the absence of electricity limits communication and connectivity, making it difficult for people to access information and stay connected with the rest of the world (Berka and Dreyfus, 2011). This hampers their ability to engage in online learning, remote work opportunities, and even basic communication with friends and family outside their immediate vicinity. Ultimately, the lack of reliable electricity exacerbates existing disparities and hinders overall development in these communities. In addition, the absence of electricity also affects healthcare services in these communities (Tadesse, 2018). Without a reliable source of power, medical facilities struggle to operate essential equipment, refrigerate vaccines and medicines, and provide adequate lighting for surgeries and examinations. This not only compromises the quality of healthcare but also puts lives at risk. The lack of electricity hampers economic growth as businesses are unable to operate efficiently, hindering job creation and income generation for individuals in these communities.

These repercussions have also highlighted the need for sustainable and resilient energy systems in Africa to mitigate future economic shocks. The pandemic has exposed the vulnerabilities of Africa's energy sector, which heavily relies on fossil fuels and lacks diversification in its energy sources. This has prompted a growing recognition of the importance of investing in renewable energy infrastructure and technologies to ensure a more stable and sustainable energy future for the continent. Investing in renewable energy infrastructure and technologies would not only reduce Africa's dependence on fossil fuels but also contribute to mitigating climate change and improving air quality. It would create new job opportunities and stimulate economic growth in the renewable energy sector, ultimately enhancing Africa's overall resilience and sustainability (Chakamera and Alagided, 2018). Furthermore, investing in renewable energy would also enhance energy access and affordability for the millions of people in Africa who currently lack reliable electricity (Possenti, 2012). This would improve their quality of life, enable better access to education and healthcare, and drive social development across the continent. By embracing renewable energy sources, Africa can position itself as a global player in clean energy innovation and attract foreign investment for further economic development. Furthermore, investing in renewable energy would reduce Africa's dependence on fossil fuels, leading to a significant decrease in carbon emissions and mitigating the impact of climate change (Proutsos, Liakatas et al., 2023). This transition to clean energy would not only benefit the environment but also create job opportunities and stimulate economic growth in the region. By embracing renewable energy, Africa can tap into its vast natural resources such as solar, wind, and hydropower, which are abundant across the continent (Nwokolo, Echeng et al., 2024). This would not only provide a sustainable and reliable source of electricity but also foster technological advancements in the clean energy sector, paving the way for export opportunities and greater energy independence for African nations. By taking a proactive approach toward clean energy adoption, Africa can showcase its commitment to sustainable development and inspire other regions to follow suit in the global fight against climate.

3.2 IMPORTANCE AND ROLE OF THE ENERGY SECTOR IN AFRICA

3.2.1 Overview of the Energy Sector's Contribution to Africa's Economy

The energy sector plays a crucial role in driving economic growth and development in Africa. It not only provides reliable and affordable electricity to power industries, businesses, and homes but also creates employment opportunities and attracts foreign investments. The expansion of renewable energy sources in Africa is not only helping to address the continent's energy needs but also contributing to reducing greenhouse gas emissions and combating climate change (Proutsos, Tigkas et al., 2023). Renewable energy sources such as solar and wind power have the potential to transform Africa's energy landscape by diversifying its energy mix and reducing its dependence on fossil fuels. This transition toward clean energy not only promotes sustainability but also positions Africa as a global player in the fight against climate change (Nwokolo, Meyer et al., 2024). The development of renewable energy infrastructure can enhance energy access in remote areas, improving the lives of millions of people and fostering inclusive growth across the continent. Investing in renewable energy can create job opportunities and stimulate economic growth in Africa. By harnessing the abundant solar and wind resources, African countries can tap into their natural potential and become self-sufficient in meeting their energy needs (Ituen et al., 2012), reducing their reliance on expensive imports. This shift toward clean energy also aligns with global efforts to mitigate greenhouse gas emissions and achieve a more sustainable future for all (Nwokolo, Proutsos et al., 2023).

According to the International Energy Agency (IEA), the energy sector's contribution to Africa's economy reached $90 billion in 2019 (International Energy Agency [IEA], 2023), and this figure is expected to rise as the continent invests more in renewable energy. For example, in 2022, the figure increased to about $100 billion, representing a significant boost to Africa's economy (IEA, 2023). This growth is driven by the increasing adoption of renewable energy sources such as solar and wind power, which not only reduce greenhouse gas emissions but also create job opportunities and attract foreign investments. The shift toward renewable energy can help Africa reduce its dependence on fossil fuels, mitigating the impact of volatile oil prices and enhancing energy security in the region. The IEA also reported that the African country with the greatest impact on its energy sector is South Africa (IEA, 2023), which has made significant progress in diversifying its energy mix by increasing its renewable energy capacity. This has not only led to a decrease in carbon emissions but has also created a thriving renewable energy industry, attracting both local and international investors. The adoption of renewable energy in South Africa has improved access to electricity in rural areas, bringing about socio-economic benefits and improving the overall quality of life for its citizens.

The agency also stated that the percentage contribution of the energy sector to South Africa's economy between 2020 and 2022 is expected to increase from 5% to 10% (IEA, 2023), demonstrating the country's commitment to transitioning to cleaner energy sources. This shift toward renewables not only helps South Africa reduce greenhouse gas emissions but also creates new job opportunities and promotes sustainable economic growth in the region. Furthermore, the adoption of renewable energy sources in South Africa can also lead to a more reliable and resilient energy system, reducing the country's dependence on fossil fuel imports. This diversification of the energy mix can enhance energy security and mitigate the risks associated with fluctuating global oil prices. Investing in clean energy technologies can attract foreign investments and position South Africa as a player in sustainable development within the African continent. The adoption of renewable energy sources can create new job opportunities and stimulate economic growth in South Africa. By investing in clean energy infrastructure and supporting local industries, the country can foster a sustainable and inclusive green economy. Transitioning to renewable energy can also contribute to reducing greenhouse gas emissions and mitigating the impacts of climate change, aligning with global efforts to combat environmental degradation.

According to the IEA, the African country with the lowest contribution from the energy sector to the economy is Burundi (IEA, 2023). Burundi's low contribution from the energy sector highlights the need for increased investment in clean energy technologies. By transitioning to renewable sources, Burundi can reduce its reliance on fossil fuels and improve its energy security. Furthermore, embracing sustainable development practices can attract foreign investments and position Burundi as a frontrunner in the region's efforts toward a greener future. Investing in clean energy technologies will not only benefit Burundi's economy but also contribute to the global fight against climate change. By reducing greenhouse gas emissions, Burundi can play a significant role in mitigating the adverse effects of climate change and protecting its natural resources. Prioritizing renewable energy sources can create job opportunities and enhance the overall well-being of Burundi's population, leading to a more sustainable and prosperous future for the country. Investing in renewable energy technologies can also improve energy access and reliability in Burundi, ensuring that all citizens have access to clean and affordable electricity. This can have a positive impact on various sectors such as healthcare, education, and agriculture, ultimately improving the quality of life for the people of Burundi. Transitioning to renewable energy can reduce the country's dependence on imported fossil fuels, enhancing energy security and reducing vulnerability to fluctuations in global oil prices.

3.2.2 Analysis of the Role of the Energy Sector in Job Creation and Poverty Reduction

The energy sector plays a crucial role in job creation and poverty reduction in Africa. With reliable access to electricity, industries can thrive, leading to the creation of new employment opportunities. The energy sector itself requires a skilled workforce, contributing to job growth in areas such as renewable energy development, power plant construction, and maintenance. Increased access to affordable and clean energy enables households and businesses to improve their productivity and generate income, thus helping to alleviate poverty across the continent. The availability of electricity allows for the expansion of education and healthcare services, as schools and medical facilities can operate more efficiently with a reliable power supply. This leads to improved access to quality education and healthcare, ultimately enhancing the overall well-being and development of communities. Increased energy access also fosters innovation and entrepreneurship, as individuals can explore new business opportunities that rely on electricity, such as e-commerce or technology-based ventures. A reliable power supply enables the use of modern technologies and communication tools, allowing for better connectivity and information sharing within communities. This can lead to improved collaboration and knowledge exchange, driving economic growth and social progress.

Increased energy access can also reduce reliance on traditional sources of fuel, promote environmental sustainability (Etuk, Agbasi et al., 2016), and mitigate the impact of climate change (Etuk, Nwokolo et al., 2016). A reliable power supply can attract investment and spur innovation in various sectors, including healthcare and education. For example, it can support the establishment of telemedicine services and online learning platforms, improving access to quality healthcare and education in remote areas. Increased energy access can empower individuals and communities by providing opportunities for entrepreneurship and income generation through the use of energy-intensive tools and machinery. A reliable power supply can also contribute to the overall economic growth of a region. Industries and businesses heavily rely on electricity to operate efficiently, and a consistent power supply ensures uninterrupted production and operations. This, in turn, creates job opportunities and stimulates economic development in the area. A stable power supply can enhance the quality of life for residents by enabling the use of modern amenities such as refrigeration, air conditioning, and lighting, improving living conditions and the overall well-being of African citizens.

A stable power supply also plays a crucial role in attracting foreign investments and fostering economic growth. Reliable electricity infrastructure is often a key factor considered by businesses when deciding to establish operations in a region, as it ensures their ability to meet production demands and remain competitive in the global market. Access to modern amenities made possible by a stable power supply can also contribute to the overall social and economic progress of African countries, empowering individuals and communities to thrive in various sectors such as education, healthcare, and technology. For example, reliable electricity enables schools to provide quality education by powering computers and other electronic devices for interactive learning. It also allows healthcare facilities to operate efficiently with uninterrupted power supply for medical equipment and refrigeration of vaccines. A stable power supply fosters the growth of technology-based industries, creating job opportunities and driving economic development in African countries.

According to the IEA, more than 50 million people in sub-Saharan Africa are expected to gain access to electricity between 2020 and 2022 (IEA, 2022). This increase in power supply will not only improve the living conditions of individuals but also enhance productivity in various sectors, such as agriculture and manufacturing. Furthermore, reliable electricity will enable communities to establish small businesses and access information through the internet, ultimately bridging the digital divide in Africa. Access to reliable electricity is crucial for economic growth and social development. It will not only create job opportunities but also attract foreign investments, leading to overall prosperity in the region. With increased power supply, sub-Saharan Africa can reduce its reliance on fossil fuels and transition toward cleaner and more sustainable energy sources, contributing to global efforts to combat climate change. By investing in reliable electricity infrastructure, sub-Saharan Africa can also improve access to education and healthcare services. With a stable power supply, schools and hospitals can operate more efficiently, providing better-quality education and healthcare to the population. This will contribute to the overall well-being and human development of the region, further enhancing its potential for growth and prosperity.

Approximately 600 million people in Africa still do not have access to reliable electricity (IEA, 2022), which the African Development Bank claims is a major barrier to the development of the continent. This not only affects the daily lives of individuals but also limits the potential for economic growth and improvement in living standards. Without access to electricity, communities are unable to power essential equipment for small businesses, such as computers or machinery. The lack of electricity hinders individuals' ability to connect to the internet and access valuable information and resources that can enhance their education and economic opportunities. Furthermore, the absence of electricity also poses significant challenges in the healthcare sector. Without reliable power sources, medical facilities struggle to operate essential equipment like refrigerators for storing vaccines or running life-saving machines. This lack of access to electricity can have severe consequences for the overall health and well-being of communities, as it hampers the delivery of proper medical care and emergency services.

In sub-Saharan Africa, more than 600 million people lack access to dependable electricity (IEA, 2022), which limits their ability to take advantage of opportunities for economic opportunity, healthcare, and education. This lack of access disproportionately affects rural communities, where infrastructure development is often limited. The report emphasizes the urgent need for investment in energy infrastructure to improve the lives of millions and promote sustainable development in the region. Investment in energy infrastructure would not only provide reliable electricity to these communities but also create job opportunities and stimulate economic growth. Access to electricity would enable the use of modern technology and improve communication, ultimately bridging the digital divide between rural and urban areas. Furthermore, the report highlights that investment in energy infrastructure would have positive environmental impacts by reducing reliance on fossil fuels and promoting the use of renewable energy sources. This transition to cleaner energy would contribute to mitigating climate change and improving air quality, thus benefiting both present and future generations in the region.

3.2.3 Importance of Reliable and Affordable Energy Supply for Economic Growth

The importance of a reliable and affordable energy supply for economic growth in Africa cannot be overstated. Access to reliable and affordable energy is crucial for powering industries, attracting investments, and driving technological advancements. It not only enhances productivity and efficiency but also improves the quality of life for millions of people by providing access to healthcare, education, and clean water. Without a dependable energy supply, Africa's economic potential remains untapped, hindering its progress toward sustainable development and poverty alleviation. Reliable and affordable energy plays a vital role in reducing greenhouse gas emissions and combating climate change. By transitioning to renewable sources of energy, Africa can not only meet its growing energy demands but also contribute to global efforts to mitigate the impacts of climate change. Access to energy can empower communities and foster entrepreneurship, creating job opportunities and driving economic growth across the continent. Access to reliable and affordable energy can also improve the quality of education and healthcare services in Africa. With electricity, schools can have proper lighting and access to technology, enhancing the learning experience for students. Similarly, healthcare facilities can operate more efficiently with electricity, ensuring the availability of life-saving equipment and refrigeration for vaccines and medicines. Overall, energy access has the potential to uplift communities and improve the overall well-being of people in Africa.

For example, statistics from the World Bank report that reliable and affordable energy supply is important for economic growth in Africa. They further report that this is capable of increasing the continent's gross domestic product (GDP) by up to 2.2% per year (World Bank, 2022). This shows, according to the World Bank reports, that access to electricity is not only crucial for economic growth but also for social development in Africa. A reliable and affordable energy supply can pave the way for improved healthcare services, as it enables hospitals and clinics to operate efficiently with proper lighting, medical equipment, and refrigeration for vaccines and medicines. This can ultimately contribute to better healthcare outcomes and a healthier population on the continent. This is where the secrets of the world's most advanced nations' constant development lie. Africa's situation is therefore ineluctably unique, and the sooner we awaken, the better for the development of the continent. Investing in reliable and sustainable energy infrastructure is crucial for Africa's progress. By prioritizing access to electricity, African countries can unlock their full potential in various sectors, such as education, agriculture, and manufacturing. This will not only improve the quality of life for its citizens but also attract foreign investments and promote economic growth. It is imperative for African leaders to recognize the significance of energy development and work toward creating a brighter future for the continent.

3.2.4 Decreased Investment in the Energy Sector

One of the consequences of the economic crisis on the energy sector in Africa is decreased investment in renewable energy projects. As countries face financial constraints, they may prioritize other sectors over clean energy initiatives, leading to a slowdown in the development and implementation of renewable technologies. This can have long-term implications for Africa's energy transition, as it may hinder the continent's ability to meet its sustainable development goals and reduce its reliance on fossil fuels. Decreased investment in the energy sector can exacerbate existing challenges such as energy poverty and lack of access to electricity, particularly in rural areas. Without sufficient investment, it becomes more difficult to expand energy infrastructure and reach underserved communities, perpetuating the cycle of inequality and hindering overall economic growth. The slowdown in renewable energy development can also impact job creation opportunities in the sector, limiting employment prospects for the local population and impeding socio-economic progress.

According to a report by the IEA, African nations have invested approximately $25 billion in the energy sector in recent years (IEA, 2022). This amount falls significantly short of the estimated $120 billion required annually to achieve universal access to electricity by 2030 (World Bank, 2022). This stark disparity highlights the urgent need for increased investment and financial support to bridge the energy gap and foster sustainable development across Africa. Despite these efforts, there is still a substantial gap in energy access and affordability, with an estimated 600 million people lacking access to electricity in Africa (World Bank, 2022). This lack of access to electricity not only hinders economic growth and development but also affects the quality of life for millions of people, particularly in rural areas. Without reliable and affordable energy, communities struggle to access basic services such as healthcare, education, and clean water, perpetuating a cycle of poverty and inequality. It is crucial for governments, international organizations, and private sector stakeholders to collaborate and invest in sustainable energy solutions that can bring electricity to those who need it most.

Nigeria invested about $20 billion in its energy sector between 2010 and 2019, according to a report by the IEA (IEA, 2022). This investment has been aimed at improving the country's electricity generation, transmission, and distribution infrastructure to meet the growing demand for reliable power supply. Despite these efforts, Nigeria still faces significant challenges in achieving universal access to electricity, with an estimated 85 million people lacking access to grid-connected electricity (International Renewable Energy Agency [IRENA, 2022). One of the main challenges Nigeria faces is the inadequate maintenance and operation of its existing power infrastructure, leading to frequent power outages and unreliable supply. The high costs of electricity tariffs and limited affordability for many Nigerians further hinder universal access to electricity.

In their energy sectors, North African nations have made sizable investments, according to a report by the African Development Bank. For instance, Morocco has invested over $9 billion in renewable energy projects, aiming to achieve 52% of its electricity generation from renewables by 2030 (IEA, 2022). According to a report by the African Development Bank, Algeria invested approximately $3 billion in its energy sector in 2019. Similarly, Egypt has invested heavily in renewable energy projects, with a total investment of $7.4 billion in 2019 alone (IRENA, 2019). Egypt has allocated around $28 billion for its energy sector, focusing on expanding its renewable energy capacity and improving infrastructure (IRENA, 2019). These investments highlight the commitment of North African nations toward developing their energy sectors and promoting sustainable sources of energy. The significant investments made by Morocco and Egypt demonstrate their recognition of the importance of transitioning to renewable energy and reducing reliance on fossil fuels. These efforts not only contribute to global efforts to combat climate change but also create opportunities for economic growth and job creation in the region.

Several nations in the area have recently made sizable investments in their energy infrastructure, which is important to note. For example, Ethiopia has invested over $13 billion in its energy sector since 2010, with a focus on hydropower projects like the Grand Ethiopian Renaissance Dam (GERD) (World Bank, 2022). Similarly, Kenya has also made substantial investments in its energy sector, particularly in renewable energy sources such as geothermal and wind power. These investments have not only helped to increase access to electricity for the population but have also attracted foreign investment and stimulated economic growth in the region. For example, the GERD in Ethiopia is expected to generate over 6,000 MW of electricity, which will not only meet domestic demand but also provide an opportunity for exporting power to neighboring countries. Kenya's investments in renewable energy have led to a significant reduction in greenhouse gas emissions, contributing to global efforts toward mitigating climate change.

It is worth noting that Central African nations have made significant investments in the energy sector to improve access to electricity. For instance, according to a report by the World Bank (2022), between 2014 and 2018, the Democratic Republic of Congo invested approximately $2.5 billion in its energy sector to enhance generation capacity and expand access to electricity. Similarly, Cameroon allocated around $1 billion to the energy sector in 2019 to improve its electricity infrastructure and increase access to power for its citizens. These investments highlight the commitment of African nations to address energy challenges and pave the way for economic growth and development. These investments demonstrate the recognition of the importance of a reliable and accessible energy supply for driving economic progress. By investing in their energy sectors, these countries are not only addressing immediate challenges but also setting the foundation for sustainable development and attracting further investment opportunities. This proactive approach to energy infrastructure will undoubtedly contribute to improved living standards and foster economic growth in the long run.

According to a report by the IEA in 2019, East African nations collectively invested around $8.4 billion in the energy sector between 2014 and 2018 (IEA, 2022). This investment aimed to improve infrastructure, increase electricity generation capacity, and promote renewable energy sources. It is important to note that specific figures may vary for each country within the region, and more detailed data may be available for individual nations. The overall investment in the energy sector reflects a significant commitment to addressing energy challenges and fostering sustainable development in East Africa. The investment in the energy sector has already shown positive results, with an increase in electricity access and a decrease in power outages. This has not only improved the quality of life for the population but has also attracted foreign investors, leading to economic growth and job creation in the region.

It is important to note that countries in Southern Africa have made sizable investments in the energy sector to increase access to electricity. For example, according to a report by the IEA, between 2010 and 2019, Southern Africa invested over $30 billion in renewable energy projects (IEA, 2022). These investments aimed to diversify the energy mix, enhance grid infrastructure, and promote sustainable development in the region. These investments have also contributed to job creation and economic growth, as well as reducing reliance on fossil fuels and mitigating the impact of climate change. Renewable energy projects in Southern Africa have not only boosted the region's electricity generation capacity but have also increased access to electricity for remote and underserved communities. This has resulted in improved living standards and opportunities for education, healthcare, and economic activities in these areas. Furthermore, the investments in renewable energy have positioned Southern Africa as a player in clean energy innovation and have attracted international partnerships and collaborations for further advancements in the sector.

In 2019, West African countries made a combined investment in the energy sector of about $25 billion, according to a report by the IEA (IEA, 2022). This investment has primarily focused on expanding electricity generation capacity, improving transmission and distribution networks, and promoting renewable energy sources. Despite these investments, the region still faces significant energy access challenges, with an estimated 175 million people lacking access to electricity (IEA, 2022). The report highlights that while the investments have made progress in increasing electricity generation, the rapid population growth and urbanization in West Africa have posed challenges in ensuring universal access to electricity. The report emphasizes the need for continued investment and policy reforms to address affordability issues and improve the reliability of energy supply in the region. The lack of access to electricity in West Africa has significant implications for economic development, education, and healthcare. Without reliable electricity, businesses struggle to operate efficiently, students are unable to study after dark, and hospitals are limited in their ability to provide life-saving treatments. It is crucial for governments and international organizations to prioritize investments in the energy sector and implement policies that promote sustainable and affordable energy solutions for all.

South Africa reportedly invested about $12 billion in the energy industry between 2010 and 2019 (IEA, 2022), according to the IEA. This investment has been directed toward various energy sources, including renewable energy projects such as solar and wind power. Despite these investments, challenges remain in ensuring universal access to affordable and reliable energy services throughout the country. These challenges include inadequate infrastructure in rural areas, limited financial resources for ongoing maintenance and upgrades, and the need for policy reforms to attract further private sector investment. Efforts are being made to address the socio-economic disparities in energy access and promote sustainable development in marginalized communities. These efforts involve implementing targeted programs and initiatives that prioritize the provision of energy services to underserved populations, such as low-income households and remote areas. Partnerships between government entities, non-profit organizations, and private sector stakeholders are being fostered to leverage resources and expertise in overcoming these challenges and achieving the goal of universal energy access.

This lack of access to electricity not only hinders economic growth and development but also affects the quality of life for millions of people, particularly in rural areas. Without reliable and affordable energy, communities struggle to access basic services such as healthcare, education, and clean water, perpetuating a cycle of poverty and inequality. It is crucial for governments, international organizations, and private sector stakeholders to collaborate and invest in sustainable energy solutions that can bring electricity to those who need it most. These solutions should prioritize renewable energy sources, such as solar and wind power, to reduce reliance on fossil fuels and minimize environmental impact. Initiatives should focus on building infrastructure and capacity in underserved areas to ensure long-term access to reliable energy services. It is crucial for these initiatives to prioritize the involvement and empowerment of local communities, as they are key stakeholders in the sustainable energy transition. By engaging with and training local residents, these projects can not only provide access to electricity but also create opportunities for economic growth and development. Efforts should be made to address affordability barriers and ensure that the cost of accessing sustainable energy remains within reach for marginalized populations.

3.2.5 Impaired Infrastructure Development

One of the consequences of the economic crisis on the energy sector in Africa is impaired infrastructure development. This crisis has hindered the ability of African countries to invest in and maintain their energy infrastructure, resulting in a lack of access to reliable and affordable energy for many communities. As a result, the development and expansion of crucial energy projects such as power plants, transmission lines, and renewable energy installations have been significantly delayed or even abandoned. The impaired infrastructure development not only exacerbates the existing energy deficit but also limits the potential for economic growth and development in these countries. Without access to reliable and affordable energy, industries struggle to operate efficiently, hindering job creation and stifling innovation. The lack of energy access disproportionately affects marginalized communities, exacerbating social inequalities and hindering progress toward sustainable development goals. In addition, the reliance on outdated and inefficient energy sources contributes to environmental degradation and worsens the impact of climate change, further jeopardizing the long-term well-being of these countries. Limited access to reliable energy sources hampers the provision of essential services such as healthcare and education, perpetuating a cycle of poverty and underdevelopment. The high costs associated with inefficient energy systems strain national budgets and limit investments in other critical sectors like infrastructure and social welfare programs.

For example, in many African nations, the economic crisis has severely impacted the energy sector, leading to a lack of investment in infrastructure development. This has resulted in inadequate power generation and distribution systems, hindering industrial growth and economic productivity. According to the World Bank, an estimated 600 million people in sub-Saharan Africa lack access to electricity, with unreliable energy infrastructure being a major barrier to economic development (World Bank, 2022). Limited financial resources due to the economic crisis have constrained governments from allocating sufficient funds toward infrastructure development. As a result, maintenance and expansion of power generation and distribution systems have been neglected, exacerbating the existing deficiencies. This lack of investment in infrastructure not only hampers economic growth but also hinders efforts to improve living standards and access to basic services for the population.

The low level of investment in infrastructure development and modernization is one instance of how the economic crisis has impacted the energy sector in African countries. Due to budget constraints, governments have struggled to allocate sufficient funds for building new power plants, transmission lines, and distribution networks. This has resulted in inadequate access to electricity for many communities, hindering economic growth and development. According to a report by the African Development Bank, as of 2019, around 600 million people in Africa still lacked access to electricity (IEA, 2022). The report also highlights that the lack of infrastructure investment has led to frequent power outages and unreliable electricity supply in many regions. This not only affects households but also hampers the productivity of businesses, making it difficult for them to operate efficiently and compete in the global market. The report emphasizes the urgent need for increased investment in renewable energy sources to address both the infrastructure gap and environmental concerns.

For instance, in many African nations, the economic crisis has resulted in a lack of funding for energy infrastructure projects. This has led to inadequate access to electricity and an unreliable power supply, hindering industrial growth and productivity. According to the IEA, around 600 million people in Africa still lack access to electricity, with rural areas being particularly affected (IEA, 2022). The World Bank estimates that Africa's infrastructure investment needs amount to around \$130–170 billion annually (World Bank, 2022). Insufficient funding for energy infrastructure projects not only hampers economic development but also exacerbates social inequalities, as marginalized communities are disproportionately affected by the lack of electricity access. This funding gap calls for urgent attention and innovative financing mechanisms to ensure sustainable and inclusive energy solutions for Africa's growing population.

3.3 CASE STUDIES OF AFRICAN COUNTRIES AFFECTED BY THE ECONOMIC CRISIS

In this section, we delve into the case studies of various African countries that have been significantly impacted by the recent economic crisis. By examining the unique challenges faced by these nations, we aim to shed light on the underlying factors contributing to their vulnerability and explore potential strategies for recovery. Through a comprehensive analysis of each country's economic policies, social dynamics, and external influences, this section offers valuable insights into the complex interplay between global economic forces and local contexts in shaping the outcomes of these crises in Africa. This analysis will also consider the role of international organizations and their response to the economic crises in these African nations. By understanding how these organizations have intervened and provided support, we can gain a better understanding of the effectiveness of such interventions and their impact on the recovery process. This section will explore any potential long-term implications of these crises on the overall economic development and stability of the African continent. It will examine the measures taken by these organizations to prevent future economic crises and promote sustainable growth in African nations. This analysis will assess the challenges faced by international organizations in coordinating their efforts with national governments and local stakeholders to ensure effective crisis management and recovery strategies.

One notable case study of an African country affected by the economic crisis is Nigeria. Despite being one of the largest oil producers in Africa, Nigeria faces significant challenges due to the sharp decline in global oil prices. This section explores how Nigeria's overreliance on oil revenue and inadequate diversification strategies exacerbated the impact of the economic crisis on its economy and population. It examines the government's response and policy measures implemented to mitigate the effects of the crisis and foster long-term economic resilience. The decline in global oil prices led to a decrease in government revenue, which in turn resulted in budget deficits and reduced public spending on essential services such as healthcare and education. Nigeria's heavy dependence on oil exports left the country vulnerable to fluctuations in global oil markets, making it susceptible to economic shocks.

The economic crisis in Nigeria had a detrimental impact on various sectors. For instance, the healthcare system suffered as funding for hospitals and medical supplies was significantly reduced. According to a report by the World Health Organization (WHO), Nigeria's healthcare spending per capita dropped from $118 in 2014 to $78 in 2016 (Rudnicka et al., 2020), leading to a decline in the quality of healthcare services and accessibility for the population. The education sector faced severe challenges due to budget cuts. The Nigerian government's allocation for education decreased by 9% in 2016 (Rudnicka et al., 2020), resulting in a shortage of teachers, inadequate infrastructure, and limited access to quality education for many children. As a result, the literacy rate in Nigeria declined, and the country struggled to meet its educational goals. Furthermore, the transportation sector also suffered as funding for road maintenance and construction was reduced. This led to deteriorating road conditions, increased traffic congestion, and hindered economic growth as goods and services faced difficulties being transported efficiently across the country.

During the economic crisis, Nigeria's employment sector experienced a significant downturn. According to the National Bureau of Statistics, the unemployment rate surged from 6.4% in 2014 to 23.1% in 2018 (National Bureau of Statistics, 2020), leaving millions of Nigerians jobless. This rise in unemployment was primarily driven by layoffs and business closures as companies struggled to stay afloat amidst the economic turmoil. The crisis also led to a decline in job creation, with new employment opportunities becoming scarce and highly competitive. As a result, job seekers faced fierce competition for limited positions, exacerbating the unemployment problem. Furthermore, the lack of economic stability and uncertainty discouraged foreign investments, further hindering job growth and exacerbating the employment crisis in Nigeria.

According to data from the Nigerian Immigration Service, the economic crisis has had a significant impact on immigration rates in Nigeria. In 2019 (Yauri, 2022), there was a noticeable decline in the number of immigrants entering the country, with a decrease of 15% compared to the previous year (Yauri, 2022). This decline can be attributed to limited job opportunities and economic instability, which deterred foreigners from seeking employment or investing in Nigeria. The uncertainty surrounding the economy has also led to a decrease in skilled workers immigrating to the country. Many professionals are hesitant to leave their current jobs or relocate to a country facing economic challenges. This decrease in skilled workers can further exacerbate the economic crisis, as Nigeria may struggle to attract the expertise needed for its development and growth.

According to a report by the International Organization for Migration (IOM), Nigeria experienced a significant decline in migration rates during the economic crisis. The report states that between 2014 and 2018, the number of Nigerians migrating to other countries decreased by 55% (Bradley, 2023). This decline can be attributed to the economic challenges faced by Nigeria, which have made it less attractive for individuals seeking better job opportunities and higher living standards. Furthermore, data from the Nigerian Immigration Service reveals that there has been a decrease in the number of work permits issued to foreign nationals during this period. This suggests that not only are Nigerians choosing to stay in the country, but there is also a decrease in the number of foreigners coming to Nigeria for employment opportunities. This decline in migration rates has had significant implications for both the Nigerian economy and the countries that traditionally receive a large number of Nigerian migrants.

One case study of an African country affected by the economic crisis is South Africa. According to a report by the World Bank, South Africa experienced a decline in its GDP growth rate from 1.3% in 2017 to 0.8% in 2018 (World Bank, 2022), largely due to weak domestic demand and declining investment. This economic slowdown has had implications for job creation and has resulted in increased unemployment rates. Another case study is Angola, which heavily relies on oil exports for its revenue. The drop in global oil prices in recent years has significantly impacted Angola's economy. The country experienced a sharp decline in government revenue, leading to budget cuts and reduced public spending. As a result, Angola faced challenges in diversifying its economy and creating alternative sources of income. This economic crisis has highlighted the vulnerability of countries heavily dependent on a single commodity for their economic stability. Countries heavily reliant on a single commodity, like Angola, are particularly susceptible to fluctuations in global markets. The decline in oil prices not only affected government revenue but also led to job losses and hindered infrastructure development. To mitigate future risks, Angola has been making efforts to diversify its economy by investing in sectors such as agriculture, manufacturing, and tourism. Achieving economic stability through diversification remains a long-term challenge for the country.

Zimbabwe is one instance of an African nation impacted by the economic crisis. Zimbabwe's GDP shrank by 12.8% in 2020, according to a study by the International Monetary Fund (IMF), as a result of the COVID-19 pandemic's effects and pre-existing structural problems (International Monetary Fund [IMF], 2020). This contraction resulted in widespread job losses and increased poverty levels in the country (IMF, 2020). Another case study is Kenya, where the economic crisis had significant effects on various sectors. The World Bank reported that Kenya's GDP growth rate declined to 0.6% in 2020 (World Bank, 2022), the lowest in over a decade. This was primarily attributed to disruptions in key sectors such as tourism, agriculture, and manufacturing, leading to job cuts and reduced incomes for many Kenyans. The country faced challenges in accessing international markets and attracting foreign investment, further exacerbating the economic crisis.

Ethiopia is one of the African nations impacted by the economic crisis. Ethiopia's economy would likely undergo a substantial downturn in 2020, with a forecast decline of 2.9%, according to the IMF (IMF, 2020). This was primarily due to disruptions in key sectors such as manufacturing and construction, as well as reduced foreign direct investment. Another case study is Ghana, where the economy contracted by 1.1% in 2020, according to the World Bank (2022). The decline was attributed to the impact of the COVID-19 pandemic, which led to a decline in global demand for Ghana's key exports, such as oil and cocoa. The country faced challenges in managing its fiscal deficit and debt levels, further exacerbating the economic crisis. Ghana also experienced disruptions in its supply chains due to lockdown measures and restrictions on international trade, which further hampered its economic performance. The government implemented various measures to mitigate the impact, including stimulus packages and support for affected industries, but the road to recovery remains uncertain as the country continues to grapple with the ongoing effects of the pandemic.

To address these challenges, the government implemented measures such as promoting non-oil sectors like agriculture and manufacturing, attracting foreign direct investment, and implementing fiscal reforms to diversify the economy and reduce reliance on oil revenue. The government focused on improving infrastructure and investing in human capital development to enhance productivity and competitiveness in non-oil sectors. These efforts aimed to create a more sustainable and resilient economy that was less affected by global oil market fluctuations. The government implemented policies to encourage entrepreneurship and innovation, supporting the growth of small and medium-sized enterprises in non-oil sectors. They also prioritized sustainable development initiatives, such as renewable energy projects and environmental conservation efforts, to further diversify the economy and reduce its environmental impact. These initiatives not only helped to reduce the country's reliance on oil but also created new job opportunities and stimulated economic growth in various sectors. By promoting a more sustainable and resilient economy, the government aimed to ensure long-term stability and prosperity for its citizens while also contributing to global efforts to combat climate change.

3.4 POTENTIAL SOLUTIONS TO MITIGATE THE CONSEQUENCES

One potential solution to mitigate the consequences of the economic crisis on the energy sector in Africa is to promote renewable energy sources. By investing in solar, wind, and hydroelectric power, African countries can reduce their dependence on fossil fuels and create a more sustainable energy system. Investing in renewable energy can reduce dependence on fossil fuels and enhance energy security. By diversifying the energy mix, countries can become less vulnerable to fluctuations in oil prices and geopolitical tensions. Transitioning to clean energy sources can contribute to reducing greenhouse gas emissions and combating climate change, thus ensuring a sustainable future for generations to come. Investing in renewable energy can create new job opportunities and stimulate economic growth. The renewable energy sector has the potential to generate a significant number of jobs across various skill levels, from manufacturing and installation to research and development. This can not only boost local economies but also provide long-term stability and prosperity for communities. Transitioning to clean energy sources can improve air quality and public health by reducing pollution from traditional fossil fuel–based power generation. This can lead to lower healthcare costs and a higher quality of life for residents. Clean energy technologies have the potential to mitigate the impacts of climate change by reducing greenhouse gas emissions. By embracing renewable energy sources, communities can contribute to global efforts to combat climate change and create a sustainable future.

According to a report by the International Renewable Energy Agency (IRENA) (IRENA, 2022), promoting renewable energy sources in African nations could not only help mitigate the consequences of the economic crisis but also drive economic growth. The report highlights that investing in renewable energy can create jobs, attract investments, and improve energy access in these countries. For instance, it states that by 2030, scaling up renewables in Africa could generate over 2 million jobs and provide electricity to an additional 600 million people (IRENA, 2022). They also reported that increasing the share of renewable energy in Africa's power sector to 67% by 2030 could generate over 2 million jobs and reduce carbon dioxide emissions by around 310 megatons annually (IRENA, 2022). The report highlights that investing in renewable energy can help African nations achieve universal electricity access, which is crucial for economic growth and development. Investing in renewable energy sources can also help African nations diversify their energy mix, reduce reliance on fossil fuels, and enhance energy security. Therefore, promoting renewable energy sources in African nations not only addresses the economic crisis but also aligns with global efforts to combat climate change and achieve sustainable development goals. It can lead to job creation, attract foreign investments, and foster technological innovation in the region. Transitioning to clean energy can improve air quality, reduce health risks associated with pollution, and mitigate the impacts of climate change on vulnerable communities.

Promoting research and development in renewable energy technologies can drive innovation and foster a competitive market, attract investments, and drive down costs for consumers. For example, a study conducted by the IRENA found that in Kenya, the promotion of research and development in renewable energy technologies led to a 70% reduction in the cost of solar power between 2010 and 2018. The study highlighted that this decrease in costs attracted significant investments, with Kenya receiving over $2 billion in renewable energy investments during the same period. These investments not only contributed to the growth of the renewable energy sector in Kenya but also created job opportunities and stimulated economic development. The decrease in solar power costs made it more accessible for consumers, leading to a wider adoption of clean energy and a reduction in greenhouse gas emissions.

For example, in South Africa, the Renewable Energy Independent Power Producer Procurement Program (REIPPPP) has been successful in attracting significant investments in renewable energy projects. According to a report by the IRENA, the program has resulted in the creation of thousands of jobs and attracted over $14 billion in investments since its inception in 2011. This demonstrates how promoting research and development in renewable energy technologies can not only drive innovation but also have positive economic impacts. Similarly, Morocco has also seen success in its renewable energy sector. The country has implemented a comprehensive strategy to increase its renewable energy capacity, resulting in the construction of several large-scale solar and wind projects. This has not only helped Morocco reduce its reliance on fossil fuels but has also created job opportunities and attracted foreign investments in the country's renewable energy sector.

The promotion of research and development in renewable energy technologies, for instance, has seen significant progress in Nigeria. According to a report by the IRENA (IRENA, 2019), Nigeria's investment in renewable energy reached $2.7 billion in 2018, showcasing the country's commitment to fostering innovation and attracting investments. Similarly, Ethiopia has also emerged as a player in renewable energy development. The country has set ambitious targets to achieve universal electricity access by 2025, with a significant focus on renewable energy sources. Ethiopia's commitment to renewable energy is evident in its development of large-scale hydropower projects, such as the GERD. The country has implemented various policies and incentives to attract private investments in solar and wind energy projects, further driving its renewable energy growth.

By way of its Renewable Energy Act of 2011, which offers incentives like tax exemptions and feed-in tariffs, Ghana, for instance, has successfully attracted private investments in renewable energy projects. According to a report by the IRENA, Ghana's renewable energy sector has experienced significant growth, with installed capacity increasing from 2 MW in 2011 to over 200 MW in 2020 (IRENA, 2022). This demonstrates how policy frameworks can play a crucial role in driving the development and expansion of renewable energy industries. By providing financial incentives and creating a favorable investment environment, Ghana's Renewable Energy Act has encouraged both local and international investors to participate in the country's renewable energy projects. This not only contributes to the reduction of greenhouse gas emissions but also stimulates economic growth and job creation in the renewable energy sector.

Implementing policies that encourage energy efficiency and conservation can help optimize energy usage and minimize waste, further mitigating the impact of the economic crisis on the energy sector. These measures can also lead to job creation and stimulate economic growth in the renewable energy industry. For example, a study conducted by the IRENA (IRENA, 2022) found that implementing energy efficiency measures in South Africa could lead to a 15% reduction in electricity demand by 2030. Another study conducted by the IRENA (IRENA, 2022) found that implementing energy efficiency measures in South Africa could result in annual energy savings of up to 22%. A case study on Kenya's energy sector by the United Nations Development Programme (UNDP) revealed that investing in renewable energy and energy efficiency could create up to 300,000 jobs by 2030 (IEA and UNEP, 2018). A report by the African Development Bank (IEA, 2022) highlighted that improving energy efficiency in Nigeria's manufacturing sector could create over 500,000 jobs and contribute to a 2% increase in GDP growth. These findings demonstrate the potential benefits of adopting energy efficiency policies in African countries, not only for reducing energy consumption and greenhouse gas emissions but also for stimulating economic growth and job creation. By investing in renewable energy sources and implementing energy efficiency measures, African countries can not only improve their energy security and reduce their reliance on fossil fuels but also tap into the growing global market for clean energy technologies. This can lead to a more sustainable and resilient economy while also addressing the urgent need to mitigate climate change.

3.5 IMPLICATIONS FOR SUSTAINABLE DEVELOPMENT AND THE ENVIRONMENT

This book section explores the multifaceted implications of the economic crisis on the energy sector in Africa, shedding light on its profound impact on sustainable development and the environment. It delves into the challenges faced by African countries in achieving their sustainable development goals amidst dwindling resources and limited investments in renewable energy. It critically examines the potential long-term consequences of this crisis on environmental degradation, emphasizing the urgent need for innovative solutions to mitigate its adverse effects and foster a more sustainable future for Africa's energy sector. The section also highlights the importance of international cooperation and financial support in assisting African countries in transitioning to cleaner and more sustainable energy sources. It explores the role of policy frameworks and regulatory mechanisms in promoting renewable energy adoption and creating an enabling environment for investment in the sector. These policy frameworks and regulatory mechanisms can include feed-in tariffs, tax incentives, and streamlined permitting processes. The section emphasizes the need for capacity building and technology transfer to ensure that African countries have the necessary knowledge and tools to effectively implement renewable energy projects. By addressing these key areas, Africa can overcome its energy challenges and pave the way for a greener and more prosperous future.

The economic downturn in Africa's energy industry has serious consequences for environmental protection and sustainable development, according to a report by the IRENA (IRENA, 2022). The report highlights that the decline in investments and job losses due to the crisis could hinder progress toward achieving renewable energy targets and addressing climate change. For instance, it states that if investments continue to decline, Africa may not be able to meet its target of generating 310 GW of renewable energy capacity by 2030 (IRENA, 2022), resulting in increased reliance on fossil fuels and higher greenhouse gas emissions. This could exacerbate the already dire consequences of climate change, such as rising temperatures, extreme weather events, and the loss of biodiversity. The report emphasizes that the energy crisis in Africa disproportionately affects marginalized communities, which rely heavily on affordable and accessible energy sources for their basic needs. Without adequate investment and support, these communities may face increased poverty, health risks, and limited opportunities for economic growth.

IRENA also reports that the economic crisis in the energy sector in Africa has significant implications for sustainable development and the environment. The report states that due to the crisis, there has been a decline in investments in renewable energy projects, resulting in slower progress toward achieving sustainable development goals. It highlights that this setback has led to an increased reliance on fossil fuels, leading to higher carbon emissions and exacerbating climate change. For instance, the report estimates that if this trend continues, Africa's carbon emissions from the energy sector could increase by 45% by 2030 (IRENA, 2022). This not only poses a threat to the environment but also hinders efforts to mitigate climate change and transition toward a low-carbon economy. Furthermore, the report emphasizes the need for urgent action and increased investments in renewable energy to reverse this trend and ensure a sustainable future for Africa.

The economic crisis in Africa's energy sector has significant ramifications for environmental sustainability, according to a report by the IRENA (IRENA, 2022). The report states that the decline in investment and slow progress in renewable energy projects due to the crisis could result in a 20% increase in greenhouse gas emissions by 2030 (IRENA, 2022). This not only undermines efforts to combat climate change but also poses risks to public health and exacerbates environmental degradation. The lack of access to clean and affordable energy in Africa hinders economic growth and social development. It disproportionately affects marginalized communities, limiting their opportunities for education, healthcare, and entrepreneurship. The reliance on fossil fuels in the absence of renewable energy alternatives perpetuates a cycle of dependence on non-renewable resources, further depleting the continent's natural resources and exacerbating environmental degradation.

REFERENCES

Asiri, M. A. Y., Liu, C., & Liu, X. (2022). MENA energy trends, global energy warming, and COVID-19 crisis: An investigation analysis from oil, gas, regional economy, and labor productivity. *Mathematical Problems in Engineering*. https://doi.org/10.1155/2022/2594422

Ballantyne, J., Curry, A., & Sumner, A. (2011). Africa 2010–2020: Poverty reduction beyond the global crisis. *Foresight*. https://doi.org/10.1108/14636681111138749

Bano, S., Liu, L., & Khan, A. (2022). Dynamic influence of aging, industrial innovations, and ICT on tourism development and renewable energy consumption in BRICS economies. *Renewable Energy*. https://doi.org/10.1016/j.renene.2022.04.134

Berka, A., & Dreyfus, M. (2021). Decentralisation and inclusivity in the energy sector: Preconditions, impacts and avenues for further research. Renewable and Sustainable Energy Reviews. https://doi.org/10.1016/j.rser.2020.110663

Bradley, M. (2023). Colonial continuities and colonial unknowing in international migration management: the International Organization for Migration reconsidered. *Journal of Ethnic and Migration Studies*. https://doi.org/10.1080/1369183X.2022.2127407

Buckland, C. E., Thomas, D. S. G., Jägermeyr, J., Müller, C., & Smith, J. A. C. (2023). Drought-tolerant succulent plants as an alternative crop under future global warming scenarios in sub-Saharan Africa. *GCB Bioenergy*. https://doi.org/10.1111/gcbb.13095

Chakamera, C., & Alagidede, P. (2018). Electricity crisis and the effect of CO_2 emissions on infrastructure-growth nexus in Sub Saharan Africa. *Renewable and Sustainable Energy Reviews*. https://doi.org/10.1016/j.rser.2018.06.062

Danso-Wiredu, E. Y., Dadson, Y. I., & Amoako-Andoh, F. O. (2016). Social, economic and environmental impacts of the recent electricity crisis in Ghana: A study of Winneba. *Journal of Social Sciences*. https://doi.org/10.1080/09718923.2016.11893621

Etuk, S., Agbasi, O., & Nwokolo, S. C. (2016). Modelling and estimating photosynthetically active radiation from measured global solar radiation at Calabar, Nigeria. *Physical Science International Journal*. https://doi.org/10.9734/psij/2016/28446

Etuk, S., Nwokolo, S. C., Agbasi, O., & John-Jaja, S. A. (2016). Analysis of photosynthetically active radiation over six tropical ecological zones in Nigeria. *Journal of Geography, Environment and Earth Science International*. https://doi.org/10.9734/jgeesi/2016/27945

Guina, J., Ferrão, J., Bell, V., & Fernandes, T. (2021). The impact of COVID-19 in sub-Saharan Africa food security and human development. *European Journal of Agriculture and Food Sciences*. https://doi.org/10.24018/ejfood.2021.3.2.267

Hakan, Ç., Hasan, M. E., & Nazmi, Ç. A. (2023). Sustainable environmental and economic relations of Turkiye and republic of South Africa—A focus study for the period 2000–2010. *E3S Web of Conferences*. https://doi.org/10.1051/e3sconf/202338303012

Howard, R., Singh, J., & Odeku, K. O. (2014). The risk and vulnerability assessment centre at the University of Limpopo: Generating new research possibilities within southern African-Nordic centre. *Mediterranean Journal of Social Sciences*. https://doi.org/10.5901/mjss.2014.v5n3p684

International Energy Agency. (2022). *World energy outlooks special report Africa energy outlook 2022*. International Energy Agency.

International Energy Agency. (2023). Global EV outlook 2023. *Geo, Geo*. International Energy Agency.

International Energy Agency (IEA) and UNEP. (2018). International energy agency and the United Nations environment programme – Global status report 2018: Towards a zero-emission, efficient and resilient buildings and construction sector. *Global Status Report*.

International Monetary Fund (IMF). (2020). World economic outlook database. www.imf.org.

International Renewable Energy Agency (IRENA). (2019). *Future of solar photovoltaic: Deployment, investment, grid integration and socio-economic aspects*. International Renewable Energy Agency (IRENA).

International Renewable Energy Agency (IRENA). (2022). World energy transitions outlook 2023: 1.5°C Pathway. *World Energy Transitions*.

Ituen, E. E., Esen, N. U., Nwokolo, S. C., & Udo, E. G. (2012). Prediction of global solar radiation using relative humidity, maximum temperature and sunshine hours in Uyo, in the Niger Delta Region, Nigeria. *Advances in Applied Science Research*, *3*(4), 1923–1937.

Khan, M. A., Khan, M. Z., Zaman, K., & Arif, M. (2014). Global estimates of energy-growth nexus: Application of seemingly unrelated regressions. *Renewable and Sustainable Energy Reviews*. https://doi.org/10.1016/j.rser.2013.08.088

Laura, C. et al. (2023). World energy outlook. In *World Energy Outlook*. International Energy Agency. www.iea.org/weo

NBS. (2020). Nigeria living standards survey a survey report by the Nigerian National Bureau of Statistics (in collaboration with the World Bank). *National Bureau of Statistics (NBS)*.

Nene, S., & Nagy, H. (2021). Legal regulations and policy barriers to development of renewable energy sources in South Africa. *Engineering for Rural Development*. https://doi.org/10.22616/ERDev.2021.20.TF049

Nwokolo, S. C. (2017). A comprehensive review of empirical models for estimating global solar radiation in Africa. *Renewable and Sustainable Energy Reviews*, *78*(October), 955–995. https://doi.org/10.1016/j.rser.2017.04.101

Nwokolo, S. C., Amadi, S. O., Obiwulu, A. U., Ogbulezie, J. C., & Eyibio, E. E. (2022). Prediction of global solar radiation potential for sustainable and cleaner energy generation using improved Angstrom-Prescott and Gumbel probabilistic models. *Cleaner Engineering and Technology*, *6*(January), 100416. https://doi.org/10.1016/j.clet.2022.100416

Nwokolo, S., Echeng, E., Obiwulu, A., Meyer, E. L., Christian, C., Ogbulezie, J. C., & Proutsos, N. (2024). A multi-model approach based on CARIMA-SARIMA-GPM for assessing the impacts of climate change on concentrated photovoltaic (CPV) potential. *Physics and Chemistry of the Earth*, *134*(January), 103560. https://doi.org/10.1016/j.pce.2024.103560

Nwokolo, S. C., Meyer, E. L., & Ahia, C. C. (2023). Credible pathways to catching up with climate goals in Nigeria. *11*(196), 1–32. https://doi.org/org/10.3390/cli11090196

Nwokolo, S. C., Meyer, E. L., & Ahia, C. C. (2024). Exploring the interactive influences of climate change and urban development on the fraction of absorbed photosynthetically active radiation. *Atmosphere*, 1–37.

Nwokolo, S. C., Obiwulu, A. U., & Ogbulezie, J. C. (2023). Machine learning and analytical model hybridization to assess the impact of climate. *Physics and Chemistry of the Earth*, *130*(February 2023), 103389. https://doi.org/10.1016/j.pce.2023.103389

Nwokolo, S. C., & Ogbulezie, J. C. (2018a). A qualitative review of empirical models for estimating diffuse solar radiation from experimental data in Africa. *Renewable and Sustainable Energy Reviews*, *92*(18), 353–393. https://doi.org/10.1016/j.rser.2018.04.118

Nwokolo, S. C., & Ogbulezie, J. C. (2018b). A quantitative review and classification of empirical models for predicting global solar radiation in West Africa. *Beni-Suef University Journal of Basic and Applied Sciences*, *7*(4), 367–396. https://doi.org/10.1016/j.bjbas.2017.05.001

Nwokolo, S. C., Ogbulezie, J. C., & Obiwulu, A. U. (2022). Impacts of climate change and meteo-solar parameters on photosynthetically active radiation prediction using hybrid machine learning with Physics-based models. *Advances in Space Research*, *70*(11), 3614–3637. https://doi.org/10.1016/j.asr.2022.08.010

Nwokolo, S. C., Ogbulezie, J. C., & Ushie, O. J. (2023). A multi-model ensemble-based CMIP6 assessment of future solar radiation and PV potential under various climate warming scenarios. *Optik*, *285*(May), 170956. https://doi.org/10.1016/j.ijleo.2023.170956

Nwokolo, S., & Otse, C. (2019). Impact of sunshine duration and clearness index on diffuse solar radiation estimation in mountainous climate. *Trends in Renewable Energy*, *5*(3), 307–332. https://doi.org/10.17737/tre.2019.5.3.00107

Nwokolo, S. C., Proutsos, N., & Meyer, E. L. (2023). Machine learning and physics-based hybridization models for evaluation of the effects of climate change and urban expansion on photosynthetically active radiation. *Atmosphere*, *14*(687), 1–36. https://doi.org/doi.org/10.3390/atmos14040687

Obiwulu, A. U., Chendo, M. A. C., Erusiafe, N., & Nwokolo, S. C. (2020). Implicit meteorological parameter-based empirical models for estimating back temperature solar modules under varying tilt-angles in Lagos, Nigeria. *Renewable Energy*, *145*. https://doi.org/10.1016/j.renene.2019.05.136

Obiwulu, A. U., Erusiafe, N., Olopade, M. A., & Nwokolo, S. C. (2020). Modeling and optimization of back temperature models of mono-crystalline silicon modules with special focus on the effect of meteorological and geographical parameters on PV performance. *Renewable Energy*, *154*. https://doi.org/10.1016/j.renene.2020.02.103

Obiwulu, A. U., Erusiafe, N., Olopade, M. A., & Nwokolo, S. C. (2022). Modeling and estimation of the optimal tilt angle, maximum incident solar radiation, and global radiation index of the photovoltaic system. *Heliyon*, *8*(6), e09598. https://doi.org/10.1016/j.heliyon.2022.e09598

Olatunji, O. O., Adedeji, P. A., Madushele, N., Akinlabi, S. A., Ishola, F., & Kalu, I. E. (2022). Reimagine renewable energy exploration in post-COVID-19 Africa. *Energy Sources, Part A: Recovery, Utilization and Environmental Effects*. https://doi.org/10.1080/15567036.2022.2074173

Pendame, S. (2021). Covid 19 and beyond – Solar energy in Africa: Powering responses, accelerating inclusive and sustainable development. *African Economic Conference*.

Pombo, D. V. (2022). Impact of the COVID-19 pandemic on energy islands: The case of Cape Verde. *Electric Power Systems Research*. https://doi.org/10.1016/j.epsr.2022.108611

Possenti, S. (2012). Rural development strategies as a path to decent work and reducing urban informal employment: The case of South Africa. *Working Paper Series*.

Pradella, L., & Cillo, R. (2021). Bordering the surplus population across the Mediterranean: Imperialism and unfree labour in Libya and the Italian countryside. *Geoforum*. https://doi.org/10.1016/j.geoforum.2020.06.019

Pretorius, I., Piketh, S. J., & Burger, R. P. (2015). The impact of the South African energy crisis on emissions. *WIT Transactions on Ecology and the Environment*. https://doi.org/10.2495/AIR150211

Proutsos, N., Liakatas, A., Alexandris, S., Nwokolo, S. C., Solomou, A. D., & Amadi, S. O. (2023). Assessing the impact of atmospheric attributes on the effectiveness of solar irradiance for photosynthesis of urban vegetation in Attica, Greece. *Theoretical and Applied Climatology*, 0123456789. https://doi.org/10.1007/s00704-023-04700-0

Proutsos, N., Tigkas, D., Tsevreni, I., Alexandris, S. G., Solomou, A. D., Bourletsikas, A., Stefanidis, S., & Nwokolo, S. C. (2023). A thorough evaluation of 127 potential evapotranspiration models in two mediterranean urban green sites. *Remote Sensing*, *15*(14), 1–41. https://doi.org/10.3390/rs15143680

Rudnicka, E., Napierała, P., Podfigurna, A., Męczekalski, B., Smolarczyk, R., & Grymowicz, M. (2020). The World Health Organization (WHO) approach to healthy ageing. *Maturitas*. https://doi.org/10.1016/j.maturitas.2020.05.018

Schulhof, V., van Vuuren, D., & Kirchherr, J. (2022). The belt and road initiative (BRI): What will it look like in the future? *Technological Forecasting and Social Change*. https://doi.org/10.1016/j.techfore.2021.121306

Sonebi, M., Abdalqader, A., Amziane, S., Dvorkin, L., Ghorbel, E., Kenai, S., Khatib, J., Lushnikova, N., & Perrot, A. (2022). Trends and opportunities of using local sustainable building materials in the Middle East and North Africa region. *RILEM Technical Letters*. https://doi.org/10.21809/rilemtechlett.2022.169

Tadesse, T. (2016). *African Drought Conference 2016 Drought Risk Management and Enhancing Resilience in Africa Strategic Framework for Drought Risk Management and Enhancing Resilience in Africa*. University of Nebraska-Lincoln.

Tadesse, T. (2018). *Strategic Framework for Drought Risk Management and Enhancing Resilience in Africa*. University of Nebraska-Lincoln.

The World Bank Group. (2022). *Risk of Global Recession in 2023 Rises Amid Simultaneous Rate Hikes*. The World Bank Group.

Watts, N. (2018). "It's not fair": Maize, money, and moral economies in the Southern Highlands of Tanzania. *Journal of Political Ecology*. https://doi.org/10.2458/v25i1.22009

Yauri, A. R. (2022). The impact of information technology on organizational performance of Nigerian Immigration Service, Kebbi State Command. *Equity Journal of Science and Technology*. https://doi.org/10.4314/equijost.v8i1.15

4 Current Energy Landscape in Africa

4.1 OVERVIEW OF THE CURRENT ENERGY LANDSCAPE IN AFRICA

As the world continues to grapple with the challenges of climate change and sustainable development, Africa's energy landscape has emerged as a crucial focal point. With its vast natural resources and growing population, the continent holds immense potential for transforming its energy sector and driving economic growth. It is also confronted with unique obstacles such as limited access to electricity, outdated infrastructure, and a heavy reliance on fossil fuels. In this chapter, we will explore the current energy landscape in Africa, examining the opportunities and barriers that exist for sustainable energy development. We will also delve into the efforts being made by governments, international organizations, and private sector entities to address these challenges and promote renewable energy sources. By understanding the complexities of Africa's energy landscape, we can identify strategies and solutions that will pave the way for a more sustainable and prosperous future for the continent. This comprehensive analysis will take into account the challenges in the current energy landscape in Africa.

We will explore the social and economic implications of sustainable energy development, including job creation and improved access to electricity for rural communities. Initiatives and policies to address the challenges in Africa's energy landscape that are crucial for achieving sustainable development will also be discussed. Governments and international organizations can play a significant role by implementing renewable energy targets, promoting investment in clean technologies, and supporting research and development in the sector. Fostering partnerships with private sector entities can facilitate the transfer of knowledge and expertise, leading to innovative solutions and increased access to affordable and reliable energy sources across the continent. The chapter concludes by highlighting the importance of creating an enabling policy and regulatory environment that encourages the growth of renewable energy markets. This includes implementing feed-in tariffs (FiTs), providing tax incentives for renewable energy projects, and establishing clear guidelines for grid integration. By addressing these key areas, Africa can unlock its vast renewable energy potential and contribute to global efforts to combat climate change while simultaneously driving economic growth and improving energy access for its citizens.

DOI: 10.1201/9781003483175-5

According to the International Energy Agency (IEA, 2022), Africa is home to abundant energy resources, including significant oil and gas reserves, vast solar potential, and ample hydroelectric capacity. The continent still faces energy poverty, with over 600 million people lacking access to electricity (IEA, 2022). To address this challenge, several African countries have been actively investing in renewable energy sources. For instance, Morocco has made remarkable progress in developing its solar power sector, with the Noor Ouarzazate Solar Complex being one of the largest solar power plants in the world. Countries like Kenya and Ethiopia have made significant strides in harnessing their geothermal energy potential, further diversifying their energy mix. These investments in renewable energy not only help to bridge the energy gap but also contribute to reducing greenhouse gas emissions and promoting sustainable development across the continent. Countries such as South Africa and Egypt have also shown commitment to renewable energy by implementing large-scale wind power projects. This demonstrates a growing recognition of the importance of transitioning to clean energy sources and reducing reliance on fossil fuels in Africa. In addition to South Africa and Egypt, other African countries have also made significant strides in renewable energy. For example, Kenya has become a leader in geothermal power generation, harnessing its abundant volcanic resources. Similarly, Morocco has invested heavily in solar energy, with the world's largest concentrated solar power plant located in the country. These efforts highlight the continent's potential to become a global renewable energy hub and inspire other nations to follow suit.

Nigeria has also made notable progress in renewable energy by increasing its wind power capacity. The country has established wind farms in several regions, taking advantage of its strong and consistent winds. Ethiopia has focused on hydropower, utilizing its vast river systems to generate clean and sustainable electricity. These diverse efforts across Africa demonstrate the continent's commitment to a greener future and showcase the various renewable energy sources that can be tapped into. In addition to wind and hydropower, Ethiopia has also invested in solar energy. The country has implemented large-scale solar projects, harnessing the abundant sunlight it receives throughout the year. These initiatives have not only helped reduce reliance on fossil fuels but have also created job opportunities and contributed to economic growth in the region. Ethiopia's dedication to renewable energy extends beyond solar power. The country has also made significant strides in geothermal energy, utilizing its geologically active landscapes to generate clean and sustainable electricity. These efforts have not only positioned Ethiopia as a leader in renewable energy within Africa but have also attracted international investments and partnerships, fostering further advancements in the green energy sector. Ethiopia's commitment to renewable energy has not only had positive environmental impacts but has also created job opportunities and stimulated economic development. The government's investment in renewable energy infrastructure has resulted in the establishment of new industries and the growth of existing ones, leading to increased employment rates and income generation for local communities. Ethiopia's focus on sustainable energy sources has reduced its reliance on fossil fuels, mitigating the negative effects of climate change and enhancing its overall resilience to energy shocks.

4.2 CHALLENGES IN THE CURRENT ENERGY LANDSCAPE IN AFRICA

In this section, the book delves into the myriad challenges faced by Africa in its current energy landscape. As the continent grapples with increasing demands for electricity and sustainable energy sources, it confronts a host of obstacles that hinder its progress. From inadequate infrastructure and limited access to modern technologies to political and economic barriers, Africa's energy sector faces a complex web of challenges that require innovative solutions and strategic interventions. These challenges are further exacerbated by the effects of climate change, such as extreme weather events and water scarcity, which pose additional risks to energy production and distribution. The lack of financial resources and investment in the energy sector also contributes to the difficulties faced by Africa in achieving its energy goals. Inadequate infrastructure and outdated technologies hinder the efficient generation and distribution of energy across the continent. Without sufficient funding and modernization efforts, Africa's energy sector struggles to meet the growing demands of its population, resulting in limited access to reliable electricity for many communities. The lack of a skilled workforce and technical expertise further exacerbates the challenges faced by Africa's energy sector. Without a sufficient number of trained professionals, it becomes even more challenging to implement and maintain sustainable energy solutions. Political instability and corruption in some African countries can hinder progress toward achieving energy goals, as they create an uncertain investment climate and undermine effective governance in the sector. The inadequate infrastructure and limited access to reliable electricity in many parts of Africa hinder the development and distribution of energy resources. This not only affects industries and businesses but also hampers the daily lives of millions of people, limiting their access to education, healthcare, and other essential services.

Around 600 million Africans, who make up more than half of the population in many countries, are said to lack access to electricity, according to a report by the IEA (2022). The African Development Bank estimates that power outages cost African economies around 2–4% of their GDP annually (African Development Bank, 2022). These figures highlight the significant challenges faced by African nations in terms of energy accessibility and reliability. These challenges have led to a growing focus on renewable energy sources in Africa. Governments and international organizations are investing in projects such as solar and wind power to increase energy access and reduce reliance on fossil fuels. The high upfront costs and lack of infrastructure remain major obstacles to the widespread adoption of renewable energy in African nations.

The IEA (2022) reported that about two-thirds of the population of Africa—600 million people—does not have access to electricity. The same report highlights that only 30% of the African population has access to modern cooking facilities, leading to health and environmental issues. These figures illustrate the significant challenges faced by African nations in terms of energy poverty and limited access to reliable energy sources. The lack of electricity and modern cooking facilities not only hinders economic development but also poses a threat to the overall well-being of the African population. It is crucial for African nations to prioritize investments in energy infrastructure and promote sustainable energy solutions to address these pressing issues. By investing in energy infrastructure, African nations can improve the living conditions of their citizens and stimulate economic growth. Sustainable energy solutions, such as renewable sources like solar and wind power, can provide reliable and clean energy that is both environmentally friendly and cost-effective in the long run.

Approximately 600 million people in Africa lack access to electricity, or about 57% of the population, according to a report by the IEA (2022). The report highlights that only 28% of sub-Saharan Africa's population has access to clean cooking facilities, resulting in health and environmental hazards. Improving access to electricity and clean cooking facilities is crucial for improving the quality of life and reducing poverty in Africa. By prioritizing investments in energy infrastructure and promoting sustainable energy solutions, African nations can not only address these pressing issues but also drive economic growth and create job opportunities. Investments in energy infrastructure can lead to increased productivity in industries such as manufacturing and agriculture, which can contribute to economic growth. The adoption of sustainable energy solutions can help mitigate climate change and reduce greenhouse gas emissions, benefiting both the environment and public health.

Nigeria's energy sector faces significant obstacles, according to the IEA (2022). Firstly, the country's electricity generation capacity is insufficient to meet the growing demand, resulting in frequent power outages and limited access to electricity for a large portion of the population. In fact, only about 60% of Nigerians have access to electricity, with rural areas being particularly underserved. Nigeria heavily relies on fossil fuels for its energy needs, with oil and gas accounting for over 80% of the country's total energy consumption. This heavy reliance on fossil fuels not only contributes to environmental degradation and climate change but also makes Nigeria vulnerable to fluctuations in global oil prices. The lack of diversification in the energy sector hinders the country's economic growth and development, as it limits opportunities for investment and innovation in renewable energy sources.

In South Africa, the current energy environment is characterized by a heavy reliance on coal, which generates about 90% of the nation's electricity (IEA, 2022). This heavy dependence on coal not only contributes to high levels of greenhouse gas emissions and air pollution but also poses challenges in terms of environmental sustainability and public health. The aging infrastructure in South Africa's energy sector presents a major challenge. The country's power grid has been strained by frequent breakdowns and blackouts, leading to disruptions in electricity supply and hindering economic growth. These challenges highlight the urgent need for South Africa to diversify its energy sources and invest in modernizing its infrastructure. Transitioning to renewable energy sources such as solar and wind power could not only reduce the country's carbon footprint but also create new job opportunities and stimulate economic development. Improving the efficiency and reliability of the power grid through investments in smart grid technology would help address the issue of frequent breakdowns and ensure a more stable and resilient energy supply. This would not only benefit households and businesses by minimizing disruptions but also attract foreign investments and promote industrial growth. Incorporating energy storage solutions like battery technology would enable South Africa to better manage its energy demand and supply, reducing reliance on fossil fuels and enhancing grid stability during peak usage periods.

In sub-Saharan Africa, there are still about 600 million people without access to electricity, according to a report by the IEA. This represents approximately two-thirds of the population in the region (IEA, 2022). The IEA highlights that unreliable and inadequate power supply hinders economic growth and development in sub-Saharan Africa, with power outages costing the region an estimated 2–4% of GDP annually (IEA, 2022). The lack of access to electricity not only affects households but also has a severe impact on industries, healthcare facilities, and educational institutions, limiting their ability to function effectively. The report emphasizes the urgent need for investment in renewable energy sources to address this issue and promote sustainable development in the region.

East Africa faces significant challenges in its energy landscape, according to the 2019 IEA report. The report states that approximately 600 million people in the region lack access to electricity, representing a staggering 57% of the population (IEA, 2022). This lack of access not only hinders household activities but also poses severe obstacles for industries, healthcare facilities, and educational institutions, limiting their ability to operate efficiently (IEA, 2022). The report highlights that the reliance on traditional biomass for cooking and heating purposes in East Africa leads to significant health and environmental issues. The use of inefficient stoves and open fires contributes to indoor air pollution, leading to respiratory diseases and premature deaths, while also contributing to deforestation and climate change (IEA, 2022). Addressing these challenges will require a comprehensive approach that includes investment in renewable energy sources, improved infrastructure, and policy reforms to promote sustainable energy access in the region.

The IEA (2022) claims that Central Africa's current energy environment is fraught with serious difficulties. Only about 20% of the population in the region has access to electricity, with rural areas being particularly underserved. The reliance on traditional biomass for cooking and heating purposes remains high, with open fires being a prevalent method of cooking. This not only contributes to indoor air pollution and health issues but also hampers efforts toward achieving sustainable development goals related to clean energy and climate action. The lack of access to clean and reliable energy sources also limits economic growth and hinders the region's ability to attract investments in industries that require a stable energy supply. Addressing these challenges requires implementing innovative solutions, such as promoting renewable energy sources and improving infrastructure for electricity distribution, especially in rural areas. Raising awareness about the benefits of clean cooking technologies and providing affordable alternatives can help reduce the reliance on traditional biomass and improve the overall quality of life for communities in the region.

In its current energy landscape, North Africa faces a number of difficulties, according to a report by the International Renewable Energy Agency (IRENA, 2022). One of the major challenges is the high dependency on fossil fuels, with around 95% of the region's electricity generation coming from non-renewable sources. This not only contributes to greenhouse gas emissions but also makes the region vulnerable to fluctuations in global fuel prices. Limited access to electricity is a significant issue in rural areas of North Africa. The World Bank estimates that around 60% of the population in rural areas of North Africa lacks access to electricity (World Bank, 2022). This lack of access hinders economic development and limits opportunities for education and healthcare in these communities. The aging infrastructure in the region poses another challenge, as it requires significant investments to modernize and improve efficiency.

The majority of those without access to electricity in sub-Saharan Africa—600 million people—live in West Africa, according to the IEA (2022). This represents a significant challenge for the region's energy landscape, hindering economic growth and social development. Inadequate energy infrastructure and unreliable power supply are major obstacles faced by businesses and industries in West Africa, resulting in high production costs and reduced competitiveness. The IEA also highlights the need for substantial investments in renewable energy sources to diversify the energy mix and reduce dependence on fossil fuels. This would not only help mitigate the environmental impact of energy production but also create new job opportunities and stimulate economic growth. Investing in renewable energy would enhance energy security by reducing vulnerability to fluctuations in global oil prices and geopolitical tensions.

The IEA released a report stating that Southern Africa's energy sector faces significant challenges. One of the key challenges is the high reliance on coal for electricity generation, which accounts for approximately 90% of the region's power production (IEA, 2022). This heavy dependence on coal not only contributes to greenhouse gas emissions and air pollution but also poses environmental and health risks. Southern Africa experiences frequent power shortages and blackouts due to aging infrastructure and inadequate investment in alternative energy sources. These power shortages have a detrimental impact on the region's economy and hinder its potential for growth and development. The lack of access to reliable electricity hampers social progress, as it limits opportunities for education, healthcare, and overall quality of life for the population in Southern Africa.

Addressing these multifaceted challenges requires a comprehensive approach that combines renewable energy initiatives, policy reforms, and international collaborations to unlock Africa's vast energy potential and promote sustainable development. In addition to these efforts, capacity-building programs can also play a crucial role in empowering local communities and governments to effectively implement renewable energy projects. By providing training and technical assistance, these programs can enhance the skills and knowledge necessary for successful project implementation, ultimately contributing to the long-term sustainability of Africa's renewable energy sector. Capacity-building programs can also help address the challenges and barriers that hinder the adoption of renewable energy technologies in Africa. These programs can support the development of policies and regulations that promote renewable energy investments, as well as facilitate partnerships between local communities, governments, and private sector entities. By strengthening institutional capacities and fostering collaboration, these programs can create an enabling environment for the growth of Africa's renewable energy sector.

Despite these obstacles, there have been notable successes in the implementation of renewable energy projects in certain African countries. For instance, countries like Kenya and South Africa have made significant strides in harnessing solar and wind power, showcasing the potential for renewable energy to transform the energy landscape in Africa. Partnerships between governments, private sector entities, and international organizations are being formed to address the financial and infrastructural challenges, paving the way for a more sustainable and reliable energy future for African nations. These partnerships aim to attract investments and provide technical expertise to support the development of renewable energy projects. Initiatives such as the African Renewable Energy Initiative have been launched to mobilize funding and promote the deployment of renewable energy technologies across the continent. These initiatives have set ambitious targets for increasing the share of renewable energy in Africa's overall energy mix, with the goal of achieving universal access to clean and affordable energy by 2030. Regional cooperation and collaboration are being fostered to facilitate knowledge sharing and capacity building in the renewable energy sector, ensuring long-term sustainability and resilience for African nations. These efforts are also attracting significant investment from both domestic and international sources, further driving the growth of renewable energy in Africa. Governments are implementing supportive policies and regulatory frameworks to create an enabling environment for the deployment of renewable energy technologies, encouraging private sector participation and innovation in the sector.

4.3 INITIATIVES AND POLICIES TO ADDRESS THE CHALLENGES

4.3.1 General Overview

In recent years, the energy landscape in Africa has been characterized by a multitude of challenges that require urgent attention and innovative solutions. To tackle these issues head-on, governments, organizations, and stakeholders across the continent have been implementing various initiatives and policies aimed at addressing the pressing energy needs of Africa's growing population. These efforts not only seek to ensure reliable access to electricity but also strive to promote sustainable development, foster economic growth, and mitigate the environmental impact of energy production and consumption in Africa.

One notable initiative is the Africa Renewable Energy Initiative (AREI), which was launched in 2015. AREI aims to increase Africa's renewable energy generation capacity by 10 GW by 2020 and mobilize investments of at least $10 billion by 2020 (Mulopo, 2022). Over the years, there has been a growing focus on off-grid solutions, such as solar home systems and mini-grids, to reach remote areas where grid extension is not feasible. These efforts are crucial to improving energy access and transforming the energy landscape in Africa. In addition to increasing renewable energy generation capacity, AREI also focuses on promoting sustainable development and reducing greenhouse gas emissions. This initiative has gained support from various international partners, including the European Union and the United Nations, further highlighting its potential to drive positive change in Africa's energy sector.

The Agenda 2063 initiative of the African Union is another noteworthy endeavor (Mulopo, 2022). It outlines a long-term vision for the continent's development and includes the target of attaining universal access to electricity by 2025 (Mulopo, 2022). This ambitious target has prompted governments and organizations to invest in renewable energy sources such as solar and wind power, as well as promote energy efficiency measures to reduce reliance on fossil fuels. Partnerships with international organizations and private sector entities have been crucial in mobilizing financial resources (Bester, 2023) and expertise to support these initiatives, ensuring their success in transforming Africa's energy landscape (Otlhogile and Shirley, 2023). These efforts have not only helped to address the energy needs of Africa's growing population (Baptista and Plananska, 2017) but have also contributed to mitigating the effects of climate change (Spalding-Fecher et al., 2000). By embracing renewable energy sources, Africa is reducing its carbon footprint (Sanoh et al., 2012) and taking steps toward a more sustainable future (Kitonga et al., 2015). The increased access to electricity has far-reaching benefits, such as improving healthcare services (Ozohu-Suleiman, 2021), enhancing education opportunities (Gromek-Broc, 2023), and fostering economic growth across the continent.

The Power Africa initiative, which was introduced by the US government in 2013 (Brower, 2016), is another noteworthy program. Power Africa aims to add 30,000 MW of new electricity generation capacity and provide access to electricity for at least 60 million new households and businesses in Africa (Brower, 2016). This initiative focuses on public–private partnerships (PPPs) and aims to attract private sector investment in the energy sector across the continent. By leveraging both renewable and conventional energy sources, these initiatives are driving sustainable development and economic growth in Africa. These initiatives are also helping to address the energy deficit in Africa, where millions of people still lack access to reliable electricity. By increasing electricity generation capacity and expanding access (Mwale and Davidson, 2014), they are improving living standards and creating opportunities for economic development (Musango et al., 2011). These initiatives promote the use of clean energy sources (Trotter and Abdullah, 2018), which can help mitigate the impact of climate change and reduce reliance on fossil fuels.

The construction of mini-grids for renewable energy is another noteworthy endeavor in Africa. Where it is not practical to expand the national grid, these mini-grids supply power to underserved and isolated populations. By harnessing renewable energy sources such as solar and wind power, these mini-grids are not only providing reliable electricity but also reducing carbon emissions and promoting sustainable development. The implementation of these initiatives has sparked innovation and entrepreneurship in the renewable energy sector, creating job opportunities and fostering economic growth in Africa. The establishment of renewable energy mini-grids in Africa has also improved access to education and healthcare services in remote areas. With reliable electricity, schools can now provide better learning environments with access to computers and internet connectivity. Healthcare facilities can operate efficiently with powered medical equipment, refrigeration for vaccines, and lighting for surgeries, even in off-grid locations. These advancements have significantly contributed to improving the quality of life and overall development in underserved communities across Africa. The availability of electricity has also opened up opportunities for entrepreneurship and economic growth in these remote areas. With access to reliable power, small businesses can now operate longer hours and utilize modern technology, leading to increased productivity and income generation. This has not only created employment opportunities but also attracted investments and improved the overall economic stability of these communities.

Despite these initiatives and policies to address the challenges of the current energy landscape in Africa, little progress has been made in improving access to reliable and affordable energy for the majority of the population. The lack of infrastructure, limited financial resources, and political instability continue to hinder the implementation and effectiveness of these initiatives. The rapid population growth in many African countries further strains the already limited energy resources, making it even more challenging to meet the increasing demand. As such, it is crucial for African governments and international organizations to prioritize and invest in sustainable energy solutions that can effectively address these challenges. This ranges from promoting renewable energy sources to reduce dependence on fossil fuels **to** developing sustainable financing mechanisms for energy projects, including innovative financing models. By implementing these measures, African countries can not only improve their energy security but also mitigate the negative impacts of climate change. Investing in sustainable energy solutions can create new job opportunities and stimulate economic growth in the region.

4.3.2 Promoting Renewable Energy Sources to Reduce Dependence on Fossil Fuels

Promoting renewable energy sources is crucial for Africa to address the challenges of the current energy landscape. By reducing dependence on fossil fuels, Africa can achieve multiple benefits. It can enhance energy security by diversifying its energy mix and reducing its vulnerability to fluctuations in fossil fuel prices. Promoting renewable energy sources can significantly contribute to mitigating climate change and reducing greenhouse gas emissions, aligning with global sustainability goals. Investing in renewable energy infrastructure will create new job opportunities and stimulate economic growth in Africa. This transition can foster innovation and technological advancements within the continent's energy sector, attracting both domestic and foreign investments. By harnessing abundant renewable resources, African countries can achieve greater energy independence while simultaneously improving access to electricity for their populations. By diversifying their energy sources and reducing reliance on fossil fuels, African countries can also mitigate the risks associated with volatile oil prices and geopolitical tensions. Transitioning to renewable energy can help address the urgent issue of climate change, as it will significantly reduce greenhouse gas emissions and contribute to a cleaner and healthier environment for future generations.

To successfully promote renewable energy sources in Africa, it is essential to develop comprehensive policies and initiatives that prioritize sustainable development. This includes establishing favorable regulatory frameworks, providing financial incentives for clean energy projects, fostering PPPs, and investing in research and development. Collaborative efforts between governments, international organizations, private sector entities, and local communities are crucial for implementing effective strategies that accelerate the adoption of renewable energy technologies across the continent. By promoting renewable energy sources such as solar, wind, and hydroelectric power, Africa can reduce its reliance on fossil fuels and mitigate the negative impacts of climate change. Educating and raising awareness among the population about the benefits of clean energy can encourage individuals to embrace sustainable practices in their daily lives.

Promoting renewable energy sources is a vital component of addressing the challenges faced by Africa's current energy landscape. By reducing dependence on fossil fuels through innovative initiatives and policies, Africa can enhance its energy security, mitigate climate change impacts, stimulate economic growth, create job opportunities, and improve access to electricity for its population. The transition toward a sustainable future requires collaborative efforts and a holistic approach that prioritizes long-term environmental sustainability alongside socio-economic development. This approach involves investing in the research and development of renewable energy technologies, promoting the use of clean energy in transportation and industry, and implementing effective regulatory frameworks to support the growth of renewable energy markets. It is crucial for African countries to foster partnerships with international organizations and leverage financial resources to accelerate the deployment of renewable energy projects across the continent.

Promoting renewable energy sources to reduce dependence on fossil fuels is a crucial step in addressing the challenges of the current energy landscape in most African countries. According to a report by the IRENA (2022), Nigeria has significant renewable energy potential, particularly in solar, wind, and biomass resources. For instance, the country receives abundant sunlight throughout the year, making solar energy a viable option for electricity generation. Nigeria's coastal regions have substantial wind resources that can be harnessed for wind power generation. Implementing initiatives and policies that promote renewable energy sources in Nigeria can help reduce the country's dependence on fossil fuels (Areo et al., 2020). This transition would not only contribute to mitigating climate change but also enhance energy security and create job opportunities in the renewable energy sector. A study conducted by the Nigerian Economic Summit Group estimated that scaling up renewable energy investments could create over 250,000 direct jobs by 2030 (Nwokolo et al. , 2023). By investing in renewable energy infrastructure and incentivizing its adoption, Nigeria can diversify its energy mix and decrease its reliance on fossil fuels. This shift would align with global efforts to combat climate change while unlocking economic growth potential through sustainable development.

According to a report by the IRENA (2023), South Africa has significant potential for renewable energy generation, particularly solar and wind power. The report highlights that harnessing just a fraction of this potential could greatly contribute to reducing the country's dependence on fossil fuels. The South African government has recognized the importance of promoting renewable energy sources and has implemented various initiatives and policies to support their development.

For instance, the Renewable Energy Independent Power Producer Procurement Programme (REIPPPP) was launched in 2011 to attract private sector investment in renewable energy projects (Eberhard and Naude, 2016). This program has been successful in driving the deployment of renewable energy technologies, with over 6,000 MW of capacity already procured (Eberhard and Naude, 2016). These efforts not only contribute to reducing greenhouse gas emissions but also have positive socio-economic impacts. The IRENA (2022) report estimates that by 2030, South Africa's renewable energy sector could create over 200,000 jobs and attract significant investments. Promoting renewable energy sources is crucial for South Africa to address the challenges of its current energy landscape. With its abundant renewable resources and supportive policies, the country has a great opportunity to reduce its dependence on fossil fuels while fostering economic growth and sustainability.

Morocco's promotion of renewable energy sources has advanced significantly, according to a report published by the IRENA (2022). The country has set a target of reaching 52% of its electricity generation from renewable sources by 2030 (IRENA, 2022). This commitment is backed by various initiatives and policies implemented by the Moroccan government. One notable initiative is the Noor Ouarzazate Solar Complex, which is one of the largest concentrated solar power plants in the world. It has a capacity of 580 MW and plays a crucial role in reducing carbon emissions and diversifying Morocco's energy mix (IRENA, 2022). The government has implemented FiTs and other incentives to attract investments in renewable energy projects. These efforts have not only reduced Morocco's dependence on fossil fuels but also created job opportunities and contributed to sustainable economic growth. The country's commitment to renewable energy has received international recognition, positioning Morocco as a leader in the region for transitioning toward a greener and more sustainable energy landscape.

Algeria exhibits considerable potential for renewable energy, especially in solar and wind resources, according to a report published by the IRENA (2022). The report states that the country's solar energy potential alone could meet its electricity demand several times over. Algeria has set ambitious targets for renewable energy deployment, aiming to generate 27% of its electricity from renewable sources by 2030 (IRENA, 2022). To support these goals, the Algerian government has implemented various initiatives and policies. For instance, they have introduced FiTs and power purchase agreements to incentivize private investment in renewable energy projects. The government has also established a regulatory framework that promotes the development of renewable energy, including simplified administrative procedures for project approval. These efforts have already shown promising results. As of 2020, Algeria had installed over 400 MW of renewable energy capacity, primarily from solar and wind projects (IRENA, 2022). This not only contributes to reducing the country's dependence on fossil fuels but also helps diversify its energy mix and improve energy security. Overall, promoting renewable energy sources is a crucial component of Algeria's initiatives and policies to address the challenges of the current energy landscape. By harnessing its abundant renewable resources and implementing supportive measures, Algeria can significantly reduce greenhouse gas emissions, create new job opportunities, and ensure a sustainable and resilient energy future.

According to a report by the IRENA (2022), Egypt has immense potential for renewable energy, particularly solar and wind power. The report states that Egypt's technical potential for solar energy is estimated at around 53 GW, while its wind energy potential is estimated at 76 GW. These figures highlight the significant opportunity for Egypt to harness renewable energy sources and reduce its dependence on fossil fuels. The Egyptian government has recognized the importance of promoting renewable energy and has implemented various initiatives and policies to support its development. For instance, the government launched the FiT program in 2014 (IRENA, 2022), which guarantees fixed electricity prices for renewable energy producers over a specified period. This program has attracted significant investment in the renewable energy sector and has helped Egypt increase its renewable energy capacity. By promoting renewable energy sources through such initiatives and policies, Egypt aims to diversify its energy mix, enhance energy security, reduce greenhouse gas emissions, and create new job opportunities in the clean energy sector. These efforts align with global goals to combat climate change and transition toward a sustainable future.

Tunisia has considerable potential for renewable energy sources, especially solar and wind power, according to a report by the IRENA (2022). The report states that the country's solar energy potential alone is estimated to be around 34,000 MW, which is more than ten times its current electricity generation capacity. To tap into this potential, Tunisia has implemented various initiatives and policies to promote renewable energy sources. For instance, the country has set a target of generating 30% of its electricity from renewable sources by 2030 (IRENA, 2022). This target is supported by the National Renewable Energy Action Plan (NREAP), which outlines specific measures and strategies to achieve this goal (Masud et al., 2017). Tunisia has introduced attractive incentives for renewable energy investments, such as FiTs and tax exemptions. These incentives have attracted both domestic and foreign investors, leading to a significant increase in renewable energy projects in the country. Overall, promoting renewable energy sources in Tunisia not only helps reduce dependence on fossil fuels but also contributes to sustainable economic growth and job creation. The country's commitment to expanding its renewable energy capacity demonstrates its recognition of the importance of transitioning toward a cleaner and more sustainable energy landscape.

Kenya has advanced its promotion of renewable energy sources significantly, according to a report published by the IRENA (2022). The country's installed renewable energy capacity has increased by 86% between 2010 and 2019, with a total of 2,712 MW being generated from renewable sources. This shift toward renewables has helped reduce Kenya's dependence on fossil fuels and contributed to a more sustainable energy landscape. One of the key initiatives driving this transition is the FiT policy implemented by the Kenyan government. The FiT policy guarantees long-term power purchase agreements for renewable energy producers, providing them with a stable and predictable income stream. This has attracted investments in wind, solar, geothermal, and biomass projects, further diversifying Kenya's energy mix. The government has set ambitious targets to increase the share of renewables in its energy mix. The Kenya National Electrification Strategy aims to achieve universal access to electricity by 2022, with at least 50% of this coming from renewable sources (IEA, 2022). These targets provide a clear roadmap for investment and development in the renewable energy sector. By promoting renewable energy sources through initiatives like FiT policies and setting ambitious targets, Kenya is taking proactive steps toward reducing its dependence on fossil fuels. This not only helps address the challenges of the current energy landscape but also contributes to mitigating climate change and fostering sustainable development in the country.

The potential for renewable energy sources, including solar, wind, and geothermal power, is enormous in East Africa, according to a report published by the IRENA (2022). For instance, Kenya's geothermal power capacity has increased significantly in recent years, reaching over 800 MW in 2020. Tanzania has been investing in wind power projects with the potential to generate up to 1,000 MW of clean energy. Promoting renewable energy sources in East Africa not only helps reduce dependence on fossil fuels but also contributes to sustainable development and mitigates climate change. The East African Community (EAC) has recognized the importance of renewable energy and has set a target to achieve 100% renewable energy generation by 2050 (O'Reilly and Heron, 2023). This commitment is aligned with global efforts to transition toward a low-carbon economy and achieve the sustainable development goals. By implementing initiatives and policies that support the development of renewable energy infrastructure and technologies, East Africa can harness its abundant natural resources while creating job opportunities and attracting investments. Promoting renewable energy can enhance energy access in rural areas where grid connectivity is limited, improving livelihoods and fostering economic growth in these regions.

West Africa offers tremendous potential for producing renewable energy, especially in the areas of solar and wind power, according to a report published by the IRENA. The region's solar energy potential alone is estimated to be around 1,600 GW, which is more than double the current global installed capacity (IEA, 2022). To harness this potential and reduce dependence on fossil fuels, several initiatives and policies have been implemented in West Africa. For instance, the Economic Community of West African States (ECOWAS) has established the ECOWAS Renewable Energy Policy (EREP) to promote renewable energy development in the region (Sanga et al., 2022). This policy aims to increase the share of renewable energy in the overall energy mix to 35% by 2020 (Sanga et al., 2022). Individual countries within West Africa have also taken steps toward promoting renewable energy sources. For example, Ghana has implemented the Renewable Energy Act, which provides incentives for renewable energy projects and sets a target of 10% renewable energy contribution to the national electricity supply by 2020 (Aboagye et al., 2021). These initiatives and policies are crucial in addressing the challenges of the current energy landscape in West Africa. By promoting renewable energy sources, the region can reduce its reliance on fossil fuels, mitigate climate change impacts, improve energy security, and foster sustainable economic development.

North Africa offers enormous potential for renewable energy sources, especially solar and wind power, according to a report published by the IRENA. The region receives abundant sunlight throughout the year and experiences strong winds in certain areas, making it an ideal location for harnessing these resources (IRENA, 2019). In recent years, several initiatives and policies have been implemented in North Africa to promote renewable energy and reduce dependence on fossil fuels. For instance, Morocco has made significant progress in developing its renewable energy sector through the Noor Solar Power Plant, which is one of the largest concentrated solar power projects in the world (IRENA, 2019). Tunisia has set a target to generate 30% of its electricity from renewable sources by 2030 (World Bank Group, 2022). These efforts not only contribute to reducing greenhouse gas emissions and mitigating climate change but also provide opportunities for job creation and economic growth in the region. By diversifying their energy mix and embracing renewables, countries in North Africa can enhance their energy security while transitioning toward a more sustainable future.

Promoting renewable energy sources is crucial in Central Africa to address the challenges of the current energy landscape. According to a report by the IRENA (2019), Central Africa has vast untapped renewable energy potential, including solar, wind, hydro, and biomass resources. For instance, the region's solar energy potential alone is estimated to be around 13 TW-h per year. By investing in and promoting these renewable energy sources, Central African countries can reduce their dependence on fossil fuels and achieve multiple benefits. Transitioning to renewables can help mitigate climate change by reducing greenhouse gas emissions and improving air quality. It can enhance energy security by diversifying the energy mix and reducing vulnerability to fluctuating fuel prices. For example, Cameroon has recognized the importance of renewable energy in its development agenda. The country has set a target to increase its share of renewable energy in the electricity mix to 25% by 2035 (IRENA, 2019). This commitment demonstrates how promoting renewables can be a part of broader initiatives and policies aimed at addressing the challenges of the current energy landscape in Central Africa.

4.3.3 Improving Access to Electricity in Rural Areas through Decentralized Energy Systems

Improving access to electricity in rural areas through decentralized energy systems is a crucial aspect of addressing the challenges faced by the current energy landscape in Africa. By implementing such initiatives and policies, African countries can bridge the energy gap, enhance economic growth, and improve the quality of life for their rural populations. Decentralized energy systems offer a viable solution as they empower local communities to generate and distribute electricity independently (Brent et al., 2022). This approach not only reduces reliance on centralized power grids but also promotes sustainability by utilizing renewable energy sources like solar or wind power. It fosters innovation and entrepreneurship within these communities, creating opportunities for job creation and economic development. By embracing decentralized energy systems, Africa can unlock its vast potential and pave the way for a brighter and more inclusive future.

Improving access to electricity in rural areas through decentralized energy systems is a crucial component of the initiatives and policies implemented to address the challenges of the current energy landscape in Africa. According to a report by the IEA, approximately 600 million people in sub-Saharan Africa lack access to electricity, with the majority residing in rural areas (IEA, 2022). To address this challenge, initiatives and policies focused on improving access to electricity through decentralized energy systems have gained momentum in Africa. Decentralized energy systems, such as mini-grids and off-grid solutions, have proven to be effective in providing electricity to remote rural communities. These systems utilize renewable energy sources like solar, wind, and hydropower, reducing reliance on fossil fuels and promoting sustainable development (World Bank, 2022). For instance, the World Bank's Lighting Africa program has successfully deployed off-grid solar solutions across several African countries. This initiative has brought electricity access to millions of households in rural areas that were previously underserved by centralized grid infrastructure (World Bank, 2022). By implementing decentralized energy systems as part of broader initiatives and policies, African countries can overcome the challenges posed by the current energy landscape. These systems not only improve access to electricity but also contribute to poverty reduction, economic growth, and environmental sustainability in rural areas (IEA and UNEP, 2018).

According to a study conducted by the IEA, approximately 11% of South Africa's population still lacks access to electricity, with the majority residing in rural areas (IEA, 2022). In response to this challenge, the South African government has implemented initiatives and policies aimed at improving access to electricity through decentralized energy systems. One such initiative is the REIPPPP, which has successfully facilitated the deployment of renewable energy projects in rural areas (Eberhard and Naude, 2016). Through this program, private sector investors are encouraged to develop and operate renewable energy projects, such as solar and wind farms, thereby increasing electricity access in remote regions (Eberhard and Naude, 2016). The government has also prioritized the establishment of mini-grids and off-grid solutions as part of its decentralized energy approach. These systems utilize a combination of renewable energy sources and storage technologies to provide reliable electricity to communities that are geographically isolated or have limited grid infrastructure (Eberhard and Naude, 2016). By focusing on decentralized energy systems and implementing initiatives like REIPPPP, South Africa aims to address the challenges posed by its current energy landscape while simultaneously improving access to electricity in rural areas. These efforts not only contribute toward achieving universal electrification but also promote sustainable development and reduce greenhouse gas emissions in line with global climate goals.

Approximately 600 million people in sub-Saharan Africa do not have access to electricity, the majority of whom live in rural areas, according to a report published by the IEA (2022). In response to this challenge, decentralized energy systems have emerged as a promising solution to improve access to electricity in these remote regions. For instance, a study conducted by the World Bank highlighted the success of decentralized energy systems in East Africa, particularly in countries like Kenya and Tanzania. These systems, which include mini-grids and off-grid solutions such as solar home systems, have proven to be effective in providing reliable and affordable electricity to rural communities. The report also emphasized that these initiatives have contributed significantly to reducing energy poverty and improving socio-economic development in these areas. Furthermore, various initiatives and policies have been implemented by governments and international organizations to support the adoption of decentralized energy systems in East Africa. For example, the Scaling Off-Grid Energy (SOGE) initiative launched by Power Africa aims to provide access to modern energy services for 20 million households across sub-Saharan Africa by 2030 (Malhotra et al., 2017). National governments have introduced supportive policies such as tax incentives and regulatory frameworks that encourage private sector investments in decentralized energy projects. Overall, improving access to electricity through decentralized energy systems is crucial for addressing the challenges faced by rural areas in East Africa. The combination of successful case studies and supportive initiatives demonstrates the potential of these solutions to bring about positive change in the region's energy landscape.

In rural Central Africa, only 23% of people have access to reliable electricity services, according to a study by the IEA (2022). This indicates that access to electricity in these areas remains a significant challenge. To address this issue, the implementation of decentralized energy systems has emerged as a promising solution. These systems involve the use of small-scale renewable energy technologies, such as solar panels and mini-grids, to provide electricity to remote communities (IEA and UNEP, 2018). By improving access to electricity through decentralized energy systems, Central Africa can enhance socio-economic development, reduce reliance on fossil fuels, and mitigate the adverse impacts of climate change. Decentralized energy systems not only provide electricity to remote communities but also offer numerous benefits. They can improve healthcare services by powering medical facilities, enhance educational opportunities through access to computers and the internet, and stimulate economic growth by enabling small businesses to operate efficiently. The shift toward renewable energy sources can help Central Africa reduce its carbon emissions and contribute to global efforts to combat climate change.

North Africa has advanced significantly in enhancing rural areas' access to electricity through decentralized energy systems, according to a report published by the IRENA (2022). The report highlights that initiatives and policies aimed at addressing the challenges of the current energy landscape have played a crucial role in this achievement. For instance, Morocco's National Electrification Program (NEP) has been successful in providing electricity access to remote rural areas through the deployment of decentralized renewable energy systems (Mowie, 2019). This program has helped increase electricity access from 18% in 1995 to over 99% in 2020, benefiting more than 2 million people. Similarly, Tunisia's Renewable Energy Program (REP) has focused on promoting decentralized renewable energy solutions (Krarti, 2020), such as solar mini-grids and standalone systems, to improve electricity access in rural areas. As a result, over 400,000 people have gained access to electricity through these initiatives (Krarti, 2020). These examples demonstrate the effectiveness of decentralized energy systems as part of broader initiatives and policies addressing the challenges of the current energy landscape in North Africa. The figures from IRENA's report emphasize the positive impact of these efforts on improving access to electricity for rural communities (IRENA, 2022).

4.3.4 Strengthening Regional Cooperation and Integration for Better Energy Planning and Resource Sharing

Strengthening regional cooperation and integration is crucial for Africa to address the challenges of its current energy landscape effectively. By working together, African nations can pool their resources and expertise to develop comprehensive energy plans that cater to the needs of the entire region. This collaborative approach will not only enhance energy security but also promote sustainable development and economic growth across Africa. Sharing resources, such as renewable energy technologies and infrastructure, will enable countries to optimize their energy production and distribution systems, leading to a more efficient and resilient energy sector for the continent as a whole. This collaborative approach will also foster knowledge sharing and capacity building among African nations, allowing them to learn from each other's experiences and best practices in the energy sector. By leveraging their collective strengths, African countries can overcome common challenges and accelerate the transition toward cleaner and more affordable energy sources. This transition will not only contribute to mitigating climate change and reducing greenhouse gas emissions, but it will also create new job opportunities and stimulate economic growth across the continent. By embracing renewable energy sources such as solar and wind power, African countries can reduce their dependence on fossil fuels and improve energy security in the long run. Investing in renewable energy infrastructure can also enhance energy access for remote and underserved communities, improve their quality of life, and promote social inclusion. The adoption of clean energy technologies can attract foreign investments and partnerships, fostering technological innovation and knowledge transfer within the region.

Significant advantages can result from regional cooperation and integration in Africa's energy sector, according to a report published by the African Development Bank (2022). For instance, the report states that interconnecting power systems across countries can increase access to electricity, improve reliability, and reduce costs. The Southern African Power Pool (SAPP) is a successful example of regional cooperation (Wright and van Coller, 2018), with its interconnected grid enabling countries to share resources and balance supply and demand. The IRENA (2022) highlights that regional cooperation can enhance renewable energy deployment in Africa. By sharing resources such as wind and solar potential across borders, countries can tap into a wider range of renewable sources and optimize their energy mix. This not only contributes to addressing the challenges of the current energy landscape but also promotes sustainable development and reduces greenhouse gas emissions. Strengthening regional cooperation and integration in Africa's energy sector is crucial for better energy planning and resource sharing. The success of initiatives like SAPP demonstrates the potential benefits of interconnecting power systems, while leveraging regional resources can accelerate renewable energy deployment. These efforts are essential for addressing the challenges faced by Africa's energy landscape and promoting sustainable development in the region.

Regional integration and cooperation in sub-Saharan Africa can have a major positive impact on resource sharing and energy planning, according to a report published by the IRENA (2022). The report highlights that by coordinating efforts and sharing resources, countries in the region can achieve economies of scale, reduce costs, and improve energy security. For instance, the SAPP is a successful example of regional cooperation in the energy sector (Wright and van Coller, 2018). By pooling together their electricity resources and infrastructure, SAPP member countries have been able to optimize power generation and transmission, resulting in improved reliability and affordability of electricity for millions of people. A study conducted by the World Bank estimates that enhancing regional integration in sub-Saharan Africa could increase access to electricity for an additional 40 million people by 2030 (World Bank, 2022). This demonstrates the potential impact of collaborative efforts in addressing the challenges faced by the current energy landscape in the region. Strengthening regional cooperation and integration is crucial for better energy planning and resource sharing in sub-Saharan Africa. By leveraging economies of scale and optimizing infrastructure, countries can enhance energy security, reduce costs, and expand access to electricity for millions of people.

One example of strengthening regional cooperation and integration for better energy planning and resource sharing in North Africa is the Desertec initiative. The Desertec Foundation proposed a project that aimed to harness solar and wind energy from the deserts of North Africa to supply electricity to Europe. According to a study by the German Aerospace Center, this initiative has the potential to generate up to 15% of Europe's electricity demand by 2050 (Hannemann et al., 2018). Regional cooperation in North Africa can also be seen through the Maghreb-Europe Gas Pipeline project. This project aims to connect the natural gas reserves of Algeria with European markets through a pipeline network. The IEA estimates that this pipeline could transport up to 25 billion cubic meters of natural gas annually, enhancing energy security and diversification for both North Africa and Europe (IEA, 2022). These regional cooperation initiatives not only have the potential to meet Europe's growing electricity demand but also contribute to reducing greenhouse gas emissions. By harnessing renewable energy sources in North Africa, such as solar and wind power, the electricity exported to Europe can help reduce reliance on fossil fuels and support the transition toward a more sustainable energy future. The Maghreb-Europe Gas Pipeline project can strengthen economic ties between North Africa and Europe, fostering mutual benefits and cooperation in the energy sector.

4.3.5 Developing Sustainable Financing Mechanisms for Energy Projects, Including Innovative Financing Models

Developing sustainable financing mechanisms for energy projects is crucial to addressing the challenges of the current energy landscape in Africa. This requires a comprehensive approach that goes beyond traditional funding sources and explores innovative financing models. By incorporating innovative financing models into initiatives and policies, African countries can attract more investments and ensure long-term financial sustainability for their energy projects. These models may include PPPs, green bonds, crowdfunding platforms (Milligom et al., 2009), or impact investment funds that prioritize renewable energy development (Odarno et al., 2017). By diversifying their financing options, African nations can accelerate the transition toward cleaner and more reliable energy sources while fostering economic growth and job creation in the region. These financing models can also help African countries overcome the financial barriers that often hinder the implementation of large-scale energy projects. By leveraging these innovative financing mechanisms, African nations can tap into international funding sources and attract foreign investors, further boosting their energy sector and overall economic development.

According to a report by the IRENA (2022), Africa has vast untapped renewable energy potential, but the lack of sustainable financing mechanisms has been a major barrier to its development. The report highlights that innovative financing models, such as green bonds and PPPs, can play a crucial role in mobilizing the necessary investments for energy projects in Africa. For instance, the African Development Bank (2022) has been at the forefront of promoting sustainable financing for energy projects in Africa. Through its Sustainable Energy Fund for Africa (SEFA), the AfDB has provided grants and concessional loans to support renewable energy initiatives across the continent. As of 2020, SEFA had mobilized over $95 million in funding and supported more than 30 projects in various African countries. These initiatives and policies are essential for addressing the challenges of the current energy landscape in Africa by ensuring access to clean and affordable energy while promoting sustainable development.

Developing sustainable financing mechanisms for energy projects is crucial in North Africa to address the challenges of the current energy landscape. According to a report by the IRENA (2022), the region has immense renewable energy potential, particularly in solar and wind resources. Limited access to financing has been a major barrier to harnessing this potential. To overcome this challenge, innovative financing models can play a significant role. For instance, green bonds have gained popularity globally as a means to raise funds for renewable energy projects. North African countries can explore the issuance of green bonds, which are specifically designed to finance environmentally friendly initiatives. This would attract investors interested in sustainable investments and provide much-needed capital for renewable energy development. PPPs can also be effective in mobilizing financing for energy projects. By partnering with private entities, governments can leverage their expertise and financial resources to accelerate the deployment of renewable energy technologies. This approach not only shares the financial burden but also promotes knowledge transfer and technology diffusion. Developing sustainable financing mechanisms and incorporating innovative models are essential steps toward addressing the challenges of North Africa's current energy landscape. Green bonds and PPPs offer promising avenues for attracting investment and accelerating the transition toward renewable energy sources in the region.

Sub-Saharan Africa has enormous potential for the development of renewable energy, with an estimated 10 TW of solar energy and 350 GW of wind energy, according to a report by the IRENA (2022). However, the lack of sustainable financing mechanisms has been a major barrier to harnessing this potential. To address this challenge, various initiatives and policies have been implemented in sub-Saharan Africa. For instance, the African Development Bank (2022) launched the SEFA in 2011, which provides grants and technical assistance to support renewable energy projects. The United Nations Development Programme (UNDP) has been working with governments and private sector partners to establish innovative financing models such as green bonds and PPPs (Soleimani et al., 2023). These efforts have shown promising results. For example, SEFA has supported over 30 projects across the region, resulting in increased access to clean energy for millions of people. Countries like Kenya and Rwanda have successfully issued green bonds to finance renewable energy projects. By developing sustainable financing mechanisms and embracing innovative financing models, sub-Saharan Africa can unlock its vast renewable energy potential and address the challenges of its current energy landscape effectively.

4.4 IMPACT OF THE CURRENT ENERGY LANDSCAPE ON ECONOMIC DEVELOPMENT IN AFRICA

4.4.1 Limited Access to Electricity Hinders the Growth of Industries and Businesses, Leading to Reduced Job Opportunities and Income Generation

Limited access to electricity in Africa has been a significant barrier to economic growth and development. Without a reliable power supply, industries and businesses struggle to operate efficiently, hindering their productivity and competitiveness in the global market. This lack of electricity also limits the adoption of modern technologies and automation, which are crucial for increasing productivity and reducing costs. Consequently, the limited access to electricity leads to reduced job opportunities as businesses are unable to expand or create new ventures, resulting in lower income generation for individuals and communities across the continent. The lack of electricity hampers the provision of essential services such as healthcare and education. Hospitals and clinics are unable to operate effectively, compromising the quality of medical care available to people. Similarly, schools and educational institutions face challenges in delivering quality education without a reliable power supply, limiting opportunities for skill development and knowledge acquisition. Overall, the limited access to electricity not only hinders economic growth but also affects the overall well-being and development of individuals and communities in Africa. Furthermore, the lack of electricity also impacts other crucial sectors, such as agriculture and manufacturing. Without a stable power supply, farmers struggle to store and process their produce, leading to significant losses in agricultural productivity. Industries face difficulties in operating machinery and maintaining production levels, hindering economic progress and job creation. Addressing the electricity deficit in Africa is essential for promoting sustainable development and improving the lives of millions of people.

To address this challenge, it is essential for African governments and international organizations to invest in infrastructure development and renewable energy sources, such as solar or wind power, to ensure sustainable economic growth and create more employment opportunities. Investing in infrastructure development and renewable energy sources would not only address the electricity deficit but also reduce reliance on fossil fuels, leading to a cleaner and more environmentally friendly energy sector. Promoting the use of renewable energy sources would also contribute to mitigating the effects of climate change and reducing greenhouse gas emissions, aligning with global sustainability goals. The implementation of renewable energy sources would diversify the energy mix, reducing the vulnerability to fluctuations in fossil fuel prices. This would create a more stable and resilient energy sector, ensuring long-term economic stability and energy security for the country. Investing in infrastructure development and renewable energy would attract foreign investments and promote technological advancements, positioning the country as a leader in sustainable development and attracting international collaborations.

4.4.2 High Energy Costs Burden Households and Discourage Consumption, Limiting Economic Growth

The high energy costs in Africa place a significant burden on households, as they struggle to afford basic necessities and meet their daily energy needs. This burden not only affects their quality of life but also discourages consumption, as households are forced to prioritize essential expenses over discretionary spending. Consequently, this limited consumption hinders economic growth by reducing the demand for goods and services, ultimately impeding Africa's overall economic development. The lack of access to reliable and affordable energy sources also hampers the productivity and competitiveness of businesses in Africa. Without a stable energy supply, industries are unable to operate at full capacity, leading to decreased production and job opportunities. This further exacerbates the cycle of poverty and inequality, hindering Africa's progress toward sustainable development. In addition, the unreliable energy supply also affects the quality of life for individuals in Africa. Without access to electricity, basic necessities such as lighting, cooking, and healthcare become challenging to fulfill, impacting the overall well-being of communities. This lack of reliable energy sources not only hampers economic growth but also perpetuates social inequalities and limits opportunities for individuals to improve their living conditions.

To overcome this challenge, innovative solutions such as investing in renewable energy sources and implementing energy efficiency measures are crucial to alleviate the burden on households and stimulate economic growth in the region. By investing in renewable energy sources, communities can have a sustainable and reliable source of electricity that is not dependent on fossil fuels. This not only reduces greenhouse gas emissions but also creates job opportunities in the renewable energy sector, promoting economic development and reducing poverty. Implementing energy efficiency measures can help reduce energy consumption and lower costs for households, making electricity more affordable and accessible for all members of the community.

4.4.3 Reliance on Fossil Fuels Contributes to Environmental Degradation, Affecting Agricultural Productivity and Food Security

Reliance on fossil fuels in Africa has significant implications for environmental degradation, which in turn has a direct impact on agricultural productivity and food security. The burning of fossil fuels releases greenhouse gases, such as carbon dioxide and methane, into the atmosphere, leading to climate change and global warming.

These changes in temperature patterns can disrupt rainfall patterns, increase the frequency of extreme weather events like droughts and floods, and negatively affect crop yields. The extraction and transportation of fossil fuels often result in habitat destruction, soil degradation, and water pollution. These environmental consequences can directly harm agricultural lands and reduce their productivity. The reliance on fossil fuels perpetuates an unsustainable energy system that diverts resources away from investing in renewable energy sources that could provide more resilient and environmentally friendly solutions for Africa's energy needs. Addressing this issue requires a shift toward sustainable alternatives such as renewable energy sources like solar, wind, hydroelectric power, or bioenergy. Transitioning to cleaner energy options not only reduces greenhouse gas emissions but also promotes economic development by creating new job opportunities in the renewable energy sector. Investing in sustainable agriculture practices that prioritize soil health conservation and water management can help mitigate the negative impacts of climate change on agricultural productivity while ensuring food security for Africa's growing population. Implementing energy-efficient technologies and practices in industries and buildings can significantly reduce energy consumption and carbon emissions. This can be achieved through the adoption of energy-efficient appliances, insulation, and smart grid systems. Promoting sustainable transportation options such as electric vehicles and public transportation can further contribute to reducing air pollution and dependence on fossil fuels.

4.4.4 Inadequate Energy Infrastructure Hampers Transportation Systems, Hindering Trade and Economic Integration

Inadequate energy infrastructure poses a significant challenge to transportation systems in Africa, ultimately impeding trade and hindering economic integration. Insufficient access to reliable and affordable energy sources limits the efficiency and capacity of transportation networks, leading to increased costs, delays, and unreliable services. This not only hampers the movement of goods and people but also undermines the potential for regional economic cooperation and integration, as countries struggle to connect and collaborate effectively due to energy-related constraints. Consequently, addressing the energy infrastructure gap becomes crucial for fostering sustainable economic development in Africa by enabling smoother transportation systems that facilitate trade and enhance regional integration. Improving energy infrastructure in Africa would not only benefit transportation systems but also have a positive ripple effect on other sectors such as manufacturing and agriculture. Access to reliable energy sources would enable industries to operate efficiently, leading to increased productivity and job creation. It would attract foreign investment and promote economic growth, ultimately contributing to poverty reduction and improved living standards in the region. The development of energy infrastructure in Africa would also address the issue of energy poverty, providing access to electricity for millions of people who currently lack it. This would have significant social and health benefits, as reliable electricity allows for better access to healthcare services and improves overall quality of life. The shift toward cleaner and more sustainable energy sources would help mitigate the impacts of climate change, benefiting not only Africa but also the global community as a whole. The development of energy infrastructure in Africa would also create job opportunities (Nwokolo, 2017) and stimulate economic growth (Nwokolo and Otse, 2019). This would help alleviate unemployment rates and contribute to the overall economic development of the region (Ituen et al., 2012). By reducing reliance on fossil fuels and promoting renewable energy sources, Africa can contribute to global efforts to combat climate change (Nwokolo and Ogbulezie, 2018a) and reduce greenhouse gas emissions (Nwokolo and Ogbulezie, 2018b). This would further enhance the continent's reputation as a responsible global citizen and attract investment in clean energy technologies.

REFERENCES

Aboagye, B., Gyamfi, S., Ofosu, E. A., & Djordjevic, S. (2021). Status of renewable energy resources for electricity supply in Ghana. *Scientific African*. https://doi.org/10.1016/j.sciaf.2020.e00660

African Development Bank. (2022). African Development Bank – Building today, a better Africa tomorrow. www.afdb.org/en/

Africa Energy Outlook 2022. *Africa Energy Outlook 2022*. https://doi.org/10.1787/2abd9ce2-en

Aniebo, I. N., & Akpan, J. S. (2022). Energy transition: Implications, considerations, and roadmap development for sub-Saharan Africa. *Society of Petroleum Engineers – SPE Nigeria Annual International Conference and Exhibition, NAIC 2022*. https://doi.org/10.2118/211990-MS

Areo, O. S., Gershon, O., & Osabuohien, E. (2020). Improved public services and tax compliance of small and medium scale enterprises in Nigeria: A generalised ordered logistic regression. *Asian Economic and Financial Review*. https://doi.org/10.18488/journal.aefr.2020.107.833-860

Avila, S. (2018). Environmental justice and the expanding geography of wind power conflicts. *Sustainability Science*. https://doi.org/10.1007/s11625-018-0547-4

Baptista, I., & Plananska, J. (2017). The landscape of energy initiatives in sub-Saharan Africa: Going for systemic change or reinforcing the status quo? *Energy Policy*. https://doi.org/10.1016/j.enpol.2017.08.006

Bester, V., & Uys, T. (2023). Artisanal mining and its drivers in the South African context. *Extractive Industries and Society*. https://doi.org/10.1016/j.exis.2023.101278

Boafo, Y., Lam, R., Juju, D., & Saito, O. (2019). The role of sustainability science in addressing Africa's sustainable development challenges and priorities: A bibliometric survey. In *Co-Designing Science in Africa: First Steps in Assessing the Sustainability Science Approach on the Ground*.

Brent, A., Chatterjee, A., Burmester, D., & Rayudu, R. (2022). Sustainable microgrids for remote communities: A practical framework for analyzing and designing. In *The Palgrave Handbook of Climate Resilient Societies: Volumes 1-2*. https://doi.org/10.1007/978-3-030-42462-6_65

Brower, T. (2016). Shango unchained? State in(capacity), urban bias, and the power Africa initiative. *Revista Brasileira de Políticas Públicas*. https://doi.org/10.5102/rbpp.v5i3.3409

Eberhard, A., & Naude, R. (2016). The South African renewable energy independent power producer Procurement programme: A review and lessons learned. *Journal of Energy in Southern Africa*. https://doi.org/10.17159/2413-3051/2016/v27i4a1483

Gebreslassie, M. G., Cuvilas, C., Zalengera, C., To, L. S., Baptista, I., Robin, E., Bekele, G., Howe, L., Shenga, C., Macucule, D. A., Kirshner, J., Mulugetta, Y., Power, M., Robinson, S., Jones, D., & Castán Broto, V. (2022). Delivering an off-grid transition to sustainable energy in Ethiopia and Mozambique. *Energy, Sustainability and Society*. https://doi.org/10.1186/s13705-022-00348-2

Gromek-Broc, K. (2023). Regional approaches to the energy transition: A multidisciplinary perspective. In *Regional Approaches to the Energy Transition: A Multidisciplinary Perspective*. https://doi.org/10.1007/978-3-031-19358-3

Hafner, M., Tagliapietra, S., & de Strasser, L. (2018). Energy in Africa: Challenges and opportunities. *SpringerBriefs in Energy*.

Hannemann, K., Martinez Schramm, J., Wagner, A., Friedl, D., & Ponchio Camillo, G. (2018). The high enthalpy shock tunnel Göttingen of the German Aerospace Center (DLR). *Journal of Large-Scale Research Facilities JLSRF*. https://doi.org/10.17815/jlsrf-4-168

IEA, & UNEP. (2018). International Energy Agency and the United Nations Environment Programme – Global status report 2018: Towards a zero-emission, efficient and resilient buildings and construction sector. *Global Status Report*.

International Energy Agency. (2022). World energy outlooks special report Africa energy outlook 2022. IEA.

International Renewable Energy Agency (IRENA). (2019). IRENA (2019), *Global energy transformation: A roadmap to 2050*. IRENA

International Renewable Energy Agency (IRENA). (2022). World energy transitions outlook 2023: 1.5°C pathway. *World Energy Transitions*. IRENA

International Renewable Energy Agency (IRENA). (2023). World energy transitions outlook 2023: 1.5°C pathway. *World Energy Transitions*. IRENA

Ituen, E. E., Esen, N. U., Nwokolo, S. C., & Udo, E. G. (2012). Prediction of global solar radiation using relative humidity, maximum temperature and sunshine hours in Uyo, in the Niger Delta Region, Nigeria. *Advances in Applied Science Research*, *3*(4), 1923–1937.

Kitonga, J. F., & Bonafé Ernesto. (2015). *Tanzanian Energy Sector under the Universal Principles of the Energy Charter*. Ministry of Energy and Minerals of the United Republic of Tanzania and Energy Charter Secretariat.

Klenam, D. E. P., Bamisaye, O. S., Williams, I. E., Van Der Merwe, J. W., & Bodunrin, M. O. (2022). Global perspective and African outlook on additive manufacturing research – An overview. *Manufacturing Review*. https://doi.org/10.1051/mfreview/2022033

Krarti, M. (2020). Multiple-benefit analysis of scaling-up building energy efficiency programs: The case study of Tunisia. *Journal of Engineering for Sustainable Buildings and Cities*. https://doi.org/10.1115/1.4045871

Lombard, A., & Ferreira, S. L. A. (2015). The spatial distribution of renewable energy infrastructure in three particular provinces of South Africa. *Bulletin of Geography. Socio-Economic Series*. https://doi.org/10.1515/bog-2015-0036

Malhotra, A., Schmidt, T. S., Haelg, L., & Waissbein, O. (2017). Scaling up finance for off-grid renewable energy: The role of aggregation and spatial diversification in derisking investments in mini-grids for rural electrification in India. *Energy Policy*. https://doi.org/10.1016/j.enpol.2017.06.037

Mas'ud, A. A., Wirba, A. V., Ardila-Rey, J. A., Albarracín, R., Muhammad-Sukki, F., Duque, Á. J., Bani, N. A., & Munir, A. B. (2017). Wind power potentials in Cameroon and Nigeria: Lessons from South Africa. *Energies*. https://doi.org/10.3390/en10040443

Mewenemesse, H. T., & Yan, Q. (2023). Policy analysis of the challenges to an effective switch to low-carbon energy in the economic community of West African states. *Energies*. https://doi.org/10.3390/en16052191

Milligan, V., Gurran, N., Lawson, J., Phibbs, P., & Phillips, R. (2009). Innovation in affordable housing in Australia: Bringing policy and practice for not-for-profit housing organisations together. *AHURI Final Report.*

MOWIE, 2019. (2019). Lighting to all: National Electrification Program 2.0: Integrated Planning for Universal Access. *Thematic Program Document.*

Mubaraq, T., Abubakar, M. T., Olatunji-Ohwovoriole, O. (2023). Investigating engagement strategies in the presidential media debate of the 2023 general election in Nigeria. *Tasambo Journal of Language, Literature, and Culture.* https://doi.org/10.36349/tjllc.2023.v02i01.011

Mukelabai, M. D., Wijayantha, U. K. G., & Blanchard, R. E. (2022). Renewable hydrogen economy outlook in Africa. *Renewable and Sustainable Energy Reviews.* https://doi.org/10.1016/j.rser.2022.112705

Mulopo, J. (2022). A mini-review of practical interventions of renewable energy for climate change in sub-Saharan Africa in the last decade (2010–2020): Implications and perspectives. *Heliyon.* https://doi.org/10.1016/j.heliyon.2022.e11296

Musango, J. K., Amigun, B., & Brent, A. C. (2011). Sustainable electricity generation technologies in South Africa: Initiatives, challenges and policy implications. *Energy and Environment Research.* https://doi.org/10.5539/eer.v1n1p124

Mwale, S., & Davidson, I. (2014). Power deficits and outage planning in South Africa. *Journal of Energy Challenges & Mechanics.*

Nwokolo, S. C. (2017). A comprehensive review of empirical models for estimating global solar radiation in Africa. *Renewable and Sustainable Energy Reviews*, *78*(October), 955–995. https://doi.org/10.1016/j.rser.2017.04.101

Nwokolo, S. C., & Ogbulezie, J. C. (2018a). A qualitative review of empirical models for estimating diffuse solar radiation from experimental data in Africa. *Renewable and Sustainable Energy Reviews*, *92*(18), 353–393. https://doi.org/10.1016/j.rser.2018.04.118

Nwokolo, S. C., & Ogbulezie, J. C. (2018b). A quantitative review and classification of empirical models for predicting global solar radiation in West Africa. *Beni-Suef University Journal of Basic and Applied Sciences*, *7*(4), 367–396. https://doi.org/10.1016/j.bjbas.2017.05.001

Nwokolo, S., & Otse, C. (2019). Impact of sunshine duration and clearness index on diffuse solar radiation estimation in mountainous climate. *Trends in Renewable Energy*, *5*(3), 307–332. https://doi.org/10.17737/tre.2019.5.3.00107

Nwokolo, S. C., Singh, R., Khan, S., Kumar, A., & Luthra, S. (2023). Scenarios that could give rise to an African net-zero energy transition. In *CSR, sustainability, ethics and governance: Vol. Part F1850.* https://doi.org/10.1007/978-3-031-44514-9_8

Odarno, L., Agarwal, A., Devi, A., & Takahashi, H. (2017). Strategies for expanding universal access to electricity services for development. *Working Paper.*

O'Reilly, P., & Heron, T. (2023). Institutions, ideas and regional policy (un-)coordination: The East African Community and the politics of second-hand clothing. *Review of International Political Economy.* https://doi.org/10.1080/09692290.2022.2062614

Otlhogile, M., & Shirley, R. (2023). The evolving just transition: Definitions, context, and practical insights for Africa. *Environmental Research: Infrastructure and Sustainability.* https://doi.org/10.1088/2634-4505/ac9a69

Ozohu-Suleiman, A. (2021). The energy crisis and environmental concerns in sub-Saharan Africa (Nigeria): A perspective on public governance for climate action. *Journal of Public Administration and Governance.* https://doi.org/10.5296/jpag.v11i2.18242

Pollet, B. G., Staffell, I., & Adamson, K. A. (2015). Current energy landscape in the Republic of South Africa. *International Journal of Hydrogen Energy.* https://doi.org/10.1016/j.ijhydene.2015.09.141

Pollet, B. G., Staffell, I., Adamson, K.-A., & SpringerLink. (2016). The energy landscape in the Republic of South Africa. *Springer Briefs in Energy*.

Sanga, D., Ahouakan, E. W. V., & Adje, N. A. F. (2022). Institutional quality and labour productivity: Evidence from the Economic Community of West African States Countries. *Journal of Economic Integration*. https://doi.org/10.11130/jei.2022.37.3.540

Sanoh, A., Parshall, L., Sarr, O. F., Kum, S., & Modi, V. (2012). Local and national electricity planning in Senegal: Scenarios and policies. *Energy for Sustainable Development*. https://doi.org/10.1016/j.esd.2011.12.005

Soleimani, A., Fahim Nia, F., Naghshineh, N., & Soleimani Nejad, A. (2023). Providing a framework for the reuse of research data based on the development dynamic framework of United Nations Development Program (UNDP). *Library Hi Tech*. https://doi.org/10.1108/LHT-01-2021-0007

Spalding-Fecher, R., Williams, A., & van Horen, C. (2000). Energy and environment in South Africa: Charting a course to sustainability. *Energy for Sustainable Development*. https://doi.org/10.1016/S0973-0826(08)60259-8

The World Bank Group. (2022). *Risk of Global Recession in 2023 Rises Amid Simultaneous Rate Hikes*. The World Bank Group.

Trotter, P. A., & Abdullah, S. (2018). Re-focusing foreign involvement in sub-Saharan Africa's power sector on sustainable development. *Energy for Sustainable Development*. https://doi.org/10.1016/j.esd.2018.03.003

Wang, X., Van Dam, K. H., Triantafyllidis, C., Koppelaar, R. H. E. M., & Shah, N. (2017). Water and energy systems in sustainable city development: A Case of sub-Saharan Africa. *Procedia Engineering*. https://doi.org/10.1016/j.proeng.2017.07.140

Wright, J. G., & van Coller, J. (2018). System adequacy in the Southern African Power Pool: A case for capacity mechanisms. *Journal of Energy in Southern Africa*. https://doi.org/10.17159/2413-3051/2018/V29I4A5581

5 Global Strategies for Achieving Different Phases of Net Zero Energy

5.1 INTRODUCTION

There is no doubt that the low-carbon energy transition has moved into a new stage of growth, and the drive toward net zero emissions is gaining momentum. Investments in renewable energy sources like wind, solar, and hydropower are being made by both public and private organizations to reduce their carbon footprint. This shift toward cleaner energy is not only good for the environment but also for the economy, as it creates new jobs and stimulates innovation. Transitioning to a low-carbon economy is not without challenges. It requires significant investment in infrastructure and technology (Sakr, 2017), as well as changes in consumer behavior (Meena et al., 2017). Governments must also ensure that vulnerable communities are not left behind in this transition. Despite these challenges, the benefits of a low-carbon future far outweigh the costs. By working together to reduce our greenhouse gas emissions, we can create a more sustainable and prosperous world for future generations.

The increasing commitment to reduce greenhouse gas emissions to net zero by governments and private firms is a positive step toward mitigating the effects of climate change. With over 120 countries representing 61% of global greenhouse gas (GHG)emissions pledging to reach net zero by mid-century or soon after (www.climatewatchdata.org/), it is clear that the world is taking this issue seriously (Morales et al., 2021). This shift toward sustainability will require significant investment in renewable energy sources and the development of new technologies to reduce emissions from industries such as transportation and agriculture. It will also require changes in individual behavior, such as reducing meat consumption and using public transportation instead of personal vehicles. The benefits of these changes will be significant, including cleaner air and water, improved public health, and a more stable climate for future generations. As we continue on this path toward net zero emissions, it is important to prioritize collaboration between nations and industries to ensure a sustainable future for all.

This is a significant milestone in the fight against climate change. The commitments made by these countries demonstrate a growing global consensus on the need to reduce greenhouse gas emissions (Tramontin et al., 2012) and limit global warming to 1.5°C above pre-industrial levels (McGraw Hill Construction, 2009). Achieving net zero emissions will require a massive transformation of our energy systems,

DOI: 10.1201/9781003483175-6

transportation, agriculture, and industry. It will also require significant investments in renewable energy, energy efficiency, and carbon capture technologies. The involvement of major emitters such as the United States and China is particularly important given their historical contributions to greenhouse gas emissions (Nwokolo, Meyer et al., 2023). Achieving net zero will not be easy and will require sustained effort from governments, businesses, and individuals alike. The fact that so many countries have committed to this goal is a cause for optimism and should inspire us all to take action on climate change.

The shift toward net zero targets has brought about a significant change in the way we approach climate change. By focusing on achieving net zero emissions, we are acknowledging the need for transformative changes across all sectors of society. This means moving beyond incremental improvements and taking bold actions to reduce greenhouse gas emissions. It requires a fundamental shift in our energy systems, transportation, agriculture, and buildings. The societal and policy discourse surrounding climate change has also shifted as a result of this focus on net zero targets. Governments, businesses, and individuals are increasingly recognizing the urgency of the situation and are taking steps to address it. Achieving net zero emissions will not be easy, and it will require collaboration across all sectors of society. We must continue to push for ambitious policies that support the transition to a low-carbon economy while also engaging with communities to build support for these changes. Achieving net zero emissions is essential if we hope to avoid the worst impacts of climate change and create a sustainable future for ourselves and future generations.

To achieve the necessary reduction in greenhouse gas emissions, it will be crucial to implement far-reaching changes across all sectors and locations. This will require a comprehensive transformation of production and consumption systems, with a focus on reducing emissions in industries that are particularly challenging to decarbonize. For example, the aviation industry will need to adopt new technologies such as electric or hydrogen-powered planes, while shipping may need to transition to low-carbon fuels or explore alternative transportation methods (Lawson, 2020). Similarly, the cement industry will need to find ways to reduce its reliance on fossil fuels in the production process. These changes will require significant investment and innovation, but they are essential if we are to meet our climate goals and avoid the worst impacts of global warming. It will be important to compensate for any remaining emissions through measures such as carbon capture and storage (CCS) or reforestation projects. By taking these steps, we can create a more sustainable future for ourselves and future generations.

The net zero energy transition is a monumental undertaking that will require significant changes in the way we produce and consume energy. The transition will involve a complex web of technological, organizational, political, and institutional changes across multiple sectors. In the electricity sector, we will need to ramp up renewable energy sources such as wind and solar power while phasing out fossil fuels. The transport sector will need to shift toward electric vehicles (EVs) and alternative modes of transportation such as cycling and walking. Buildings will need to be retrofitted with energy-efficient technologies, while industry will need to adopt cleaner production processes. Achieving net zero emissions will also require changes in consumer behavior, including reducing energy consumption and shifting toward low-carbon diets (Chakravarty and Tavoni, 2013). The transition will not be easy, but it is essential if we are to avoid the worst impacts of climate change. Governments, businesses, and individuals all have a role to play in driving this transformation forward. By working together, we can create a cleaner, more sustainable future for ourselves and future generations.

The field of transition studies has revealed a great deal about the complex processes that underlie societal change. Researchers have identified a range of factors that interact to shape transitions, including technological innovation, social norms and values, political institutions, and economic incentives. In the domain of transport, for example, shifts toward sustainable modes of transportation have been driven by a combination of environmental concerns (Amor et al., 2011), changing consumer preferences (Boucher and Reddy, 2008), and government policies aimed at reducing emissions (Rajas-Vallejos and Lastuka, 2020). Similarly, the transition toward renewable energy sources in the electricity sector has been influenced by factors such as declining costs of solar and wind power, public support for clean energy, and regulatory frameworks that encourage investment in renewables. By understanding these factors and how they interact with one another, researchers can help policymakers design effective strategies for managing transitions toward more sustainable and equitable societies.

But the bulk of this body of work has concentrated on particular industries or the appearance of a single, significant innovation. The transition from propeller to jet engines in ships marked a significant milestone in the history of transportation. This shift was driven by technological advancements and a growing need for faster and more efficient modes of transportation. Key frameworks such as the multi-level perspective and technological innovation systems have been developed to explain transformation processes around a focal innovation or sector, but they fall short when it comes to addressing the complexity of multiple innovations and transforming sectors. The transition from propeller to jet engines involved not only changes in technology but also changes in organizational structure, business models, and regulatory frameworks. These changes had far-reaching implications for the shipping industry, including increased speed, reduced costs, and improved safety. This transition also presented challenges such as environmental concerns and job displacement (Clora and Yu, 2022). As we continue to witness rapid technological advancements across various sectors, it is essential to develop new frameworks that can capture the complexity of these transitions and their impacts on society as a whole.

As we continue to grapple with the challenges of transitioning to a net zero energy economy, it is clear that we still have much to learn. Despite the progress that has been made, we lack a comprehensive framework for understanding and managing the complex processes involved in this transition. This is particularly true as the transition enters new phases of development and change processes accelerate and broaden in scope. One of the key challenges that we face is developing policies that can keep pace with these changes and ensure that they are sustainable over the long term. This will require a deep understanding of the social, economic, and political factors that drive energy transitions, as well as innovative approaches to policy design and implementation. Our success in achieving a net zero energy future will depend on our ability to navigate these challenges effectively and build a resilient, sustainable energy system for generations to come.

In the first phase of the net zero energy transition, the focus is on building a strong foundation for the transition. This involves developing and implementing policies that encourage investment in renewable energy technologies and infrastructure, as well as promoting energy efficiency measures. The second phase is characterized by a shift toward more integrated and systemic approaches to energy management. This includes efforts to optimize energy use across different sectors and promote greater coordination between different stakeholders. In the third phase, attention turns toward addressing the challenges associated with scaling up renewable energy production and integrating it into existing energy systems. Finally, in the fourth phase, policymakers must grapple with issues related to ensuring long-term sustainability and resilience in a net zero energy system. This includes addressing potential trade-offs between different sustainability goals and developing strategies for managing risks associated with climate change impacts. Overall, our model highlights the importance of taking a comprehensive and phased approach to policymaking in order to successfully navigate the complex challenges of the net zero energy transition.

As we move toward a net zero energy transition, it is important to consider the phases and policy challenges that lie ahead. In order to effectively decarbonize our economy, we must explore a range of strategies. Low-carbon electrification is one such strategy, which involves transitioning our energy sources to renewable options such as wind and solar power. Another option is the development of new types of fuels, such as hydrogen or biofuels, which can be used in transportation and industry. Lifestyle changes can play a significant role in reducing carbon emissions, such as reducing meat consumption or increasing public transportation use. A mix of approaches will likely be necessary to successfully tackle decarbonization challenges. Policymakers must work together with industry leaders and communities to create a comprehensive framework that addresses the unique needs and challenges of each sector while also prioritizing sustainability and environmental responsibility. Along with the stages and challenges of policy, we talk about different decarbonization strategies.

The field of sustainability transition research provides valuable insights into the complexity of large-scale transformations. These transformations are systemic, meaning that they involve multiple interconnected systems and actors. They are also non-linear, which means that small changes can have significant and unpredictable effects. These transformations involve both socio-political and techno-economic processes, which makes them even more complex. To navigate this complexity, it is important to have a clear understanding of the phases and strategies involved in the transformation process. The phases we discuss will be broad and partly stylized to provide general guidance, but they will be informed by empirical evidence from previous sustainability transitions. The strategies we discuss will also be informed by this evidence as well as by insights from other fields such as innovation studies and institutional theory. Our goal is to provide a framework that can help policymakers, businesses, and civil society organizations navigate the challenges of large-scale sustainability transformations in a way that is effective, efficient, and equitable.

Guidance is essential for countries seeking to transition to a low-carbon economy. Germany, the United Kingdom, and Denmark have all made strides in this direction, despite their reliance on fossil fuels in the past. One commonality among these countries is a commitment to investing in renewable energy sources like wind and solar power. Another is a focus on energy efficiency measures, such as building retrofits and smart grid technologies. These efforts have not only reduced carbon emissions but also created new jobs and stimulated economic growth. As other countries seek to follow their lead, it will be important to prioritize collaboration and knowledge sharing between nations. With the right guidance and support, we can build a sustainable future for generations to come.

The net zero transition is a complex and multifaceted process that is influenced by a variety of contextual factors. For example, the transition may look different in emerging economies compared to industrialized ones due to differences in available resources and industry structure. Political priorities and societal preferences also play a significant role in shaping the transition. Despite these contextual differences, there are overarching patterns that can be observed across all contexts. One such pattern is the need for collaboration and cooperation between various stakeholders, including governments, businesses, and individuals. Another pattern is the importance of innovation and investment in new technologies that can help reduce emissions and increase energy efficiency. The net zero transition requires a fundamental shift in how we approach energy production and consumption, as well as a willingness to adapt to changing circumstances and emerging challenges. By recognizing these patterns and working together toward a common goal, we can successfully navigate this complex transition toward a more sustainable future.

Sustainability transitions are crucial for ensuring a sustainable future. In Section 5.2, we focus on the transition to net zero energy and provide a brief overview of the sustainability transitions. Section 5.3 is the core of this chapter, where we delve into the particularities of this transition. We discuss four main phases that are essential for achieving net zero energy, including planning, implementation, monitoring, and evaluation. We also highlight key policy challenges that need to be addressed to make this transition successful. These challenges include financing, governance, social acceptance, and technological innovation. We present five generic decarbonization strategies that can be applied in various sectors to achieve net zero energy. Section 5.4 generates lessons learned from our analysis and offers concluding remarks. Our findings suggest that a combination of policy measures and technological innovations is necessary for achieving net zero energy. It is crucial to involve stakeholders in decision-making processes to ensure social acceptance and support for sustainability transitions. Overall, our book contributes to the ongoing debate on sustainability transitions by highlighting the importance of a holistic approach that considers both policy and technological solutions, as well as the social dimensions of sustainability. It is important to note that achieving net zero energy will require significant changes in our current energy systems and lifestyles. This will involve transitioning away from fossil fuels and toward renewable energy sources, as well as improving energy efficiency in buildings and transportation. There will need to be a shift toward more sustainable consumption patterns and lifestyles, such as reducing meat consumption and using public transportation or cycling instead of driving.

global strategies for achieving different phases of net zero energy are crucial to addressing the pressing issue of climate change. These strategies require a comprehensive approach that encompasses renewable energy sources, energy efficiency measures, and policy frameworks to incentivize sustainable practices. By implementing these strategies, we can pave the way toward a more sustainable future and mitigate the adverse effects of greenhouse gas emissions on our planet. Collaboration between governments, businesses, and individuals is essential in order to successfully implement these global strategies. This includes sharing knowledge, resources, and technologies to accelerate the transition to a net zero energy future. Continuous monitoring and evaluation of the effectiveness of these strategies will be necessary to ensure their long-term success and make any necessary adjustments along the way.

5.2 SHIFTING PERSPECTIVES ON SUSTAINABILITY AND THE ENVIRONMENT

Crises that resist conventional policy approaches are a growing concern in today's world. These crises are often highly complex, spanning multiple sectors and places, which makes them difficult to tackle. To effectively address these crises, a high degree of coordination is required among various stakeholders, including government agencies, NGOs, and the private sector (Shindhal et al., 2021). Conflicting views and interests regarding problem definition and potential solutions can hinder progress. These crises are not static; they change over time and require ongoing attention and adaptation to remain effective. The impact of these crises can be far-reaching and can undermine the very basis of our coexistence as a society (Manavalan and Jayakrishna, 2019). It is crucial that we develop innovative approaches to address these challenges and work collaboratively to find solutions that are sustainable in the long term. This requires a deep understanding of the root causes of these crises and a willingness to engage in open dialogue with all stakeholders to find common ground. Only by working together can we hope to overcome the challenges posed by these complex crises and build a sustainable future for generations to come. Achieving sustainability requires a shift in mindset and a commitment to taking action (Sutcliffe et al., 2023). This means adopting practices that prioritize environmental protection, social responsibility, and economic viability. It also means recognizing the interconnectedness of these three pillars of sustainability and working toward solutions that address all three simultaneously. In order to achieve this, we must invest in renewable energy sources, reduce waste and pollution, promote fair labor practices, and support local communities. We must also prioritize education and awareness-raising efforts to ensure that everyone understands the importance of sustainability and is empowered to take action in their own lives. By taking these steps, we can create a world that is not only sustainable in the long term but also just, equitable, and prosperous for all.

The field of sustainability transitions has emerged as a promising approach to addressing the pressing challenges of climate change and environmental degradation. With a growing evidentiary base, this research has shown that simply tweaking existing systems is not enough. Rather, fundamental changes are needed in socio-technical systems such as electricity, transportation, and buildings (Schan et al., 2011). This requires a shift away from fossil fuels and toward renewable energy sources, as well as more efficient and sustainable building designs. The field also offers valuable lessons for accelerating transition processes, such as the importance of stakeholder engagement and collaboration across sectors. While there are certainly challenges to overcome in implementing these changes, the sustainability transitions approach provides a roadmap for creating a more sustainable future for all.

By studying these systems, researchers can gain insights into how they function and evolve, as well as identify opportunities for improvement. Large-scale socio-technical systems are particularly important because they provide essential services to society, such as energy supply, transportation, and housing. These systems involve a range of actors, including firms and NGOs, as well as institutions like policies and societal norms. Technologies and infrastructures also play a critical role in these systems. For example, energy supply systems may rely on power plants, transmission lines, and distribution networks. Transportation systems may involve vehicles, roads, and traffic management technologies. Housing systems may include building materials, construction techniques, and zoning regulations. By understanding the interplay between these different elements of socio-technical systems, researchers can develop strategies to improve their performance and sustainability over time (Le et al., 2022). The concept of lock-in is a significant challenge for transitioning to sustainable energy systems. Sunk costs and vested interests create a barrier to change, as actors in the energy sector are reluctant to abandon their investments and established practices. This resistance to change is particularly evident in the fossil fuel industry, where powerful corporations have built their wealth and influence on the exploitation of non-renewable resources. The result is a socio-technical system that is highly resistant to radical changes, making it difficult for alternative energy sources to gain traction. There are promising signs of progress, with increasing investment in renewable energy technologies and growing public awareness of the urgent need for action on climate change. To overcome lock-in, it will be essential to address the underlying economic and political structures that reinforce the status quo and incentivize unsustainable practices. Only then can we hope to build a truly sustainable energy system that meets our needs without compromising the future of our planet.

Socio-technical systems are complex and dynamic, with various components that can, and often do, change over time. The term "socio-technical transition" refers to the fundamental adjustments made to these systems in response to changing circumstances. These transitions can take many forms, from the adoption of new technologies to the restructuring of social institutions. One example of a socio-technical transition is the shift toward renewable energy sources, which involves not only the development of new technologies but also changes in how energy is produced, distributed, and consumed (Shamsuddoha and Woodside, 2022). Another example is the move toward more sustainable forms of agriculture, which requires changes in farming practices as well as shifts in consumer behavior. The shift from a transport system based on horse-drawn carriages to one anchored around internal combustion automobiles and the use of oil brought about significant changes in infrastructure, business models, practices, and rules. The development of better roads and ports facilitated faster and more efficient transportation of goods and people. Business models evolved to include more accurate shipping times and automotive maintenance services, leading to greater reliability in the transport industry. The advent of automobiles also led to new practices such as recreational travel, which was previously limited to the wealthy (Schwedhelm et al., 2023). New rules were put in place to regulate speed limits and ensure safety on the roads. These changes were not without their challenges, but they ultimately led to a more interconnected world where people and goods could be transported faster and more efficiently than ever before. As we continue to evolve our transportation systems, it is important that we consider the impact of these changes on our environment and work toward sustainable solutions for future generations. Successful socio-technical transitions require collaboration across multiple sectors and stakeholders, as well as a willingness to experiment and adapt to new challenges. These transitions offer opportunities for creating more equitable and sustainable societies that are better equipped to meet the needs of future generations.

Socio-technical transitions are critical to addressing the grand challenges that face society today. These transitions involve the adoption of new technologies and non-technical novelties such as changes in policies, practices, or lifestyles. Innovations in these areas are essential for sustainable development and ensuring a better future for all. For instance, the adoption of renewable energy sources such as solar and wind power is an innovation that can help address climate change and reduce carbon emissions (Azila and Alias, 2015). The implementation of policies that promote sustainable practices in agriculture and industry can lead to more efficient resource use and a reduction in waste. Changes in lifestyle, such as adopting plant-based diets or reducing meat consumption, can also have a significant impact on the environment. In conclusion, socio-technical transitions require a holistic approach that involves not only technological innovations but also non-technical novelties to address the grand challenges facing society today.

In addition to technical innovations, there are also non-technical novelties that can address the grand challenges facing society today. These non-technical changes can often be intertwined with technical innovations, which is why the term socio-technical configurations is often used. For example, solar PV is a technical innovation that can help address climate change, but it also requires changes in business models, regulations, and practices to be fully effective. One such change is the self-consumption of electricity generated at home. This allows individuals to use the energy they produce themselves rather than relying solely on the grid. Other non-technical changes that can address grand challenges include changes in social norms and behaviors, such as reducing meat consumption to combat deforestation and biodiversity loss. Changes in policies and governance structures can also play a crucial role in addressing grand challenges. For instance, implementing carbon pricing mechanisms or investing in public transportation infrastructure can help reduce greenhouse gas emissions and mitigate climate change. Overall, addressing grand challenges requires a multifactorial approach that involves collaboration between governments, businesses, and individuals. When it comes to deforestation and biodiversity loss, there are several solutions that can be implemented. One approach is to promote sustainable forestry practices that prioritize the conservation of biodiversity. This can involve setting aside protected areas for endangered species and implementing measures to prevent illegal logging. Another solution is to promote reforestation efforts in areas where forests have been destroyed. This can involve planting new trees and restoring degraded ecosystems. In addition to these solutions, it is important to address the root causes of deforestation, such as agricultural expansion and infrastructure development. This requires a shift toward more sustainable land-use practices and greater investment in renewable energy sources. Addressing deforestation and biodiversity loss requires a comprehensive approach that takes into account the economic, social, and environmental factors that contribute to these issues. By working together, we can create a more sustainable future for ourselves and future generations.

By working together toward a more sustainable future, we can ensure that our planet remains habitable for ourselves and future generations. One way to achieve this is by embracing emerging socio-technical configurations that promote sustainability. For instance, renewable electricity production is gaining momentum as a viable alternative to traditional energy sources. This shift toward low-carbon electricity has the potential to create more sustainable socio-technical systems that reduce our carbon footprint and mitigate the effects of climate change (Legutko-Kobus et al., 2023). By investing in renewable energy infrastructure and encouraging its adoption, we can create a cleaner, greener future for all. We can adopt sustainable practices in our daily lives, such as reducing waste and conserving resources. It is up to us to take action and work together toward a more sustainable future.

The S-curve is a useful model for understanding the diffusion of innovation. At the early stage, progress is often slow and incremental, with only minor changes occurring in niches where multiple innovations are being developed. As these innovations begin to take hold and gain traction, they enter the take-off phase, where they start to diffuse more widely. This diffusion then stimulates further improvements and refinements to the innovation, which in turn accelerates its uptake and adoption. This positive feedback loop can lead to exponential growth in the popularity and impact of an innovation as it becomes increasingly integrated into society or industry. Of course, not all innovations follow this trajectory perfectly; some may stall or fail to gain traction altogether, but understanding the S-curve can help innovators and policymakers anticipate and manage the diffusion process more effectively.

The process of transitioning from one socio-technical system to another is complex and multifaceted. It typically involves a period of experimentation and innovation as new technologies and practices are developed and tested. Over time, these innovations begin to gain traction and spread throughout the system, gradually transforming many of its key elements. As this process unfolds, we often see a series of feedback loops emerge, in which changes in one area of the system lead to further changes in other areas. Eventually, these feedback loops can reach a point where they trigger a more fundamental reconfiguration of the system as a whole. This can be a turbulent period as old structures are dismantled and new ones take their place. Over time, the dynamics slow down again as the new system stabilizes and becomes more established. While this pattern is most commonly associated with transitions within single sectors or industries, we also see similar dynamics at play when transitions span across multiple sectors or domains (Sarkawi et al., 2016). In these cases, the process may be more complex and require greater coordination among various stakeholders. For example, when transitioning to a low-carbon economy, it is not just the energy sector that needs to change but also transportation, agriculture, and other sectors that contribute to greenhouse gas emissions. This requires collaboration between government agencies, businesses, and civil society organizations to develop policies and strategies that support a sustainable transition. The process may be influenced by external factors such as technological advancements or changes in global markets. Despite these challenges, successful transitions can lead to significant benefits, such as improved environmental outcomes and new economic opportunities. It is important for stakeholders to remain committed to the transition process and adapt their strategies as needed in order to achieve these goals.

The context of climate change and other pressing sustainability challenges demands a comprehensive approach to transitions that involves multiple innovations and multiple socio-technical systems. This means that transitioning to a sustainable future requires not just technological advancements but also changes in social norms, policies, and cultural practices. One example of such a transition is the shift toward renewable energy sources like solar and wind power. This transition involves not only the development of new technologies but also changes in the way we produce, distribute, and consume energy. Similarly, transitioning to a circular economy requires innovations in waste management, product design, and business models. It also involves changes in consumer behavior and policy frameworks that incentivize sustainable practices. Successful transitions require collaboration across sectors and stakeholders to create a shared vision for a sustainable future.

The transition to a low-carbon economy requires a multifaceted approach that involves major changes in various sectors. The automobile sector is undergoing a significant transformation with the rise of EVs and the phasing out of traditional combustion engines. The success of EVs depends on innovations in battery technology that can improve their range and reduce charging times. Fortunately, the battery industry is also making strides toward more efficient and sustainable batteries, such as solid-state batteries and recycling initiatives. To support this transition, a robust charging infrastructure is necessary to ensure convenient access to charging stations for EV owners. This requires public infrastructure investments in urban areas and along highways. The energy sector must shift toward low-carbon electricity sources to power these EVs. This includes expanding renewable energy sources like wind and solar power while phasing out fossil fuels. Progress in one technology can stimulate further progress in another, creating a virtuous cycle of innovation and sustainability. This transition will not only reduce carbon emissions and combat climate change, but it will also lead to a more resilient and decentralized energy system. By relying on a diverse mix of renewable energy sources, we can reduce our dependence on imported fossil fuels and improve energy security. In addition, the shift toward low-carbon electricity sources will create new job opportunities in the renewable energy sector, driving economic growth and development. This transition will require significant investments in infrastructure and technology, as well as policies that incentivize the adoption of clean energy solutions. Governments, businesses, and individuals all have a role to play in driving this transition toward a more sustainable future. By working together, we can build a cleaner, more resilient energy system that benefits everyone.

The co-evolution of different sectors toward renewable energy sources can lead to a cascade of changes that reinforce the transition toward a sustainable future. For instance, as utility companies invest in vehicle charging stations, this encourages the adoption of EVs, which in turn reduces greenhouse gas emissions and air pollution. This can lead to further investments in renewable energy sources such as wind and solar power, which can then be used to power more EVs and other forms of transportation. As more sectors transition toward sustainability, they become increasingly interconnected, creating a positive feedback loop that accelerates the pace of change. This can lead to the development of new technologies and business models that are more aligned with sustainability goals, such as distributed energy systems and circular economies. This co-evolutionary process can lead to the build-up of an even larger transition that involves multiple socio-technical systems, creating a more resilient and sustainable future for all.

As we look toward a future that is decarbonized, it is important to have an overarching strategy or paradigm in place. This could be in the form of meta-rules that guide our decision-making processes. One such paradigmatic feature that we are currently observing is the approach to electrifying as many energy applications as possible. This strategy is being driven by the rapidly falling costs of renewable electricity generation, making it a viable option for a wide range of applications. We must also consider other factors such as energy storage and grid stability to ensure that this approach is sustainable in the long run. We need to focus on reducing energy consumption through energy efficiency measures and shifting toward low-carbon transportation options. Finally, we must prioritize research and development of new technologies that can further accelerate decarbonization efforts. By adopting a holistic approach and implementing these strategies in tandem, we can achieve a sustainable future powered by clean energy sources.

Several different technologies are used in the two main transitions that interact. The relationship between electric mobility and low-carbon electricity is complex. On the one hand, the growth of EVs can drive up demand for additional low-carbon electricity, which could incentivize greater investment in renewable energy sources. This would be a complementary interaction, as both electric mobility and renewable energy are key components of a sustainable future. This increased demand for electricity could also lead to tensions among different energy end-uses, such as mobility and indoor temperature regulation, as they compete for limited electricity supplies. This competition could potentially slow down the transition to a low-carbon economy if not managed effectively. Therefore, it is important to carefully balance the growth of electric mobility with investments in renewable energy infrastructure to ensure that both complement each other and contribute to a sustainable future. Policies that encourage energy efficiency and demand-side management can help alleviate some of these tensions by reducing overall electricity consumption and optimizing its use across different end-uses. Achieving a sustainable energy system will require a holistic approach that takes into account not only the technical aspects of energy production and consumption but also the social, economic, and environmental dimensions. This means that policymakers must work closely with stakeholders from different sectors to develop integrated solutions that address the complex challenges of the energy transition. For example, promoting renewable energy sources such as solar and wind power can help reduce greenhouse gas emissions while also creating new job opportunities in the clean energy sector. At the same time, efforts to improve energy efficiency in buildings and transportation can help reduce energy demand and lower costs for consumers. Finally, it is important to recognize that achieving a sustainable energy system will require long-term investments in research and development as well as public education campaigns to raise awareness about the benefits of clean energy. By working together toward a common goal, we can create a more sustainable future for ourselves and future generations.

5.3 APPROACHING NET ZERO ENERGY

The net zero energy transition is undoubtedly one of the most challenging transformations of our time. It requires the integration of multiple socio-technical systems, such as energy, transportation, and buildings. Achieving net zero emissions will require significant changes in almost every sector and part of society. This transition will not be easy, and there will be many interactions between different systems that could lead to unexpected developments and setbacks. It is crucial that we take on this challenge because the consequences of not doing so are dire. The world is already experiencing the impacts of climate change, from rising sea levels to more frequent extreme weather events. By transitioning to a net zero energy system, we can mitigate these impacts and ensure a sustainable future for generations to come. It will require cooperation across all sectors of society and a willingness to embrace new technologies and ways of living. But with determination and collective action, we can achieve a net zero energy future that benefits both people and the planet.

The transition to a net zero economy is a radical shift that requires the active participation of many key societal actors, including governments, businesses, and individuals. This transition is not only necessary to address the urgent threat of climate change but also presents an opportunity for innovation and growth. Governments have a central role to play in this transition, setting ambitious targets and implementing policies that encourage sustainable practices and investment in renewable energy. This transition cannot be achieved by governments alone. Businesses must also step up and take responsibility for their carbon footprint, investing in sustainable technologies and practices. Individuals can also contribute by making conscious choices about their consumption habits and reducing their own carbon footprint. While past transitions may have taken decades to unfold, we only have about 30 years left to complete the net zero transition. This requires coordination between all actors involved and a sense of urgency to act now before it's too late. By working together toward a common goal, we can create a more sustainable future for ourselves and generations to come.

To unfold in parallel, rather than one at a time, is to recognize that the challenges we face in achieving net zero emissions are interconnected and require a systems-level approach. Incremental improvements will not suffice, as they only address individual pieces of the puzzle without considering how they fit into the larger picture. Instead, we need radical changes that can bring about new socio-technical configurations capable of achieving net zero emissions. This will require a fundamental shift in our energy systems, transportation networks, and industrial processes. We must transition to renewable energy sources like wind and solar power while also implementing energy-efficient technologies across all sectors. We must also rethink our transportation systems to prioritize public transit and active transportation options like biking and walking. Finally, we must transition away from fossil fuel–based industries and toward circular economies that prioritize resource efficiency and waste reduction. These changes will not be easy, but they are necessary if we hope to achieve a sustainable future for ourselves and future generations.

Achieving a sustainable future for ourselves and future generations requires a concerted effort to transition to net zero energy. This transition involves not only implementing readily available innovations but also developing new ones that may not even be known yet. This presents a challenge as it requires rapid action in the face of deep uncertainty. To overcome this challenge, we must prioritize investment in research and development of new technologies while simultaneously implementing existing ones. This includes investing in renewable energy sources such as solar, wind, and geothermal, as well as improving energy efficiency in buildings and transportation. We must adopt sustainable practices such as reducing waste and conserving natural resources. By taking these steps, we can create a more sustainable future that benefits both ourselves and future generations. It is crucial that we act now to ensure a livable planet for all.

To truly understand the net zero energy transition, it is important to view it as a complex system of interdependent and complementary transitions that occur across various sectors and locations. This process is not a one-size-fits-all solution but rather a cumulative effort involving multiple socio-technical transitions that work together to move society toward complete decarbonization. From the adoption of renewable energy sources in the power sector to the electrification of transportation and the implementation of energy-efficient buildings, each transition plays a crucial role in reducing greenhouse gas emissions and achieving a net zero future. These transitions are not isolated events but rather dynamic processes that evolve over time with varying degrees of success. As such, it is important to approach this challenge with a holistic perspective and recognize the interconnectedness of these transitions in order to achieve our ultimate goal of creating a sustainable future for generations to come.

Achieving a sustainable future for generations to come requires a complex and interconnected web of socio-technical transitions. Each transition involves the development and diffusion of multiple innovations, which can have ripple effects across various sectors. For example, a disruptive business model that emerges in one industry may inspire and be adapted to another sector, leading to new innovations and transitions. It is important to ensure that these innovations are sustainable and do not cause harm to the environment or society. This requires careful consideration of the entire life cycle of products and services, from raw material extraction to disposal. Collaboration between different stakeholders, such as businesses, governments, and communities, is crucial for achieving a sustainable future. By working together and fostering innovation that is mindful of social and environmental impacts, we can create a world that is not only prosperous but also equitable and resilient for generations to come.

To come to a net zero carbon future, it is essential to recognize that transitions in multiple sectors must occur simultaneously and complement each other. For instance, EVs can only be truly sustainable if they are powered by low-carbon electricity. Similarly, renewable energy sources like wind and solar power can only be fully utilized if the grid infrastructure is modernized to accommodate them. But these transitions cannot happen in isolation; they must be part of a larger, systemic change that involves everything from urban planning to agricultural practices. This means that governments, businesses, and individuals all have a role to play in achieving net zero emissions. It will require significant investment in research and development, as well as political will and public support. But the benefits of a net zero future are immense: cleaner air and water, more resilient communities (Nwokolo and Otse, 2019), and a stable climate for future generations (Nwokolo, Obiwulu et al., 2023). We must act now to ensure that these multiple transitions occur in a coordinated way so that we can achieve our goal of net zero emissions within the necessary timeframe.

This statement highlights the importance of understanding the complexity of transitions that occur across multiple systems and innovations (Nwokolo, 2017). It suggests that a simple model can be used to represent these transitions, but it is important to recognize that this model is just a small piece of a much larger puzzle. To fully understand and navigate these transitions, it is crucial to consider the interconnectedness of different systems and innovations, as well as their potential impacts on each other. This requires a holistic approach that takes into account not only the technical aspects of innovation but also the social, economic, and political factors that shape its adoption and diffusion. By recognizing the complexity of transitions and embracing a multidisciplinary perspective (Nwokolo, Proutsos et al., 2023), we can better navigate these challenges and create more sustainable and equitable outcomes for all stakeholders involved (Ituen et al., 2012).

The success of a transition toward a low-carbon economy depends on the complementarity of individual transitions and their cumulative effects. For instance, if the second and third transitions use low-carbon electricity, they already start at a higher level, which can lead to faster progress. Innovations, policy approaches, business models, or practices that work in one transition can be transferred to another through horizontal black arrows. This cross-over can accelerate progress toward a low-carbon economy. It is important to note that these transitions are not independent of each other and may interact in complex ways. Therefore, policymakers need to carefully consider the interactions between different transitions and design policies that support complementarity and cross-over between them. This will require collaboration across sectors and stakeholders to ensure a coordinated approach toward achieving a sustainable future.

Collaboration across sectors and stakeholders is essential to achieving a sustainable future. This requires a coordinated approach that involves governments, businesses, civil society, and individuals working together toward common goals. The transition toward sustainability will not be easy, and it will require policymakers to continually adapt and develop new solution strategies as the transition unfolds. To achieve this, we need to move from easier targets to more difficult ones, prioritizing the most urgent issues first. For example, reducing carbon emissions is a critical priority that requires immediate action. We can do this by investing in renewable energy sources such as wind and solar power while phasing out fossil fuels. We must prioritize sustainable agriculture practices that reduce environmental degradation while ensuring food security for all (Proutsos, Liakatas et al., 2023; Proutsos, Tigkas et al., 2023). By working together across sectors and stakeholders, we can create a sustainable future that benefits everyone on the planet for generations to come.

5.4 A PROGRESSION TOWARD NET ZERO

We identify four qualitatively distinct phases of development in order to capture the growing complexity of the net zero energy transition. The first phase of the net zero energy transition was characterized by a wide sectoral scope with a focus on reducing emissions across all sectors. During this phase, policymakers and industry leaders explored a variety of decarbonization approaches, including renewable energy deployment, energy efficiency improvements, and CCS. In the second phase, the focus shifted to the electricity sector, as it was seen as the most promising area for rapid decarbonization. This phase saw significant growth in renewable energy deployment and grid modernization efforts. The third phase has seen a renewed focus on a wider sectoral scope, with an emphasis on the electrification of transportation and building heating. Finally, in the fourth phase, there is an increasing recognition that achieving net zero emissions will require a range of decarbonization approaches tailored to specific sectors and regions. This includes not only renewable energy deployment but also investment in low-carbon technologies such as hydrogen fuel cells and advanced nuclear reactors. As we move through these phases, it is clear that achieving net zero emissions is a monumental task that requires a multifaceted approach (Nwokolo, Meyer et al., 2024). Renewable energy deployment is a crucial component, but it is not the only solution. Low-carbon technologies, such as hydrogen fuel cells and advanced nuclear reactors, must also be invested in to achieve this goal (Nwokolo, Echeng et al., 2024). Implementing these solutions will require tailored approaches for specific sectors and regions. For example, transportation may require a shift toward EVs or biofuels, while industrial processes may require CCS technology. Building efficiency improvements and sustainable land-use practices will also play important roles in reducing emissions. It's clear that the necessary changes will not happen overnight, but with continued effort and investment in these decarbonization approaches, we can work toward a sustainable future for generations to come.

As we work toward a sustainable future for generations to come, decarbonization approaches are becoming increasingly important. One such approach is renewable energy electrification, which has emerged as a dominant method for reducing carbon emissions. This approach is being used to decarbonize applications in the building and transport sectors, and it has the potential to be applied in other areas as well. In addition to electrification, other approaches such as CCS, hydrogen fuel cells, and bioenergy are also being explored (Nwokolo, Obiwulu et al., 2023). These approaches offer promising solutions for reducing greenhouse gas emissions and mitigating the effects of climate change. By implementing these decarbonization approaches on a large scale, we can create a cleaner and more sustainable future for ourselves and future generations (Nwokolo, Proutsos et al., 2023). It is important that we continue to invest in the research and development of these technologies so that they can become even more efficient and cost-effective over time.

As businesses strive to remain competitive in today's fast-paced economy, they are constantly seeking ways to become more efficient and cost-effective. This requires a careful analysis of their operations and processes, identifying areas where improvements can be made. One approach that has gained popularity in recent years is the use of phased implementation strategies. These strategies involve breaking down complex projects into smaller, more manageable phases, each with its own set of goals and objectives. By doing so, businesses can focus their resources on specific areas of improvement, while also building momentum and achieving quick wins along the way. As these phases are implemented, they often reinforce one another, leading to even greater efficiencies and cost savings over time. Of course, every business is unique, and there may be specific challenges or opportunities that require a different approach. By adopting a phased implementation strategy and continually seeking out ways to improve their operations, businesses can position themselves for long-term success in an ever-changing marketplace.

As the world moves toward a net zero energy future, each phase of the transition brings with it new challenges and opportunities. From the initial push toward renewable energy sources to the development of more efficient technologies (Nwokolo and Ogbulezie, 2018a), the complexity of this transition continues to grow (Nwokolo and Ogbulezie, 2018b). As a result, political dynamics are also changing, with new stakeholders emerging and old power structures being challenged. The need for coordinated policy action is becoming increasingly urgent as countries seek to balance economic growth with environmental sustainability. This requires innovative approaches to governance that can navigate the competing interests of different groups and ensure that everyone benefits from the transition. In addition, there is a growing recognition that social equity must be at the heart of any successful net zero energy strategy, as vulnerable communities are often disproportionately affected by environmental degradation and climate change. Achieving a net zero energy future will require sustained effort and collaboration across all sectors of society, as we work together to build a more sustainable and equitable world for future generations.

5.4.1 First Progression

Renewed energy technology made significant strides during the first phase of the net zero energy transition, which lasted from the 1980s to the early 2000s. During this time, solar and wind power systems became more efficient and cost-effective, making them increasingly viable alternatives to traditional fossil fuel–based energy sources. The net zero energy transition was a critical turning point in the history of renewable energy. The advancements in solar and wind power systems during this period led to a significant reduction in greenhouse gas emissions, which had been a major contributor to climate change.

The net zero energy transition has been a crucial development in the fight against climate change. It has been characterized by two major factors: the emergence of climate change as a topic of societal concern and the development of low-carbon technologies in key sectors such as electricity, buildings, and transport. The foundation of the Intergovernmental Panel on Climate Change (IPCC, 2021) and its first report in 1990 marked a turning point in global awareness of the issue. During this period, the world witnessed a remarkable shift toward a more sustainable future. The significant reduction in greenhouse gas emissions was achieved through a combination of factors, including the adoption of low-carbon technologies and the implementation of policies aimed at promoting renewable energy sources. The first phase of the net zero energy transition saw the emergence of innovative solutions in sectors such as electricity, buildings, and transport. In the electricity sector, there was a significant increase in the deployment of renewable energy sources such as solar and wind power. Buildings became more energy-efficient through better insulation and smart technology, while EVs became increasingly popular in the transport sector.

As the world becomes more conscious of the need to reduce carbon emissions, the use of insulation and smart technology has become increasingly popular in the construction industry. Insulation materials such as fiberglass, cellulose, and foam can help reduce energy consumption by keeping buildings warm in the winter and cool in the summer. Smart technology, on the other hand, allows for greater control over energy usage through automated systems that adjust heating and lighting based on occupancy or time of day. In addition to these innovations, EVs have also gained popularity in the transport sector due to their lower carbon footprint compared to traditional gasoline-powered cars. The development of power generation technologies such as wind or solar PV, new heating technologies like solar heating (Obiwulu, Chendo et al., 2020) or heat pumps, building efficiency technologies (Obiwulu, Erusiafe et al., 2020), biofuels, and first-generation electric and fuel cell vehicles have all contributed to a more sustainable future. With continued innovation and investment in these areas, we can hope to achieve a cleaner and greener planet for generations to come.

The first phase of this journey has already seen significant changes in regulations, such as grid access and feed-in regulations for independent power producers, stricter building codes, and fuel efficiency standards for vehicles. These regulations have incentivized businesses to adopt more sustainable practices and encouraged consumers to make more environmentally conscious choices. There is still much work to be done. We must continue to invest in the research and development of new technologies that will allow us to transition entirely away from fossil fuels. We must also prioritize the protection of our natural resources by implementing policies that promote conservation and sustainable land-use practices. By working together toward these goals, we can create a brighter future for ourselves and the planet.

During the first stage of the net zero energy transition, a number of policies were put in place to encourage innovation, technology development, and diffusion. These policies and actors have paved the way for a brighter future for ourselves and the planet. The net zero energy transition has created new opportunities for innovation and technology development, which in turn has led to the emergence of new businesses and industries. Project developers, energy service providers, and technology producers have all played a critical role in this transition, working together to create sustainable solutions that benefit both the environment and society as a whole. As we move forward, it is important to continue supporting these efforts through policies that encourage investment in renewable energy sources and sustainable technologies. By doing so, we can ensure that we are creating a world that is not only cleaner and healthier but also more prosperous and equitable for all. With continued dedication to this cause, we can build a future that is truly sustainable and secure for generations to come.

Despite the initial debates over the seriousness of climate change and the potential for new innovations to replace existing ones, we have made significant progress toward a net zero energy future. The development of renewable energy sources, such as wind and solar power, has grown exponentially in recent years, while advances in battery technology have made it possible to store excess energy for later use. Governments around the world have also taken steps to reduce their carbon emissions through policies and regulations aimed at promoting clean energy and reducing reliance on fossil fuels. As we continue on this path toward a sustainable future, it is important to remember that our actions today will impact generations to come. By investing in renewable energy and reducing our carbon footprint, we can ensure a secure and sustainable future for ourselves and future generations.

The drive for more sustainable systems persisted despite the difficulties presented by lock-ins and incumbent resistance. Governments and organizations around the world began to invest heavily in research and development, leading to breakthroughs in areas such as energy storage and grid management. As a result, wind and solar power became increasingly cost-competitive with traditional fossil fuels (Obiwulu et al., 2022b), leading to a surge in adoption. Meanwhile, new transportation solutions emerged, such as ride-sharing services and electric bikes, that helped reduce reliance on personal vehicles. The food industry also saw significant progress toward sustainability, with more consumers opting for plant-based diets and companies implementing eco-friendly practices. While there is still much work to be done, these advancements demonstrate that change is possible even in the face of entrenched systems. As we continue to prioritize sustainability in all aspects of our lives, we can build a more resilient future for ourselves and our planet.

5.4.2 Second Progression

In 2010, the rapid global spread of renewable energy sources initiated the second phase of the net zero transition. This phase marked a significant shift in the global energy landscape, as countries and industries recognized the urgent need to reduce carbon emissions and mitigate the impacts of climate change. Governments implemented ambitious renewable energy targets, incentivizing the deployment of clean technologies and fostering innovation in the sector. As a result, wind farms sprouted across vast stretches of land and offshore, harnessing the power of nature's gusts. Solar panels adorned rooftops and covered fields, converting sunlight into electricity on an unprecedented scale. Simultaneously, advancements in battery storage technology unlocked the potential for reliable renewable energy supply, enabling a more stable grid. The momentum behind this transition was further fuelled by increased public awareness and support for sustainable practices. Communities embrace decentralized energy systems, with microgrids empowering local resilience and self-sufficiency. Corporations also joined the movement, recognizing that embracing renewables not only aligned with their environmental commitments but also made economic sense in the long run.

For example, this significant growth in solar PV installations over the past decade showcases its immense potential as a renewable energy source. With a staggering increase from 40 GW in 2010 to an impressive 700 GW in 2020, solar PV has experienced a remarkable 30% annual growth rate (IEA, 2023). This surge in solar power capacity has not only demonstrated its effectiveness but also its growing popularity among individuals, businesses, and governments worldwide. Conversely, onshore wind witnessed a contrasting trend during this period, experiencing a decline from 178 GW to 698 GW (IEA, 2023). While this decrease may seem puzzling at first glance, it could be attributed to various factors such as shifting priorities, regional variations in resource availability, or the emergence of alternative renewable energy technologies. Nonetheless, the remarkable rise of solar PV remains a testament to its promising future and its ability to contribute significantly to global efforts toward sustainable and clean energy solutions.

This shift was crucial in driving the adoption of renewable energy sources such as solar and wind power. As a result, the electricity sector witnessed a significant reduction in carbon emissions, leading to a cleaner and more sustainable energy landscape. This momentum paved the way for advancements in energy storage technologies, enabling better integration of intermittent renewable sources into the grid. The increased emphasis on electrification also spurred innovations in EV technology, further reducing reliance on fossil fuels and contributing to overall decarbonization efforts. The net zero transition fostered collaborations between governments, industries, and research institutions to develop cutting-edge solutions for efficient energy use and distribution. These collective efforts have not only accelerated the low-carbon transition but have also created new economic opportunities and jobs in the renewable energy sector. Overall, the momentum generated during this phase has been instrumental in propelling the electricity sector toward a sustainable future while mitigating climate change impacts.

This transition has been driven by various factors, including government policies and incentives, technological advancements, and increasing public awareness of the need to address climate change. Government policies such as renewable energy targets and feed-in tariffs have encouraged the deployment of renewable energy technologies. These policies have created a favorable investment environment for renewable energy projects, attracting private sector investments and driving down the costs of renewable technologies. Technological advancements in solar and wind power have also played a crucial role in the transition. The efficiency and cost-effectiveness of these technologies have improved significantly over the years, making them increasingly competitive with fossil fuels. Public awareness of the environmental impacts of fossil fuel use has grown, leading to a greater demand for clean energy sources. This increased demand has further incentivized the shift toward renewables. As a result of these combined efforts, the electricity sector has witnessed a remarkable transformation toward net zero emissions, setting an example for other sectors to follow suit.

These changes in power supply, system balancing, sector-specific policies, and the emergence of new entrants and business models have revolutionized the energy landscape. Distributed generation has become increasingly prevalent with a shift toward renewable sources such as solar and wind power. This decentralization of power production has not only reduced reliance on traditional centralized power plants but has also empowered individuals and communities to generate their own electricity. As a result, the grid has become more resilient and less vulnerable to disruptions. System balancing has also undergone a significant transformation with the integration of advanced technologies like smart grids and energy storage systems. These innovations enable real-time monitoring and optimization of energy flows, ensuring efficient utilization of resources and minimizing waste. Sector-specific policies have played a crucial role in promoting sustainability and encouraging the adoption of clean energy solutions. Governments around the world have implemented supportive regulations, incentives, and subsidies to accelerate the transition toward a low-carbon economy. These policies have fostered innovation by creating a conducive environment for new entrants to develop and scale up renewable energy technologies.

These contests have revealed the deep divisions within society regarding the role of traditional energy sources and the urgency of transitioning to cleaner alternatives. Stakeholders from various sectors, including government, industry, and environmental organizations, have engaged in heated debates and lobbying efforts to shape policies and regulations that favor their interests. The incumbent actors, aware of the potential disruption to their established assets and business practices, have employed political strategies aimed at slowing down the pace of transformation. They have sought to influence public opinion, fund research studies that cast doubt on the viability of renewable energy sources, and leverage their political connections to maintain their dominance in the energy sector. Despite their efforts, the momentum for change continues to build as more countries commit to ambitious renewable energy targets and as public awareness of climate change impacts grows. The transition away from fossil fuels toward a sustainable energy future is inevitable, and it is crucial for all stakeholders to collaborate and find common ground in order to navigate this complex and challenging process. Only through collective action can we ensure a smooth and successful transition.

This transition toward renewable electrification is driven by the urgent need to combat climate change and reduce greenhouse gas emissions. The advancements in wind, solar-PV, and battery technology have made them more efficient and affordable, making it increasingly feasible to rely on these sources for our energy needs. As a result, we are witnessing a shift away from fossil fuels toward clean and sustainable energy sources. This not only helps in decarbonizing the electricity sector but also paves the way for electrification in other sectors such as transportation and heating. The ongoing efforts to fully decarbonize electricity are crucial in achieving our goal of net zero emissions and creating a sustainable future for generations to come.

These interdependencies can be both beneficial and challenging. On the one hand, they highlight the interconnectedness of various elements within the socio-technical system, emphasizing the need for a holistic approach to sustainability transitions. This understanding allows for more effective coordination and collaboration among stakeholders as they recognize the importance of aligning their efforts toward a common goal. On the other hand, these interdependencies can also create vulnerabilities and bottlenecks within the system. Changes in one component can have unintended consequences for others, potentially leading to disruptions or delays in the transition process. Therefore, it becomes crucial to carefully analyze and anticipate these interdependencies to minimize risks and maximize positive outcomes. Declines in certain sectors during transitions pose additional challenges. As old technologies or practices become obsolete, industries may face economic downturns or job losses. Managing these declines requires proactive strategies such as retraining programs or supporting alternative industries to ensure a just transition for affected communities. Overall, phase 2 of sustainability transitions provides valuable insights into navigating complex systems and promoting long-term resilience.

This transition from high-carbon to low-carbon technologies requires careful planning and coordination among governments, industries, and communities. It involves investing in renewable energy sources like solar and wind power as well as improving energy efficiency in buildings, transportation, and industrial processes. It necessitates the development of sustainable infrastructure that supports clean energy generation and distribution. Alongside these efforts, there is a growing recognition of the importance of nature-based solutions in mitigating climate change. Restoring and protecting ecosystems such as forests, wetlands, and mangroves can help sequester carbon dioxide from the atmosphere while providing numerous co-benefits like biodiversity conservation and water resource management. Transitioning to a low-carbon economy also presents opportunities for job creation and economic growth through the development of green industries and sustainable practices. It is crucial to ensure a just transition that considers the social implications of phasing out carbon-intensive technologies. This includes supporting affected workers with retraining programs and providing assistance to communities heavily dependent on fossil fuel industries.

5.4.3 Third Progression

The third stage of the net zero energy transition started in or around 2015, and it is characterized by an increase in electrification to further reduce carbon emissions across a variety of sectors. This stage witnessed a significant shift toward renewable energy sources, such as solar and wind power, to meet the growing electricity demand. Innovations in battery storage technology also played a crucial role in enabling the integration of intermittent renewable energy into the grid. There was a notable focus on improving energy efficiency in buildings and transportation through advanced technologies and smart systems. The adoption of EVs gained momentum, with governments and industries investing heavily in charging infrastructure and incentivizing their use. Industries like manufacturing and agriculture embraced cleaner technologies and sustainable practices to minimize their environmental impact. The third stage of the net zero energy transition marked a turning point in global efforts to combat climate change as electrification became a key strategy for achieving carbon neutrality across diverse sectors.

Developments in low-carbon electricity supply, as well as ancillary innovations that use electricity for things like heating and transportation, support the phase's directionality. This shift toward electrification has been instrumental in achieving significant reductions in carbon emissions. With the advancement of technology, heat pumps have become more efficient and widely adopted, replacing traditional heating systems that heavily rely on fossil fuels. Similarly, the widespread adoption of EVs has revolutionized the transportation sector, reducing reliance on gasoline-powered cars and significantly lowering emissions. The integration of renewable energy sources into the electricity grid has also played a crucial role in this phase, ensuring that the increased demand for electricity is met with clean and sustainable power generation. As a result, sectors such as residential heating and transportation have witnessed a substantial decrease in their carbon footprint, contributing to the overall goal of achieving net zero energy. This phase marks a pivotal moment in our transition toward a sustainable future, where electrification becomes the cornerstone for decarbonizing various sectors and paving the way for a greener and more environmentally friendly world.

Renewable electrification has taken on a preponderant role as the third stage of the net zero energy transition and the corresponding carbon-neutral approach of "electrifying everything." This approach involves shifting away from fossil fuel–based energy sources and transitioning to renewable electricity for all sectors of the economy. By electrifying everything, we can reduce greenhouse gas emissions and achieve a sustainable future. This means replacing gas-powered vehicles with electric cars, using electric heat pumps for heating and cooling, and relying on electric appliances for cooking and other household needs. The goal is to maximize the use of clean energy generated from renewable sources such as solar, wind, and hydroelectric power. Renewable electrification not only helps reduce carbon emissions but also improves air quality by eliminating pollutants associated with burning fossil fuels. It promotes energy independence by reducing reliance on imported oil and gas. As we continue to advance in this direction, it is crucial to invest in the development of efficient battery storage systems to ensure a reliable supply of electricity even during times of intermittent renewable generation. With renewable electrification at the forefront of our energy transition, we can pave the way toward a sustainable and carbon-neutral future.

In this phase, complementary innovations play a crucial role in driving progress toward a sustainable future. These innovations encompass various areas, such as renewable energy technologies, smart grid systems, and energy storage solutions. Advancements in transportation, agriculture, and industrial sectors are also pivotal in achieving the goals set by the Paris Climate Agreement. As electrification gains momentum across diverse sectors, it unlocks new opportunities for decarbonization and reduces reliance on fossil fuels. The integration of EVs into transportation networks and the adoption of clean energy sources in manufacturing processes exemplify the transformative potential of electrification. The development of advanced battery technologies enables efficient energy storage and utilization, ensuring a reliable and resilient power supply. The third phase represents a critical juncture where these complementary innovations converge to maximize their collective impact on mitigating climate change and fostering sustainable development worldwide. As we continue to push boundaries and explore new frontiers, this phase will mark a significant milestone in our journey toward a greener and more sustainable future for generations to come. With the integration of renewable energy sources such as solar and wind power, we are witnessing a paradigm shift in the way we generate electricity. Advancements in battery technology have revolutionized energy storage, allowing us to harness and store excess energy during peak production periods for use during times of high demand. This not only ensures a reliable power supply but also reduces our dependence on fossil fuels, leading to a significant reduction in greenhouse gas emissions.

One of the key challenges is the need for effective coordination and collaboration among different sectors and systems. This involves not only aligning goals and objectives but also integrating diverse technologies, policies, and stakeholders. The complexity arises from the interdependencies and interconnections that emerge when sectors such as energy, transportation, and water intersect. For example, in the case of cross-sectoral balancing of electricity supply and demand, it requires close coordination between power grid operators, renewable energy providers, and consumers to ensure a reliable and sustainable energy system. It necessitates the development of advanced monitoring and control systems that can dynamically adjust electricity generation and consumption based on real-time data. These multi-sector interactions also raise issues related to governance, regulation, and financing. As different sectors converge, there is a need for new institutional frameworks that can effectively govern these complex systems while ensuring fairness, transparency, and accountability. Innovative financing mechanisms are required to support cross-sectoral projects that may involve multiple stakeholders with diverse interests.

This trade-off is not unique to Norway but is a common challenge faced by many countries with abundant renewable energy resources. The decision to prioritize either exporting hydroelectricity or utilizing it for domestic decarbonization depends on various factors, such as economic considerations, environmental goals, and the overall energy demand. While exporting hydroelectricity can bring significant revenue and contribute to global decarbonization efforts, focusing on driving domestic decarbonization of transport and industry can have long-term benefits for the country's energy independence and sustainability. By leveraging their hydroelectric base to power EVs and electrify industrial processes, Norwegian utilities can reduce greenhouse gas emissions within their own borders, enhance energy security, and foster the growth of a green economy. Finding the right balance between export opportunities and domestic decarbonization is crucial to ensuring a stable energy supply while simultaneously reducing the carbon footprint. Strategic planning and collaboration between stakeholders are essential to making informed decisions that align with Norway's broader sustainability objectives.

In this evolving landscape, incumbents in the energy sector are grappling with complex decisions and uncertainties. They find themselves at a crossroads, facing the challenge of balancing their existing infrastructure with the need to adapt to a rapidly changing market. One key dilemma they encounter is whether to prioritize the expansion of transmission lines for exporting hydroelectricity or leverage their hydroelectric base for driving domestic decarbonization efforts in transportation and industry. This decision carries significant implications for their future success and relevance. The traditional strategies that incumbents have relied upon may no longer be viable as shifting consumer preferences, policy changes, and the emergence of new competitors reshape industry dynamics. As a result, incumbents must carefully navigate this transformative period, exploring innovative approaches that align with evolving market demands while ensuring long-term sustainability and competitiveness. By embracing change and seizing opportunities in this dynamic environment, incumbents can position themselves as leaders in the transition toward a cleaner and more sustainable energy future.

This shift toward sustainable practices is not limited to the automotive industry. In fact, various sectors are starting to embrace the need for change. The building sector, although still facing resistance, is slowly recognizing the importance of adopting environmentally friendly practices. Architects and developers are incorporating energy-efficient designs and materials into their projects, aiming to reduce carbon emissions and minimize waste. They are exploring innovative technologies such as solar panels and smart home systems to create more sustainable living spaces. Governments and organizations are providing incentives and regulations to encourage green building practices. This collective effort is paving the way for a greener future in the construction industry. While challenges remain, progress is being made. On the other hand, in the transportation sector, automakers are taking significant steps toward a greener future. Many renowned companies have announced ambitious targets to phase out gasoline cars entirely. They are investing heavily in EV technology and infrastructure, aiming to make EVs more accessible and convenient for consumers. Automakers are also exploring alternative fuel options such as hydrogen and biofuels. This diversification of the transportation sector is crucial to reducing greenhouse gas emissions and mitigating climate change.

This shift toward electrification will lead to a transformation in the energy landscape, disrupting traditional industries and infrastructure. As conventional fueling and automotive servicing stations become obsolete, new charging stations and maintenance facilities will emerge to cater to the growing demand for EVs. Similarly, the existing natural gas distribution networks will need to be reimagined to accommodate changing energy needs. This transition presents an opportunity for innovation and investment in renewable energy sources and smart grid technologies. The erosion of these old assets will pave the way for a more sustainable and efficient energy system, reducing reliance on fossil fuels and promoting environmental stewardship. It is crucial for stakeholders to adapt and embrace this change, as it holds the potential to revolutionize transportation and reshape our cities.

These disruptions can lead to significant power shifts and reshape the political landscape. As new players enter the market, traditional service providers may find themselves facing unprecedented challenges. The battle for dominance becomes more fierce as resources become scarcer, and political struggles become inevitable. Incumbents must adapt quickly to keep up with the changing dynamics or risk being left behind. This intensified competition forces all parties involved to innovate and improve their services, ultimately benefiting consumers with better options and lower costs. This also creates a volatile environment where alliances are formed and broken, and political power constantly shifts hands. The future will witness a relentless cycle of disruption and adaptation as new entrants continue to disrupt established industries, forever altering the political struggles that accompany such transformations.

These tensions arise from the complex interplay between various actor networks, business models, and social practices. The future prospects of these entities are crucial in understanding the underlying dynamics. Actor networks, comprising individuals, organizations, and institutions, play a pivotal role in shaping the direction of industries and economies. Their ability to adapt to changing circumstances and leverage emerging technologies will determine their long-term success. Similarly, business models need to evolve continuously to remain competitive in an ever-changing landscape. Innovation, diversification, and strategic partnerships become essential for sustainable growth. Social practices influence the behavior of actors and organizations within a given context. As societal norms shift and values change, businesses must align their practices with evolving expectations to maintain relevance and legitimacy. Therefore, understanding the long-term prospects of actor networks, business models, and social practices is crucial for navigating the complexities of today's interconnected world. By anticipating future trends and adapting accordingly, stakeholders can position themselves for success while mitigating potential tensions that may arise along the way.

A growing right-wing populism serves as a backdrop to many of these debates, which have become more and more polarized. This surge in populism has fueled a divisive political climate, where opposing viewpoints are met with hostility rather than constructive dialogue. The once nuanced discussions on important issues have now devolved into echo chambers, where individuals only seek validation for their pre-existing beliefs. This polarization has hindered progress and hindered the search for common ground, as people are more focused on defending their own ideologies than finding solutions that benefit society as a whole. The rise of right-wing populism has also exacerbated social divisions, pitting different groups against each other based on race, religion, and nationality. This toxic environment fosters fear and animosity, further deepening the divide between communities. It is crucial to address this growing polarization by promoting empathy, understanding, and respectful discourse. By encouraging open-mindedness and actively seeking out diverse perspectives, we can bridge the gap between different ideologies and work toward finding common solutions to the challenges we face as a society.

5.4.4 Fourth Progression

This ambitious goal aims to completely eliminate greenhouse gas emissions by a certain deadline, typically 2050 (IEA, 2022). The introduction of net zero emission targets by the European Union has sparked global momentum toward achieving a sustainable and carbon-neutral future (International Renewable Energy Agency [IRENA], 2022). As other countries and regions follow suit, we are witnessing a significant shift in the way we approach energy production, transportation, and consumption. This transition requires not only technological advancements but also changes in policies, regulations, and public awareness. Governments are now implementing measures to promote renewable energy sources, invest in green infrastructure, and incentivize sustainable practices across industries. Simultaneously, businesses are recognizing the economic opportunities that come with embracing clean technologies and reducing their carbon footprint. The fourth phase of the net zero transition is characterized by a collective effort to mitigate climate change and ensure a habitable planet for future generations. It signifies a turning point in our commitment to address the urgent environmental challenges we face today. With each step taken toward achieving net zero emissions, we move closer to creating a more sustainable and resilient future.

In this phase, the focus shifts toward addressing the sectors and applications that pose significant challenges in terms of decarbonization. These are the areas where electrification is not a straightforward solution due to various factors. One such challenge is the limited power storage capacity, which hampers the effective utilization of electricity in these sectors. The remoteness of grid interconnection poses a hurdle as it becomes difficult to establish a reliable and efficient connection to the main power grid. Cost-efficiency considerations also play a crucial role in determining the feasibility of electrification in these difficult-to-decarbonize sectors and applications. Therefore, during phase four of the net zero transition, concerted efforts are directed toward finding innovative solutions that go beyond traditional electrification methods to overcome these obstacles and achieve decarbonization targets. This involves exploring alternative technologies, developing advanced energy storage systems, and implementing cost-effective strategies tailored to each specific sector's requirements.

For example, the decarbonization of the transportation sector is significantly hampered by the aviation, shipping, and long-distance trucking industries for a variety of reasons. In aviation, the challenge lies in the lack of efficient power storage capacity, making it difficult to rely solely on electrification. The remoteness of grid interconnection further complicates the transition toward net zero emissions in this sector. Similarly, shipping and long-distance trucking face similar obstacles, with insufficient power storage capacity and cost-efficiency considerations hindering progress. To successfully address these difficulties, innovative solutions beyond electrification need to be explored. This may involve advancements in alternative fuels or propulsion systems that can effectively reduce emissions without relying solely on electric power. By focusing on these challenging sectors and applications, we can pave the way for a comprehensive net zero transition that encompasses all aspects of our economy and ensures a sustainable future for generations to come.

Phase 4 is expected to feature low-carbon innovations, including hydrogen, a versatile general-purpose energy carrier. This innovative technology can be produced in various ways and used for various applications, making it a significant focus in the industry. Hydrogen has the potential to revolutionize various sectors, including transportation, industry, and power generation. Its versatility stems from its ability to be produced through diverse methods such as electrolysis, natural gas reforming, or biomass gasification. This ensures a sustainable and clean energy source that can significantly reduce greenhouse gas emissions. In transportation, hydrogen fuel cells offer an alternative to conventional internal combustion engines, providing longer ranges and shorter refueling times for EVs. Hydrogen can be utilized in industrial processes like steel manufacturing or chemical production, replacing carbon-intensive practices with greener alternatives. Hydrogen power plants have the capacity to generate electricity while emitting only water vapor as a byproduct. As we enter phase 4, the integration of hydrogen into our energy systems holds great promise for a more sustainable future. With ongoing research and development efforts focused on enhancing efficiency and reducing costs, hydrogen is poised to play a crucial role in the global transition toward a low-carbon economy.

The integration of competing alternative fuels such as biofuels and hydrogen and the expansion of electrification through full battery trucks, catenary trucks, and fuel cell trucks present additional challenges in the pursuit of sustainable transportation (Mäkitie et al., 2023). These advancements hold the potential to revolutionize the industry, but their implementation may encounter struggles and obstacles. As efforts to achieve net zero emissions align with post-pandemic recovery initiatives, there is hope for accelerated climate action. It is still uncertain how political dynamics will shape the future trajectory of these endeavors. The interplay between technological advancements, environmental goals, and political factors will ultimately determine the success and pace of our transition toward a greener future.

In addition to supply-side innovations, phase four may also bring about substantial demand-side changes, such as increased engagement in demand reduction and lifestyle adjustments. These changes in lifestyles could involve a shift toward more sustainable and environmentally friendly practices. People might start adopting minimalist lifestyles, focusing on reducing their consumption and waste generation. This could lead to a decrease in demand for products that have a high environmental impact, such as single-use plastics or fast fashion. There might be an increased demand for products and services that promote sustainability, such as renewable energy sources or zero-waste stores. Governments and businesses might also play a crucial role in shaping these changes by implementing policies and practices that encourage sustainable consumption patterns. Overall, phase 4 could witness a transformation in the way people perceive and engage with consumption, leading to a more sustainable and responsible society.

In order to address the limitations of supply-side innovations and avoid the need for extreme options, a comprehensive approach to sustainability is crucial. This entails focusing not only on reducing emissions from aviation through measures like decreased overseas travel but also considering the broader picture. By adopting a holistic perspective, we can explore alternative solutions that mitigate environmental impact while also promoting social and economic well-being. For instance, investing in renewable energy sources and improving energy efficiency across various sectors can significantly reduce carbon emissions without compromising land use or depleting resources. Prioritizing sustainable practices such as circular economy models and responsible consumption can help minimize waste and pollution while fostering long-term sustainability. Emphasizing these multifaceted strategies will enable us to tackle climate change effectively without relying solely on extreme options that may have unintended consequences.

5.5 POLICY ROADBLOCKS ON THE PATH TO NET ZERO EMISSIONS

There is a growing need for comprehensive and forward-thinking policies that address the multifaceted challenges of achieving net zero energy. Policymakers must consider various factors such as technological advancements, economic implications, social equity, and environmental sustainability. It requires a delicate balance between incentivizing renewable energy sources, promoting energy efficiency measures, and ensuring a just transition for affected communities. International collaboration and coordination are crucial to tackling global climate change effectively. Policymakers must navigate through intricate negotiations, considering the diverse interests and priorities of different countries. They need to foster innovation and research to develop breakthrough technologies that can accelerate the transition toward a net zero energy future. The complexity of policymaking in this context necessitates interdisciplinary approaches that involve experts from various fields such as energy, economics, sociology, and environmental science. By fostering collaboration between policymakers, scientists, industry leaders, and civil society organizations, we can develop robust policies that drive the net zero energy transition while minimizing unintended consequences. Effective policymaking will require a deep understanding of the interconnectedness between energy systems, economic dynamics, societal behaviors, and environmental impacts. This holistic approach will enable policymakers to identify potential trade-offs and synergies across different sectors. For instance, policies promoting renewable energy sources must consider not only their technical feasibility but also their economic viability and social acceptance.

As society evolves and technology advances, governments are faced with the task of adapting policies to address emerging issues. One such challenge is the regulation of artificial intelligence (AI) and its impact on various sectors. Policymakers must grapple with the ethical implications of AI, ensuring that it is developed and used responsibly to prevent potential harm. The rise of automation raises concerns about job displacement and income inequality, prompting policymakers to explore solutions such as universal basic income or retraining programs. Another pressing policy challenge is climate change and environmental sustainability. Governments worldwide are under pressure to implement measures that reduce carbon emissions, promote renewable energy sources, and protect natural resources. Balancing economic growth with environmental conservation requires innovative policies that encourage sustainable practices across industries. In an increasingly interconnected world, cybersecurity has become a critical concern for policymakers. Safeguarding sensitive data from cyber threats and ensuring privacy protection are paramount in the digital age.

This evolution in problem framing, conceptual frameworks, and approaches for innovation policy has been a dynamic and transformative journey since the 1950s. Initially, the focus was primarily on R&D policy, emphasizing research and development as the driving force behind innovation. As our understanding of innovation deepened, so did our approach. The shift toward systemic innovation policy recognized that innovation is not limited to isolated activities but rather a complex interplay of various actors and factors within a system. This broader perspective acknowledged the importance of fostering collaboration, knowledge exchange, and the integration of different stakeholders to drive innovation forward. Building upon this foundation, we have now entered an era of transformative innovation policy. This approach recognizes that true innovation goes beyond incremental improvements; it seeks to address grand challenges and fundamentally transform systems for a sustainable future. It emphasizes disruptive technologies, social change, and systemic shifts that can reshape entire industries and societies. By embracing this transformative mindset, we can unlock new opportunities and tackle pressing global issues in ways that were previously unimaginable.

Policymakers prioritized fostering innovation as the cornerstone of achieving net zero emissions. They recognized that ground-breaking technological advancements were crucial to revolutionizing industries and transforming our energy systems. Investments were directed toward research and development, encouraging scientists and engineers to push the boundaries of what was possible. Collaborative partnerships between governments, academia, and the private sector flourished, creating an ecosystem of knowledge exchange and resource sharing. This phase witnessed a surge in breakthroughs such as efficient renewable energy technologies, CCS solutions, and sustainable transportation alternatives. These innovations not only paved the way for cleaner energy production but also fostered sustainable practices across various sectors. Policymakers understood that without disruptive innovation, the transition to net zero emissions would be arduous. Therefore, they championed policies that incentivized companies to invest in clean technologies while simultaneously phasing out fossil fuel subsidies. The first phase set the stage for a transformative journey toward a decarbonized future, where innovation became synonymous with progress in the fight against climate change.

The policy challenge was to stimulate technology development and the formation of early niche markets. To tackle this challenge, policymakers implemented various strategies. Firstly, they focused on fostering collaboration between research institutions and industry players to accelerate technology development. This involved establishing innovation hubs and providing grants to encourage joint projects and knowledge sharing. They incentivized private sector investment in emerging technologies by offering tax breaks and funding opportunities. Policymakers recognized the importance of creating a supportive regulatory environment. They worked closely with industry experts to develop flexible frameworks that would encourage experimentation and entrepreneurship while ensuring consumer safety and privacy. This approach allowed innovative startups to navigate regulatory hurdles more easily, fostering a vibrant ecosystem of technology development. In parallel, policymakers also prioritized building a skilled workforce capable of driving technological advancements. They invested heavily in education and training programs that focused on science, technology, engineering, and mathematics (STEM) fields. By equipping individuals with the necessary skills for emerging industries, they aimed to create a talent pool. The emphasis was on research, development, and demonstration of alternative socio-technical configurations, and scholars as well as policymakers highlighted the importance of systems approaches such as technological innovation systems.

These efforts led to significant advancements in renewable energy technologies and a shift toward sustainable practices. Governments implemented various incentives and subsidies to encourage the adoption of low-carbon innovations, fostering collaboration between academia, industry, and government agencies. This multidimensional approach not only stimulated economic growth but also paved the way for a greener future. As a result, Denmark became a global leader in wind energy, harnessing its abundant resources to generate clean electricity. Similarly, Germany's commitment to solar power propelled it to the forefront of photovoltaic technology, revolutionizing the way we harness the sun's energy. These success stories inspired other nations to follow suit, sparking a global movement toward decarbonization and mitigating climate change. The lessons learned from these localized initiatives laid the foundation for broader international cooperation on low-carbon innovation, with countries sharing best practices and collaborating on research and development.

These approaches are crucial for fostering economic growth, promoting sustainability, and addressing societal challenges. By implementing systemic innovation policies, governments can create an enabling environment that encourages collaboration between different stakeholders, such as businesses, academia, and civil society. This collaborative approach allows for the sharing of knowledge and resources, leading to the development of breakthrough technologies and solutions. Systemic innovation policies help identify and remove barriers that hinder innovation, such as outdated regulations or a lack of access to funding. They also promote the adoption of new business models that prioritize sustainability and social impact. In the future, as economies become increasingly interconnected and complex, systemic innovation policy approaches will play a vital role in driving transformative change across sectors and ensuring a prosperous and inclusive society.

This significant investment is crucial to driving innovation and accelerating the development of new technologies that can effectively reduce greenhouse gas emissions. It is evident that relying solely on existing technologies will not be sufficient to achieve the ambitious goal of net zero emissions by 2050 (IRENA, 2019). The IEA's recognition of the need for novel solutions highlights the importance of fostering research and development in clean energy sectors. By allocating public funding toward R&D and demonstration projects, governments can actively support the creation and implementation of breakthrough technologies. These investments will not only facilitate the transition to a sustainable future but also stimulate economic growth and job creation in emerging industries. Collaboration between governments, academia, and private sector entities will be essential to leverage expertise, share knowledge, and accelerate the deployment of these innovative technologies. By embracing this forward-thinking approach, we can pave the way for a greener and more resilient planet while simultaneously driving technological advancements that benefit society as a whole.

The second phase of net zero emissions shifts from innovation to deployment policies for technology improvement and diffusion acceleration. These deployment policies aim to support the widespread adoption of clean energy technologies and ensure their integration into various sectors. Governments implemented measures such as feed-in tariffs, tax incentives, and grants to incentivize businesses and individuals to invest in renewable energy sources like solar and wind power. Regulations were put in place to encourage the development of energy-efficient buildings and transportation systems. The focus was on creating an enabling environment that would facilitate the transition toward a low-carbon economy. International collaborations also played a crucial role in sharing best practices and knowledge transfer, allowing countries to learn from each other's experiences and avoid reinventing the wheel. As a result of these efforts, clean energy technologies became more affordable and accessible, leading to their widespread adoption across industries and communities worldwide. The deployment phase marked a significant milestone in achieving net zero emissions by harnessing the power of existing technologies and leveraging supportive policies for their rapid diffusion.

These decline policies play a crucial role in accelerating the transition toward a low-carbon future. By specifically targeting sectors that contribute significantly to greenhouse gas emissions, such as coal-fired power plants, incandescent bulbs, and internal combustion engines, these policies effectively address the root causes of climate change. The phased-out approach ensures a gradual reduction in the use of these high-emission sources, allowing for alternative and cleaner technologies to take their place. This not only reduces carbon dioxide emissions but also encourages innovation and investment in renewable energy sources. Decarbonization policies create a sense of urgency and accountability among industries and individuals alike, fostering a collective commitment toward decarbonization.

Due to their broad implications for the targeted industries and the communities they are associated with, these new issues for policy development and implementation present significant challenges. As we navigate this complex landscape, it becomes crucial to strike a delicate balance between promoting economic growth and ensuring social and environmental sustainability. Policymakers must carefully consider the potential consequences of their decisions, taking into account the diverse needs and perspectives of various stakeholders. This requires fostering open dialogue and collaboration among industry leaders, community representatives, and government officials to develop inclusive policies that address the concerns of all parties involved. Effective implementation strategies should be devised to ensure that these policies are not only enacted but also effectively enforced. This may involve investing in capacity-building initiatives, providing necessary resources, and establishing robust monitoring mechanisms to track progress and identify areas for improvement. By adopting a comprehensive approach that takes into account the unique characteristics of each industry and community, policymakers can pave the way for sustainable development while minimizing any adverse impacts.

In order to successfully transition toward a sustainable energy future, it is crucial to consider the impact on affected regions, such as those around coal mining. These regions may require support or compensation to mitigate the economic and social consequences of the transition. The shift in policy focus is now toward embracing entire sectors or whole systems, recognizing the interconnectedness of various elements. This approach entails incorporating a wide range of system components, including transmission grids and storage technologies, in the electricity sector. By addressing potential bottlenecks that may arise during the transition, such as infrastructure limitations or supply–demand imbalances, we can ensure a smoother and more efficient pathway toward a sustainable energy future. This comprehensive approach not only facilitates the integration of renewable energy sources but also supports the development of resilient and adaptable energy systems that can withstand future challenges.

In order to successfully navigate phase three of the net zero energy transition, effective cross-sectoral policy coordination and multi-system interaction are paramount. This entails bringing together stakeholders from various sectors, including government bodies, industries, and communities, to collaborate and align their efforts toward a common goal. By fostering strong partnerships and encouraging information sharing, policymakers can ensure that policies are integrated across sectors and avoid siloed approaches. Multi-system interaction plays a crucial role in optimizing energy efficiency and reducing carbon emissions. This involves leveraging technological advancements to enable seamless communication and integration between different energy systems, such as electricity grids, transportation networks, and buildings. By creating interconnected systems that can efficiently exchange energy and resources, we can maximize the overall effectiveness of the net zero energy transition. Therefore, robust cross-sectoral policy coordination and multi-system interaction serve as the foundation for a successful phase three, enabling us to overcome challenges and accelerate progress toward a sustainable future.

Prior to the third phase, interactions across sectors—possibly involving competition—emerged, as opposed to the earlier net zero energy transition phases, where decarbonization policies were independent. This necessitates a holistic approach to decarbonization, where policies are designed with a comprehensive understanding of the interconnectedness between sectors. For instance, while transitioning to low-carbon electricity is crucial for reducing emissions in multiple sectors, it is important to carefully manage the allocation of this limited resource. Prioritizing sectors that have fewer alternative decarbonization options, such as heavy industry or transportation, can maximize the overall impact of limited low-carbon electricity availability. Fostering collaboration and knowledge sharing between sectors becomes essential to identifying synergies and avoiding conflicts. This can involve initiatives like integrated planning processes that bring together stakeholders from different sectors to collectively develop strategies that optimize resource utilization and minimize competition. As the net zero energy transition progresses, it becomes increasingly vital to consider the broader implications of sectoral policies on society as a whole. Ensuring equitable access to clean energy resources and addressing potential social disparities arising from sectoral competition should be key considerations in policy formulation.

This shift toward integrated energy system planning marks a significant departure from the traditional approach of treating each sector in isolation. By breaking down the barriers between electricity, transport, buildings, and hydrogen, we can unlock a multitude of synergies and opportunities. For instance, by coordinating the generation and consumption of electricity with the transportation sector, we can optimize charging infrastructure for EVs and leverage vehicle-to-grid technologies to enhance grid stability. Simultaneously, integrating buildings into the equation allows for smarter energy management systems that reduce overall demand and enable efficient use of renewable energy sources. The inclusion of novel energy carriers like hydrogen opens up avenues for long-term energy storage and decarbonization efforts. This holistic approach to energy planning not only maximizes efficiency but also paves the way for a more sustainable and resilient future.

As phase 3 of the net zero transition unfolds, the demand for new infrastructure investments will intensify. One crucial aspect will be the development of public charging stations to support the growing number of EVs on the road. This expansion requires substantial funding, and public financing will be instrumental in making it a reality. In this context, the state is expected to assume a more prominent role, leading to a potential increase in contestation and debate surrounding its involvement. As governments strive to facilitate the transition toward net zero energy, they will need to navigate complex discussions around funding allocation and ensure that public resources are effectively utilized to drive sustainable infrastructure development.

To complete phase 4 of the net zero transition, it will be necessary to continue to support innovation, implement sectoral transition policies, and implement cross-sectoral policies. These measures will play a crucial role in accelerating the shift toward a carbon-neutral economy. Innovation is key to developing new technologies and solutions that can reduce greenhouse gas emissions across various sectors. By investing in research and development, governments can foster the creation of breakthrough innovations that will drive the net zero transition forward. Sectoral transition policies are essential for guiding industries toward cleaner and more sustainable practices. By setting targets, providing incentives, and implementing regulations, governments can encourage sectors such as energy, transportation, and manufacturing to adopt low-carbon alternatives. It is important to recognize that achieving net zero requires collaboration across sectors. Cross-sectoral policies that promote coordination and cooperation between different industries are vital for maximizing efficiency and minimizing emissions. These policies can encourage knowledge sharing, resource optimization, and the adoption of best practices across sectors.

These new technologies and infrastructures require significant investment and collaboration between governments, industries, and research institutions. Governments will need to implement policies that encourage the development and deployment of these technologies, such as providing financial incentives and regulatory support. Sectoral transition policies will be crucial in ensuring a smooth shift from traditional energy sources to these novel technologies. This may involve phasing out fossil fuel–based industries and supporting the growth of renewable energy sectors. Cross-sectoral policies are also essential to address the interconnectedness of different industries and ensure a coordinated approach to the adoption of these new technologies. For example, the transportation sector will need to work closely with the energy sector to develop efficient ways to transport hydrogen or captured CO_2 on a large scale. Collaboration between stakeholders is key to overcoming challenges related to infrastructure development, as it requires expertise from various fields such as engineering, logistics, and environmental science. International cooperation will be necessary to share best practices and knowledge in order to accelerate the transition toward a more sustainable future.

As we enter the fourth phase, it becomes evident that tackling major changes in lifestyles and consumption patterns is a paramount policy challenge. The world has witnessed a rapid transformation in how people live and what they consume. It is imperative to recognize and adapt to these shifts to ensure a sustainable future. This requires innovative approaches that encourage responsible consumption and promote environmentally friendly choices. Policymakers must prioritize education and awareness campaigns to empower individuals to make informed decisions about their lifestyles. Fostering collaboration between industries, governments, and communities can lead to the development of sustainable alternatives that minimize the negative impact on our planet. Embracing technological advancements can also play a pivotal role in revolutionizing consumption patterns by offering eco-friendly solutions for everyday needs.

To address industries and applications that are challenging to decarbonize, a new set of low-carbon innovations must be developed and disseminated, along with the infrastructure needed to support them. This requires a multifaceted approach that encompasses research, development, and implementation of breakthrough technologies. One potential solution lies in the advancement of CCS technologies, which can effectively capture and store carbon emissions from industries such as cement production and steel manufacturing. Promoting the adoption of renewable energy sources, such as solar and wind power, can significantly reduce the carbon footprint of energy-intensive sectors like transportation and heating. Investing in sustainable transportation systems, such as EVs and efficient public transit networks, can help decarbonize the transportation sector. Exploring alternative materials with lower carbon footprints for construction and manufacturing purposes can further contribute to reducing emissions in difficult-to-decarbonize industries.

In addition, the transition to low-carbon energy carriers will inevitably lead to economic shifts and challenges. Industries heavily reliant on fossil fuels may face significant disruptions, necessitating careful planning and support for affected communities. The adoption of new technologies and infrastructure for low-carbon energy may require substantial investments and international cooperation. It is crucial for governments and policymakers to navigate these complexities with strategic foresight, ensuring a just and equitable transition for all stakeholders involved. As the demand for low-carbon energy carriers grows, it is essential to prioritize research and development efforts to enhance their efficiency and scalability. This includes advancements in renewable energy sources such as solar, wind, and hydroelectric power, as well as exploring innovative solutions like CCS. Collaborative efforts between governments, academia, and industry will be vital in driving these technological advancements forward.

5.6 TECHNIQUES FOR ACHIEVING NET ZERO IN VARIOUS STAGES

Net zero strategies aim to achieve a balance between greenhouse gas emissions and removals, ultimately resulting in no net emissions. These strategies encompass a wide range of measures across various sectors. In addition to low-carbon electrification, another key approach is the adoption of renewable energy sources such as solar and wind power. By transitioning from fossil fuels to clean energy, we can significantly reduce carbon emissions. Improving energy efficiency in buildings and industries plays a crucial role in decarbonization efforts. This involves implementing advanced technologies and practices that minimize energy waste. Sustainable transportation solutions like EVs and public transit systems can help curb emissions from the transportation sector. Alongside these measures, reforestation and afforestation initiatives are essential for carbon sequestration, as trees absorb CO_2 from the atmosphere (Acheampong et al., 2023). Promoting circular economy principles that prioritize recycling and waste reduction can contribute to lowering emissions associated with resource extraction and production processes.

Below, we identify five main strategies that encompass a wide range of options for addressing the challenges of reducing carbon emissions and transitioning to a sustainable future. The first strategy focuses on reducing demand by promoting energy efficiency measures and implementing conservation practices across sectors such as transportation, buildings, and industry. This involves encouraging the adoption of smart technologies, promoting behavioral changes, and implementing policies that incentivize energy-saving practices. The second strategy revolves around substituting fuels by transitioning from fossil fuels to cleaner alternatives such as renewable energy sources like solar, wind, and hydropower. This includes investing in research and development to enhance the efficiency and affordability of these technologies. The third strategy involves changing lifestyles by promoting sustainable consumption patterns and encouraging individuals to adopt eco-friendly behaviors in their daily lives. This can be achieved through education campaigns, public awareness programs, and providing incentives for sustainable choices (Ju et al., 2021). The fourth strategy focuses on enhancing resilience through adaptation measures that prepare communities for the impacts of climate change. This includes investing in infrastructure improvements, implementing disaster risk reduction strategies, and promoting the use of renewable energy sources.

These untapped approaches hold great potential for further reducing carbon emissions and combating climate change. One such approach is the implementation of behavioral interventions, which aim to change individual and societal behaviors toward more sustainable practices. This could include initiatives such as promoting public transportation, encouraging the use of bicycles or walking for short distances, and raising awareness about the environmental impact of certain consumer choices. Another untapped approach is the integration of renewable energy sources into existing infrastructure on a larger scale. By utilizing solar panels on rooftops, wind turbines in urban areas, and harnessing the power of tidal or geothermal energy, we can significantly reduce our reliance on fossil fuels. Exploring innovative technologies like CCS can play a crucial role in achieving negative emissions. These technologies involve capturing carbon dioxide from industrial processes or directly from the atmosphere and storing it underground or converting it into useful products.

The best approach to reducing energy demand is through the adoption of smart grid technologies. Smart grids enable more efficient and effective management of electricity distribution, allowing for real-time monitoring and control of energy consumption. This not only helps to optimize energy usage but also enables the integration of renewable energy sources into the grid, further reducing reliance on fossil fuels. The implementation of demand response programs can incentivize consumers to adjust their energy usage during peak demand periods, thus alleviating strain on the grid and reducing overall energy demand. Another strategy involves promoting sustainable transportation options such as EVs and public transit systems. By encouraging the use of these alternatives, we can significantly decrease the energy consumed by traditional gasoline-powered vehicles (Rahnama Mobarakeh and Kienberger, 2022). Investing in research and development for advanced energy storage technologies can enhance the efficiency and reliability of renewable energy sources like solar and wind power. These innovations allow for better utilization of intermittent renewable resources, ensuring a more stable and sustainable energy supply.

To address this residual demand, it is crucial to explore renewable energy sources and integrate them into the energy mix. Solar power, wind turbines, and hydroelectric plants offer sustainable alternatives that can help offset the remaining energy needs. Implementing smart grid systems can optimize energy distribution and reduce wastage. By monitoring consumption patterns and adjusting supply accordingly, we can minimize inefficiencies and maximize the utilization of available resources. Promoting energy-conscious behavior among individuals and businesses through education and incentives can contribute to overall energy savings. Encouraging practices such as proper insulation, efficient appliances, and responsible usage can make a significant difference in reducing energy consumption. Achieving comprehensive energy efficiency requires a multifaceted approach that combines technological advancements, renewable energy integration, smart grid implementation, and widespread behavioral changes. Only through these combined efforts can we strive toward a more sustainable future with minimized residual demand.

Efficiency is a key aspect that should be prioritized in all phases of any process or project. It is unfortunate that this strategy has not received the attention it deserves and remains largely untapped. By focusing on efficiency, organizations can streamline their operations, reduce costs, and maximize productivity. From the initial planning stage to execution and evaluation, incorporating efficiency measures can lead to significant improvements. By identifying bottlenecks, eliminating redundancies, and optimizing resources, businesses can achieve better outcomes while minimizing waste. Embracing efficiency as a core principle will not only enhance overall performance but also contribute to sustainable growth and success in the long run. It is crucial for decision-makers to recognize the importance of this strategy and actively implement it throughout all phases of their endeavors. Only by doing so can they unlock the full potential of their operations and stay ahead in today's competitive landscape.

In order to achieve net zero, the second strategy, which focuses on low-carbon electrification, has become the most popular. This strategy focuses on reducing carbon emissions by transitioning to renewable energy sources for electricity generation. It involves increasing the deployment of solar, wind, and hydroelectric power systems while phasing out fossil fuel–based power plants. Energy efficiency measures are implemented to optimize electricity usage and minimize waste (Chapungu et al.,

2022). This approach also promotes the electrification of various sectors, such as transportation, heating, and industrial processes. By replacing fossil fuel-powered vehicles with EVs, adopting electric heating systems, and utilizing electric-based technologies in industries, the reliance on carbon-intensive practices is reduced significantly. This strategy encourages the development of smart grids and energy storage solutions to enhance the integration of intermittent renewable energy sources into the grid. By prioritizing low-carbon electrification, economies can achieve substantial reductions in greenhouse gas emissions while simultaneously fostering sustainable economic growth and creating new job opportunities in the clean energy sector.

The first sub-strategy, the substitution of carbon-intensive power generation, aims to replace the existing fossil fuel–based power generation with cleaner alternatives. This can be achieved through the widespread adoption of renewable energy sources such as solar, wind, and hydroelectric power. Advancements in energy storage technologies can help overcome the intermittent nature of renewable sources and ensure a reliable and stable power supply (Delbeke et al., 2019). Exploring the potential of nuclear energy as a low-carbon option could be considered, although its acceptance would depend on socio-political factors. The second sub-strategy focuses on extending electrification to sectors that traditionally relied on fossil fuels. This involves transitioning buildings and transportation systems to electric power. In buildings, this can be achieved by promoting energy-efficient designs and retrofitting existing structures with smart technologies that optimize energy consumption. The electrification of transportation involves the widespread adoption of EVs and the development of charging infrastructure. Alternative modes of transport such as electric buses and trains can contribute to reducing carbon emissions in the transportation sector.

To address these challenges, researchers and policymakers are exploring alternative solutions. One promising avenue is the development of advanced technologies that can enable a more flexible power system. These technologies include extensive interconnections between different energy sources and regions, allowing for the efficient transfer of electricity. Energy storage capabilities play a crucial role in balancing supply and demand fluctuations, ensuring a reliable and resilient power grid (Kichere, 2020). Rethinking the way we value energy is essential for achieving sustainability goals. Traditional metrics such as cost per kilowatt-hour may not fully capture the environmental and social impacts of different energy sources. Therefore, incorporating broader considerations like carbon emissions, air pollution, and ecosystem impacts into energy valuation frameworks can provide a more comprehensive understanding of the true costs and benefits. Addressing difficult-to-decarbonize sectors presents another challenge on the path to sustainability. While some sectors have successfully transitioned to cleaner energy sources, others remain heavily reliant on fossil fuels due to technological limitations or a lack of viable alternatives. To overcome this hurdle, innovative solutions and investments in research and development are crucial.

The first sub-strategy involves the development and promotion of biofuels. This entails utilizing organic matter, such as crops or agricultural waste, to produce renewable fuels that emit significantly lower carbon emissions compared to traditional fossil fuels. Biofuels offer a promising alternative for various sectors, including transportation and aviation, reducing their carbon footprint and contributing to a more sustainable future. The second sub-strategy centers around the advancement of hydrogen as a low-carbon fuel option. Hydrogen has gained attention as a versatile energy carrier that can be produced through renewable sources like electrolysis. By utilizing hydrogen fuel cells, vehicles can operate with zero emissions, offering an efficient and environmentally friendly transportation solution (Guo et al., 2023). Hydrogen can also be utilized in industries such as manufacturing and power generation, replacing carbon-intensive processes with cleaner alternatives. Both sub-strategies under the low-carbon fuels approach demonstrate the commitment to reducing greenhouse gas emissions while addressing the limitations of electrification. By diversifying our energy sources and promoting the adoption of biofuels and renewable energy technologies, we can significantly decrease our reliance on fossil fuels.

Despite the initial promise of biofuels in reducing greenhouse gas emissions and creating new opportunities in agriculture, several unintended consequences have emerged. One such consequence is the conflict between land use for food production and biofuel production, leading to concerns about food security. Limitations in feedstock availability have posed challenges to the widespread adoption of biofuels. The cultivation of energy crops in monocultures has also raised environmental concerns, as it can lead to biodiversity loss and soil degradation. Previously overlooked emissions of greenhouse gases, such as nitrous oxide from soils, have further complicated the sustainability of biofuels. These issues highlight the need for careful consideration and comprehensive assessments when implementing biofuel policies to ensure their overall effectiveness and minimize negative impacts on both the environment and food systems.

To truly achieve a net zero future, we must explore alternative solutions that go beyond the incremental improvements offered by biofuels (Chisale and Lee, 2023). While they may have their place in certain niche applications, relying solely on biofuels could prove to be a dead-end pathway. It is crucial to acknowledge the risk that biofuels might inadvertently prolong the lifespan of internal combustion vehicles, merely mitigating emissions without catalyzing the transformative shift required for reaching net zero goals. Therefore, we must prioritize investing in more revolutionary technologies and sustainable energy sources that can drive us toward a truly carbon-neutral society. By embracing innovation and pursuing diverse avenues for decarbonization, we can break free from the limitations of biofuels and pave the way for a cleaner and greener future.

Despite the initial hype, hydrogen has faced several challenges that have hindered its widespread adoption. One major obstacle is the lack of infrastructure for hydrogen production, storage, and distribution. Unlike traditional fuels, such as gasoline or diesel, hydrogen requires a completely different infrastructure to be built from scratch. This includes hydrogen production facilities, transportation pipelines or tankers, and refueling stations. The high cost and complexity of developing this infrastructure have been major barriers to the widespread use of hydrogen as a fuel source. Another challenge is the efficiency of hydrogen production. Currently, most hydrogen is produced through a process called steam methane reforming (SMR), which involves reacting natural gas with steam to produce hydrogen gas. This process is energy-intensive and relies on fossil fuels as a feedstock, which undermines the environmental benefits of using hydrogen as a clean fuel alternative. Developing more efficient and sustainable methods for producing hydrogen, such as electrolysis powered by renewable energy sources like solar or wind power, could help overcome this challenge and make hydrogen a truly green fuel option.

Electrolysis involves splitting water molecules into hydrogen and oxygen using an electric current. By utilizing renewable energy sources like solar or wind power to generate this electricity, the entire process becomes carbon-neutral and sustainable. Advancements in electrolysis technology have made it more efficient and cost-effective over the years. Implementing large-scale electrolysis facilities powered by renewable energy could significantly reduce our reliance on fossil fuels for hydrogen production. Research is underway to explore alternative methods, such as biological hydrogen production through the use of microorganisms or algae. These innovative approaches hold great promise for a future where hydrogen can be produced without any negative environmental impact. By investing in these sustainable methods, we can unlock the full potential of hydrogen as a clean and versatile fuel that can power various sectors, including transportation, industry, and even homes.

Advancements in hydrogen production methods, such as electrolysis using renewable energy sources, are reducing energy losses and making it a more sustainable option. Research and development efforts are focused on improving the efficiency of hydrogen conversion technologies to minimize costs. The potential of hydrogen as an energy carrier lies in its versatility and ability to be stored and transported easily. It can play a crucial role in decarbonizing sectors that are challenging to transition to renewable energy sources alone. The emergence of hydrogen-powered vehicles has already demonstrated its viability as a clean transportation solution. As governments and industries increasingly commit to net zero targets, the demand for hydrogen is expected to grow exponentially, driving further innovation and investment in this sector. With continued political support and ongoing technological advancements, hydrogen has the potential to revolutionize our energy systems and contribute significantly to a sustainable future.

To truly achieve a net zero future, it is crucial to carefully consider the risks associated with hydrogen production from natural gas. While hydrogen may seem like a promising alternative, there is a danger of perpetuating our dependence on fossil fuels. This is why the success of hydrogen in achieving net zero emissions relies heavily on two factors: carbon capture utilization and sequestration, as well as the widespread adoption of low-carbon electricity for electrolysis. Carbon capture, utilization, and sequestration technology holds promise in mitigating the environmental impact of natural gas-derived hydrogen by capturing and storing carbon emissions. Its effectiveness and scalability remain uncertain. Alternatively, expanding the use of low-carbon electricity for electrolysis can provide a cleaner method of producing hydrogen. By leveraging renewable energy sources such as wind or solar power, we can ensure that hydrogen production aligns with our goals of reducing greenhouse gas emissions.

"Negative emission technologies" and carbon dioxide removal make up a fourth strategy. It is predicated on the notion that CO_2 removal from the atmosphere must be proactive because other methods will not be effective in time to achieve net zero goals. This approach involves the deployment of various techniques, such as direct air capture, enhanced weathering, and afforestation. Direct air capture involves using large-scale machines to extract CO_2 directly from the air, which can then be stored underground or used for other purposes. Enhanced weathering aims to accelerate natural processes by exposing minerals that react with CO_2 to the atmosphere, thus capturing and storing it in solid form. Afforestation focuses on planting trees on a massive scale to absorb CO_2 through photosynthesis and store it in their biomass. These negative emission technologies offer promising solutions to combat climate change by actively reducing atmospheric CO_2 levels. Their implementation requires significant investment, research, and careful consideration of potential environmental impacts. It is crucial to prioritize reducing emissions at their source while simultaneously pursuing carbon dioxide removal strategies to achieve a sustainable and carbon-neutral future.

CCS-based approaches involve capturing CO_2 emissions from power plants and industrial facilities and storing them underground or utilizing them for enhanced oil recovery. This technology has the potential to significantly reduce CO_2 emissions and prevent them from entering the atmosphere. It requires extensive infrastructure development and faces challenges such as high costs, limited storage capacity, and potential leakage risks. On the other hand, other negative emission strategies focus on removing CO_2 directly from the atmosphere. These include afforestation and reforestation efforts, which involve planting trees to absorb CO_2 through photosynthesis. Bioenergy with CCS (BECCS) combines biomass energy production with CCS technology to remove CO_2 from the atmosphere. These strategies have the advantage of being natural-based solutions that can provide additional benefits such as biodiversity conservation and sustainable land use. To effectively address climate change, a combination of these strategies is necessary.

CCS technology plays a crucial role in mitigating greenhouse gas emissions by capturing and storing CO_2 from fossil fuel sources. By implementing different methods such as pre-combustion, post-combustion, and oxyfuel, CCS effectively prevents the release of CO_2 into the atmosphere. The potential of CCS extends beyond mere carbon capture. When combined with other technologies like bioenergy, it offers a pathway toward achieving negative emissions. This is evident in the case of BECCS, where biomass serves as a carbon sink during its growth phase and further contributes to carbon sequestration when converted into bioenergy. The integration of CCS with bioenergy not only helps reduce CO_2 emissions but also provides an opportunity to utilize renewable energy sources sustainably. By maximizing the synergies between these technologies, we can make significant strides toward addressing climate change and transitioning to a low-carbon future.

Recent advancements in DAC technology offer a promising solution. By utilizing large-scale fans and filters, DAC systems can extract CO_2 from ambient air and store it underground through CCS. This process not only directly tackles the issuc of atmospheric CO_2, but also provides a means to effectively sequester and store the captured carbon. While there are still challenges to overcome, such as improving the efficiency and scalability of DAC systems, ongoing research and development efforts are driving progress in this field. The combination of DAC and CCS holds significant potential for addressing the urgent need to reduce greenhouse gas emissions. With DAC's ability to directly capture CO_2 from the atmosphere, it offers a complementary approach to traditional emission reduction strategies that primarily focus on limiting emissions at their source. This means that even existing infrastructure and industries with high carbon footprints can be retrofitted with DAC-CCS systems to actively remove CO_2 from the atmosphere. DAC-CCS technology has the advantage of being adaptable to various locations and environments, making it a versatile solution for combating climate change. By strategically placing DAC-CCS facilities near areas with high CO_2 concentrations or in proximity to industrial hubs, we can effectively target and capture emissions at their highest points. This not only helps reduce the overall carbon footprint but also creates opportunities for job growth and economic development in these regions.

Another important approach to mitigating climate change is through reforestation and afforestation. These methods involve the strategic planting of trees in areas where forests used to exist or where there were previously none. By doing so, we can restore lost ecosystems and create new ones, which in turn help to absorb and store carbon dioxide from the atmosphere. Innovative techniques like enhanced weathering or ocean alkalinity enhancement offer promising solutions. These methods involve spreading fine basalt rock over large land or sea areas to accelerate chemical weathering reactions that remove CO_2 from the air and store it in solid carbonate minerals. This not only reduces greenhouse gas emissions but also provides a valuable source of carbon storage. Changing agricultural practices is another effective way to sequester carbon. Moving away from conventional farming toward practices like agroforestry allows for greater carbon capture in the soil and vegetation, contributing to overall climate resilience. By implementing these diverse strategies, we can make significant strides in combating climate change and safeguarding our planet for future generations.

Major lifestyle changes play a crucial role in achieving a net zero future. By encouraging sustainable habits such as reducing meat consumption, adopting renewable energy sources, and promoting public transportation, individuals can significantly contribute to carbon emissions reduction. Radical substitution of products offers an innovative approach to decarbonization. This involves replacing carbon-intensive materials and processes with more sustainable alternatives, such as utilizing biodegradable materials instead of plastics or implementing circular economy practices. Restricting new carbon-intensive practices is essential to prevent the development of environmentally harmful industries. Implementing stringent regulations and policies can deter the establishment of new ventures that contribute to greenhouse gas emissions, ensuring a more sustainable and low-carbon future for generations to come. These untapped approaches complement existing solutions and provide a comprehensive framework for achieving the ambitious goal of net zero emissions by the mid or end of the century.

These lifestyle changes can have a significant impact on reducing energy demand and achieving net zero emissions. Car-free housing projects, for example, encourage residents to rely on public transportation, walking, or cycling instead of owning cars, leading to a decrease in carbon emissions from transportation. This approach can be further enhanced by incorporating new approaches in city planning that prioritize pedestrian-friendly infrastructure and efficient public transportation systems. Promoting telework and virtual conferencing can greatly reduce the need for business travel, resulting in decreased energy consumption associated with transportation. Adopting low-carbon diets that prioritize plant-based foods and reduce meat consumption can also contribute to lowering greenhouse gas emissions from the agricultural sector. Encouraging local vacations instead of long-distance travel not only reduces carbon emissions from air travel but also supports local economies. Lastly, extensive changes in consumption patterns, such as shifting away from fast fashion toward sustainable and durable clothing options, can significantly reduce the environmental impact of the fashion industry.

To combat these challenges, it is crucial to implement effective policies and regulations that prioritize sustainability and discourage the proliferation of energy-intensive vehicles like pickup trucks and SUVs. Encouraging the adoption of EVs through incentives and infrastructure development can help mitigate the increased energy demand in transportation. Investing in research and development for alternative energy sources, such as hydrogen fuel cells or advanced biofuels, can provide viable options for reducing emissions in the transportation sector. As new energy uses continue to emerge, it is essential to anticipate their potential environmental impacts and proactively address them. For instance, with the rise of space tourism, stringent regulations should be put in place to ensure that emissions from rocket launches are minimized or offset. By taking a proactive approach and considering the long-term consequences of emerging technologies, we can navigate these challenges and work toward achieving our net zero emission targets.

These proposals faced significant backlash from conservative groups and individuals who viewed them as an infringement on personal freedoms and an attack on traditional values. Despite these challenges, it is crucial to push forward with innovative policies that promote sustainability and address pressing environmental concerns. One potential avenue is to incentivize the adoption of renewable energy sources by offering tax breaks or subsidies to homeowners and businesses that invest in solar panels or wind turbines. Implementing stricter regulations on carbon emissions for industries and promoting the development of clean technologies can help reduce our reliance on fossil fuels. Fostering a culture of sustainable transportation through the expansion of public transit systems, bike lanes, and EV infrastructure can contribute to significant emissions reductions. Embracing circular economy principles, where resources are reused and recycled rather than discarded, can also play a pivotal role in minimizing waste and conserving natural resources. Promoting education and awareness campaigns about the benefits of sustainable living can empower individuals to make environmentally conscious choices in their daily lives.

As the need for decarbonization becomes more pressing, Switzerland has taken steps to address the issue. A policy initiative aimed at banning SUVs and off-road vehicles was proposed in the past, but it was ultimately rejected by parliament and the government. This setback did not deter the country from implementing stricter emission regulations instead. This pragmatic approach demonstrates Switzerland's commitment to reducing carbon emissions without alienating certain sectors of society. While political feasibility remains a challenge for more radical decarbonization approaches, there is growing pressure to explore a wider range of solutions. It is clear that Switzerland recognizes the urgency of the situation and is willing to adapt its strategies to meet the demands of a changing world. By embracing innovative approaches and considering various options, Switzerland aims to pave the way toward a sustainable future while balancing political realities.

5.7 CONCLUSION

According to sustainability research, there are four major stages of change that must occur in order to achieve net zero GHG emissions. The first phase involves raising awareness and understanding of the urgent need to reduce GHG emissions. This includes educating the public about the detrimental effects of climate change and the benefits of transitioning to a low-carbon economy. It also involves engaging with policymakers, businesses, and communities to create a shared sense of responsibility and commitment toward achieving net zero emissions. The second phase focuses on implementing policy measures and regulations that incentivize sustainable practices and discourage carbon-intensive activities. This includes setting ambitious targets for renewable energy deployment, implementing carbon pricing mechanisms, and promoting energy efficiency measures across sectors. It involves supporting research and development in clean technologies to accelerate their adoption. The third phase centers around transforming key sectors such as transportation, industry, and agriculture toward low-carbon alternatives. This requires investing in infrastructure for EVs, promoting circular economy principles in manufacturing processes, and encouraging sustainable farming practices. It also entails fostering innovation and collaboration between industries to develop sustainable solutions at scale. Finally, the fourth phase revolves around ensuring long-term sustainability by implementing policies and regulations that support the transition to a low-carbon economy.

In order to address the challenge of decarbonization in difficult-to-decarbonize sectors such as aviation, shipping, and cement, alternative strategies must be explored. One potential solution lies in the development and implementation of advanced biofuels for aviation and shipping, which can significantly reduce carbon emissions compared to traditional fossil fuels. The use of CCS technology in cement production can help mitigate its environmental impact. Exploring innovative technologies like hydrogen fuel cells and synthetic fuels could offer promising alternatives for these sectors. It is crucial to invest in research and development to accelerate progress in these areas and ensure a comprehensive approach toward achieving a sustainable future.

These radical strategies may require significant societal shifts and policy interventions. For instance, promoting sustainable transportation options such as EVs and public transit can reduce carbon emissions from the transportation sector. Investing in research and development for innovative technologies like CCS can help mitigate emissions from industries that are challenging to decarbonize. Adopting circular economy principles that prioritize recycling and reusing materials can minimize the demand for carbon-intensive products. Encouraging international cooperation and agreements can ensure a global effort toward achieving net zero emissions. Lastly, empowering individuals through education and awareness campaigns can drive behavioral changes toward more sustainable lifestyles. By combining these approaches with existing strategies, we can accelerate our progress toward a net zero future while minimizing the impact of climate change on our planet.

To ensure a successful transition toward net zero emissions, it is crucial to explore a range of strategies, both engineered and natural, that can effectively reduce historical emissions or offset ongoing emissions. It is important to acknowledge that net zero pathways will differ significantly depending on the context. For instance, a jurisdiction with an already near-decarbonized electric power system will have different challenges compared to one with mixed or carbon-intensive electricity sources. Therefore, tailored approaches must be adopted to address the specific circumstances of each region. While striving for net zero emissions, there is a need to remain cautious of potential pitfalls and undesirable outcomes. It is possible for society to encounter dead-end pathways where strategies prove unsuccessful in achieving the desired results. These setbacks can hinder progress and delay the attainment of net zero goals. There is a risk of dystopian futures if society fails to reach net zero emissions altogether. Such scenarios could have severe consequences for the environment and human well-being.

First, it is crucial to prioritize decarbonization efforts across all sectors of the economy. This means implementing ambitious targets and regulations that encourage the rapid adoption of renewable energy sources, such as solar and wind power. Investing in research and development for clean technologies can drive innovation and create new opportunities for sustainable growth. Second, a holistic approach is necessary to address the interconnectedness of environmental, social, and economic factors. Policies should not only focus on reducing greenhouse gas emissions but also consider the conservation of natural resources, the protection of biodiversity, and the promotion of social equity. This requires collaboration between governments, businesses, and civil society to develop integrated strategies that tackle multiple sustainability challenges simultaneously. Third, fostering international cooperation is essential for a successful transition to net zero. Climate change is a global issue that requires collective action and shared responsibility. Governments must work together to establish international agreements that set clear targets and facilitate the transfer of clean technologies to developing countries. Collaboration on research and knowledge sharing can also accelerate progress toward a sustainable future.

Fourth, investing in renewable energy sources and improving energy efficiency is crucial for achieving net zero emissions. Transitioning to clean and sustainable energy systems will not only reduce greenhouse gas emissions but also create new job opportunities and stimulate economic growth. Governments, businesses, and individuals must prioritize the development and deployment of renewable energy technologies such as solar, wind, and hydropower. Adopting energy-efficient practices in industries, transportation, and buildings can significantly contribute to reducing carbon footprints. Fifth, promoting sustainable agriculture and land use is vital for mitigating climate change. Deforestation, intensive farming practices, and land degradation are major contributors to greenhouse gas emissions. Encouraging sustainable farming methods, protecting forests and natural habitats, and restoring degraded lands can help sequester carbon dioxide from the atmosphere while ensuring food security for future generations. Finally, educating and empowering individuals to make sustainable choices is key to achieving a net zero future. Public awareness campaigns on climate change impacts and the importance of sustainable lifestyles can drive behavioral change and promote the adoption of renewable energy sources.

REFERENCES

Acheampong, A. O., Opoku, E. E. O., & Aluko, O. A. (2023). The roadmap to net-zero emission: Do geopolitical risk and energy transition matter? *Journal of Public Affairs*. https://doi.org/10.1002/pa.2882

Africa Energy Outlook 2022. (2022). *Africa Energy Outlook 2022*. https://doi.org/10.1787/2abd9ce2-en

Amor, M. B., Pineau, P. O., Gaudreault, C., & Samson, R. (2011). Electricity trade and GHG emissions: Assessment of Quebec's hydropower in the Northeastern American market (2006–2008). *Energy Policy*. https://doi.org/10.1016/j.enpol.2011.01.001

Azila, A. S., & Alias, A. (2015). The concept of sustainability from the Islamic perspectives. *International Journal of Business, Economics and Law*.

Boucher, O., & Reddy, M. S. (2008). Climate trade-off between black carbon and carbon dioxide emissions. *Energy Policy*. https://doi.org/10.1016/j.enpol.2007.08.039

Chakravarty, S., & Tavoni, M. (2013). Energy poverty alleviation and climate change mitigation: Is there a trade off? *Energy Economics*. https://doi.org/10.1016/j.eneco.2013.09.022

Chapungu, L., Nhamo, G., Chikodzi, D., & Maoela, M. A. (2022). BRICS and the race to net-zero emissions by 2050: Is COVID-19 a barrier or an opportunity? *Journal of Open Innovation: Technology, Market, and Complexity*. https://doi.org/10.3390/joitmc8040172

Chisale, S. W., & Lee, H. S. (2023). Evaluation of barriers and solutions to renewable energy acceleration in Malawi, Africa, using AHP and fuzzy TOPSIS approach. *Energy for Sustainable Development*. https://doi.org/10.1016/j.esd.2023.101272

Clora, F., & Yu, W. (2022). GHG emissions, trade balance, and carbon leakage: Insights from modeling thirty-one European decarbonization pathways towards 2050. *Energy Economics*. https://doi.org/10.1016/j.eneco.2022.106240

Delbeke, J., Runge-Metzger, A., Slingenberg, Y., & Werksman, J. (2019). The paris agreement. In *Towards a Climate-Neutral Europe: Curbing the Trend*. https://doi.org/10.4324/9789276082569-2

Guo, Q., Abbas, S., AbdulKareem, H. K. K., Shuaibu, M. S., Khudoykulov, K., & Saha, T. (2023). Devising strategies for sustainable development in sub-Saharan Africa: The roles of renewable, non-renewable energy, and natural resources. *Energy*. https://doi.org/10.1016/j.energy.2023.128713

Intergovernmental Panel on Climate Change (IPCC). (2021). Climate change 2021—The physical science basis. *Chemistry International*. https://doi.org/10.1515/ci-2021-0407

International Energy Agency. (2023). Global EV Outlook 2023. *Geo*, *Geo*.

International Renewable Energy Agency (IRENA). (2019). Future of Solar Photovoltaic: Deployment, investment, grid integration and socio-economic aspects. *International Renewable Energy Agency (IRENA)*.

International Renewable Energy Agency (IRENA). (2022). World Energy Transitions Outlook 2023: 1.5°C pathway. *World Energy Transitions.*

Ituen, E. E., Esen, N. U., Nwokolo, S. C., & Udo, E. G. (2012). Prediction of global solar radiation using relative humidity, maximum temperature and sunshine hours in Uyo, in the Niger Delta Region, Nigeria. *Advances in Applied Science Research*, *3*(4), 1923–1937.

Ju, Y., Xi, C., Wang, S., Mao, L., Wang, K., & Zhou, H. (2021). 3-D fracture evolution and water migration in fractured coal under variable stresses induced by fluidized mining: In situ triaxial loading and CT imaging analysis. *Energy Reports*. https://doi.org/10.1016/j.egyr.2021.05.036

Kichere, C. E. (2020). Performance audit report on the management of accessibility and reliability of electricity supply services. *National Audit Office.*

Laura, C. et al. (2023). World Energy Outlook. In *World Energy Outlook*. International Energy Agency. www.iea.org/weo

Lawson, L. A. (2020). GHG emissions and fossil energy use as consequences of efforts of improving human well-being in Africa. *Journal of Environmental Management*. https://doi.org/10.1016/j.jenvman.2020.111136

Le, T. T., Kieu, X. H., Behl, A., & Pereira, V. (2022). Building up more sustainable food supply chains: Implications for sustainable development. *Journal of Cleaner Production*. https://doi.org/10.1016/j.jclepro.2022.134650

Legutko-Kobus, P., Nowak, M., Petrisor, A. I., Bărbulescu, D., Craciun, C., & Gârjoabă, A. I. (2023). Protection of environmental and natural values of urban areas against investment pressure: A case study of Romania and Poland. *Land*. https://doi.org/10.3390/land12010245

Mäkitie, T., Hanson, J., Damman, S., & Wardeberg, M. (2023). Digital innovation's contribution to sustainability transitions. *Technology in Society*, 73. https://doi.org/10.1016/j.techsoc.2023.102255

Manavalan, E., & Jayakrishna, K. (2019). A review of Internet of Things (IoT) embedded sustainable supply chain for industry 4.0 requirements. *Computers and Industrial Engineering*. https://doi.org/10.1016/j.cie.2018.11.030

McGraw Hill Construction. (2009). Green Building Retrofit and Renovation. In *Green Building Smartmarket Report*. Bedford, MA: … .

Meena, R. S., Gogoi, N., & Kumar, S. (2017). Alarming issues on agricultural crop production and environmental stresses. *Journal of Cleaner Production*. https://doi.org/10.1016/j.jclepro.2016.10.134

Morales, P. K., Roslan, N., Haas, A., & Lykins, A. D. (2021). Citizen science engagement: Lessons learned from the ClimateWatch "Scientist for a Day" program. *Applied Environmental Education and Communication*. https://doi.org/10.1080/1533015X.2021.1922102

Nwokolo, S. C. (2017). A comprehensive review of empirical models for estimating global solar radiation in Africa. *Renewable and Sustainable Energy Reviews*, *78*(October), 955–995. https://doi.org/10.1016/j.rser.2017.04.101

Nwokolo, S. C., Echeng, E., Obiwulu, A., Meyer, E. L., Christian, C., Ogbulezie, J. C., & Proutsos, N. (2024). A multi-model approach based on CARIMA-SARIMA-GPM for assessing the impacts of climate change on concentrated photovoltaic (CPV) potential. *Physics and Chemistry of the Earth*, *134*(January), 103560. https://doi.org/10.1016/j.pce.2024.103560

Nwokolo, S. C., & Ogbulezie, J. C. (2018a). A qualitative review of empirical models for estimating diffuse solar radiation from experimental data in Africa. *Renewable and Sustainable Energy Reviews*, *92*(18), 353–393. https://doi.org/10.1016/j.rser.2018.04.118

Nwokolo, S. C., & Ogbulezie, J. C. (2018b). A quantitative review and classification of empirical models for predicting global solar radiation in West Africa. *Beni-Suef University Journal of Basic and Applied Sciences*, *7*(4), 367–396. https://doi.org/10.1016/j.bjbas.2017.05.001

Nwokolo, S. C., Meyer, E. L., & Ahia, C. C. (2023). Credible pathways to catching up with climate goals in Nigeria. *Climate*, *11*(196), 1–32. https://doi.org/org/10.3390/cli11090196

Nwokolo, S. C., Meyer, E. L., & Ahia, C. C. (2024). Exploring the interactive influences of climate change and urban development on the fraction of absorbed photosynthetically active radiation. *Atmosphere*, 1–37.

Nwokolo, S. C., Obiwulu, A. U., & Ogbulezie, J. C. (2023). Machine learning and analytical model hybridization to assess the impact of climate. *Physics and Chemistry of the Earth*, *130*(February 2023), 103389. https://doi.org/10.1016/j.pce.2023.103389

Nwokolo, S. C., & Otse, C. (2019). Impact of sunshine duration and clearness index on diffuse solar radiation estimation in mountainous climate. *Trends in Renewable Energy*, *5*(3), 307–332. https://doi.org/10.17737/tre.2019.5.3.00107

Nwokolo, S. C., Proutsos, N., & Meyer, E. L. (2023). Machine learning and physics-based hybridization models for evaluation of the effects of climate change and urban expansion on photosynthetically active radiation. *Atmosphere*, *14*(687), 1–36. https://doi.org/doi.org/10.3390/atmos14040687

Obiwulu, A. U., Chendo, M. A. C., Erusiafe, N., & Nwokolo, S. C. (2020). Implicit meteorological parameter-based empirical models for estimating back temperature solar modules under varying tilt-angles in Lagos, Nigeria. *Renewable Energy*. https://doi.org/10.1016/j.renene.2019.05.136

Obiwulu, A. U., Erusiafe, N., Olopade, M. A., & Nwokolo, S. C. (2020). Modeling and optimization of back temperature models of mono-crystalline silicon modules with special focus on the effect of meteorological and geographical parameters on PV performance. *Renewable Energy*, *154*, 404–431. https://doi.org/10.1016/j.renene.2020.02.103

Obiwulu, A. U., Erusiafe, N., Olopade, M. A., & Nwokolo, S. C. (2022). Modeling and estimation of the optimal tilt angle, maximum incident solar radiation, and global radiation index of the photovoltaic system. *Heliyon*, *8*(6), e09598. https://doi.org/10.1016/j.heliyon.2022.e09598

Proutsos, N., Liakatas, A., Alexandris, S., Nwokolo, S. C., Solomou, A. D., & Amadi, S. O. (2023). Assessing the impact of atmospheric attributes on the effectiveness of solar irradiance for photosynthesis of urban vegetation in Attica, Greece. *Theoretical and Applied Climatology*, *155*, 1415–1427. https://doi.org/10.1007/s00704-023-04700-0

Proutsos, N., Tigkas, D., Tsevreni, I., Alexandris, S. G., Solomou, A. D., Bourletsikas, A., Stefanidis, S., & Nwokolo, S. C. (2023). A thorough evaluation of 127 potential evapotranspiration models in two mediterranean urban green sites. *Remote Sensing*, *15*(14), 1–41. https://doi.org/10.3390/rs15143680

Rahnama Mobarakeh, M., & Kienberger, T. (2022). Climate neutrality strategies for energy-intensive industries: An Austrian case study. *Cleaner Engineering and Technology*. https://doi.org/10.1016/j.clet.2022.100545

Rojas-Vallejos, J., & Lastuka, A. (2020). The income inequality and carbon emissions trade-off revisited. *Energy Policy*. https://doi.org/10.1016/j.enpol.2020.111302

Sakr, D. (2017). Sustainability and innovation: The next global industrial revolution. *Journal of Cleaner Production*. https://doi.org/10.1016/j.jclepro.2016.10.135

Sarkawi, A. A., Abdullah, A., & Dali, N. M. (2016). Reviewing the concept of sustainability from the Islamic perspectives. *Kuala Lumpur International Islamic Studies and Civilisations (KLiISC) 2016.*

Schau, E. M., Traverso, M., Lehmannann, A., & Finkbeiner, M. (2011). Life cycle costing in sustainability assessment—A case study of remanufactured alternators. *Sustainability*. https://doi.org/10.3390/su3112268

Schwedhelm, M. C., Wilhelm, A. K., Bigelow, M., Bates, N., Vibar, T. M., Ortega, L., & Allen, M. L. (2023). What is sustainable participatory research? Insights from a school-university partnership. *Educational Action Research*. https://doi.org/10.1080/09650792.2021.2007783

Shamsuddoha, M., & Woodside, A. G. (2022). Achieving radical process innovations by applying technology-mindset transformations via second-order system-dynamics engineering. *Journal of Business Research*. https://doi.org/10.1016/j.jbusres.2022.04.006

Shindhal, T., Rakholiya, P., Varjani, S., Pandey, A., Ngo, H. H., Guo, W., Ng, H. Y., & Taherzadeh, M. J. (2021). A critical review on advances in the practices and perspectives for the treatment of dye industry wastewater. *Bioengineered*. https://doi.org/10.1080/21655979.2020.1863034

Sutcliffe, C., Knox, J., & Hess, T. (2023). 'I think this is where this lovely word "sustainability" comes in': Fruit and vegetable growers' narratives concerning the regulation of environmental water use for food production. *Sociologia Ruralis*. https://doi.org/10.1111/soru.12437

Tramontin, V., Loggia, C., & Trois, C. (2012). Strategies for sustainable building design and retrofit in developing countries. New goals for green buildings in South Africa. *Journal of Construction*.

6 Foreign Investment in Africa for a Net Zero Economy

6.1 INTRODUCTION

Foreign investment in Africa for a net zero economy is becoming increasingly important as African countries seek to attract investment while also transitioning to a more sustainable and environmentally friendly economy. It requires a delicate balance between economic growth and environmental responsibility. Many African countries have set ambitious targets for achieving net zero emissions, and foreign investment can play a crucial role in helping them achieve these goals. It is important for investors to prioritize sustainable and socially responsible investments that support the development of local communities and protect the environment. This can be achieved through partnerships with local businesses and organizations that prioritize sustainable practices and community development.

Investors should also consider the long-term impacts of their investments on the environment (African–EU: Investment promises, 2022) and society rather than solely focusing on short-term financial gains in African nations (Africanacity of international investment law, 2021). By promoting sustainable practices and community development, businesses can also attract socially conscious consumers who prioritize ethical and environmentally responsible products and services in African nations (Amendolagine and Prata, 2021). This can lead to increased customer loyalty and brand reputation, ultimately resulting in long-term financial benefits for businesses in African nations. Adopting sustainable practices can also reduce operational costs for businesses in African nations as they become more efficient and environmentally friendly. This can result in increased profitability (Asiedu, 2004) and competitiveness in the market (Baltag, 2009). In addition, sustainable practices can also attract socially responsible investors (Capital investment in Africa: Its course and effects, 1939) and customers who are willing to pay a premium for eco-friendly products and services. This can lead to long-term financial stability and growth for businesses in Africa.

Sustainable practices can also create job opportunities and improve the livelihoods of local communities (Cheng, 2023), ultimately contributing to poverty reduction (Cheung et al., 2012) and economic development in Africa. Therefore, adopting sustainable practices is not only beneficial for the environment but also for the social (Clark, 1997) and economic well-being of the continent. For instance, sustainable

DOI: 10.1201/9781003483175-7

agriculture practices can lead to increased crop yields and income for farmers in African nations while also preserving soil fertility and reducing the use of harmful pesticides. Additionally, renewable energy projects can provide access to electricity in remote areas (Nwokolo and Otse, 2019), creating new business opportunities and improving the overall quality of life for Africans. These sustainable practices can also contribute to mitigating climate change by reducing greenhouse gas emissions and promoting carbon sequestration.

They can enhance biodiversity and ecosystem services, supporting the conservation of natural resources and wildlife habitats. Implementing sustainable practices can also have economic benefits, such as reducing costs and increasing efficiency, while promoting social equity and improving community well-being in Africa. These practices can create new job opportunities and stimulate local economies, particularly in rural areas of African nations. For example, investing in renewable energy sources like solar power can reduce the reliance on expensive imported fossil fuels and create jobs in the installation and maintenance of solar panels (Nwokolo and Ogbulezie, 2018a). Additionally, sustainable agriculture practices can improve crop yields and support local food systems, providing economic opportunities for farmers and reducing food insecurity in African communities (Nwokolo and Ogbulezie, 2018b).

These sustainable practices not only have economic benefits but also have a positive impact on the environment by reducing greenhouse gas emissions and preserving natural resources. Implementing these solutions can lead to a more resilient and sustainable future for African communities. By promoting sustainable practices, African communities can also improve their quality of life by reducing pollution and improving access to clean water and air. Additionally, these practices can create new job opportunities and support local economies. Sustainable practices can also lead to the preservation of natural resources and biodiversity, which are essential for the long-term survival of these communities. Furthermore, by reducing their carbon footprint, African communities can contribute to global efforts to mitigate climate change (Nwokolo, Obiwulu, et al., 2023) and protect the planet for future generations (Nwokolo, Ogbulezie, et al., 2023).

Sustainable practices can also improve the health and well-being of individuals within these communities by providing access to clean water, nutritious food, and better sanitation in African nations. This can lead to increased productivity and economic growth, ultimately benefiting the entire community. Sustainable practices can help mitigate the effects of climate change, which disproportionately affect vulnerable communities in Africa. By reducing greenhouse gas emissions and promoting renewable energy sources, these communities can become more resilient to the impacts of climate change. Sustainable practices can also lead to economic benefits for these communities by creating job opportunities and reducing their dependence on expensive fossil fuels in African nations. This can ultimately improve their overall quality of life and promote long-term sustainability.

Sustainable practices can also attract eco-tourism and foreign investment, which can further boost the local economy and provide additional sources of income for the community in African nations. This can lead to a more diversified and resilient economy that is less vulnerable to external shocks in African nations. Implementing sustainable practices can also contribute to the preservation of natural resources and biodiversity, which are crucial for the long-term well-being of both local communities and the planet as a whole. This can help ensure that future generations in African nations have access to these resources and can benefit from them in a sustainable way. Sustainable practices can also lead to economic benefits, such as increased efficiency and cost savings, as well as improved marketability for businesses that prioritize sustainability. By investing in sustainable practices, African nations can create a more resilient and prosperous future for their citizens.

It can also attract foreign investment and support from international organizations, which can provide additional resources and expertise to help African nations achieve their sustainability goals. This can lead to further economic growth and development in the region. In addition, sustainable practices can also enhance the reputation of African nations on the global stage, leading to increased tourism and trade opportunities. This can create a positive cycle of economic growth and sustainability for the continent. Furthermore, the adoption of sustainable practices can also lead to the preservation of natural resources and biodiversity, which are crucial for the long-term survival of African ecosystems and the species that inhabit them. This can also have positive impacts on the health and well-being of local communities that rely on these resources for their livelihoods.

It can help mitigate the effects of climate change, such as droughts and floods, which are becoming increasingly common in many parts of Africa. By reducing greenhouse gas emissions and promoting ecosystem resilience, sustainable practices can help communities adapt to these challenges and build a more sustainable future. For instance, promoting agroforestry and sustainable land management practices can help improve soil health and water retention, reducing the risk of erosion and flooding. Additionally, investing in renewable energy sources such as solar and wind power can reduce reliance on fossil fuels (Obiwulu et al., 2022) and lower carbon emissions (Obiwulu, Chendo, et al., 2020). These measures can contribute to mitigating the impacts of climate change and ensuring a more sustainable future for generations to come.

It is crucial for governments, businesses, and individuals to take action toward implementing these solutions. By reducing greenhouse gas emissions, investing in renewable energy, and promoting sustainable practices, we can create a healthier planet for ourselves and future generations. It is our responsibility to act now and make a positive impact on the environment before it's too late. The consequences of not taking action will be catastrophic, leading to more frequent natural disasters, rising sea levels, and irreversible damage to ecosystems in African nations. Let's take action today and work toward a sustainable future for all in African countries. It is crucial that we prioritize sustainable development in African nations by investing in renewable energy, reducing carbon emissions, and promoting eco-friendly practices (Obiwulu, Erusiafe, et al., 2020). By doing so, we can mitigate the negative impacts of climate change and ensure a better future for generations to come. Sustainable development will not only benefit the environment but also the economy and social well-being of African nations. It is essential to work toward a sustainable future for the continent by implementing policies that prioritize environmental protection and economic growth. By promoting sustainable practices such as renewable energy, waste reduction, and conservation of natural resources, African nations can create new job opportunities and improve the standard of living for their citizens.

Sustainable development can attract foreign investment and enhance international cooperation for a greener future. Sustainable practices can also lead to a reduction in greenhouse gas emissions and mitigate the impacts of climate change, which disproportionately affect African countries. By adopting sustainable development strategies, African nations can contribute to global efforts to address the climate crisis while simultaneously improving their own economic and social well-being. This can also lead to a reduction in poverty and inequality, as well as contribute to mitigating the effects of climate change. It is important for African governments to prioritize sustainable development and work toward achieving the United Nations Sustainable Development Goals (SDGs). This can be achieved through implementing policies that promote renewable energy, sustainable agriculture, and responsible natural resource management. Additionally, partnerships with international organizations and the private sector can provide resources and expertise to support sustainable development efforts in Africa. By prioritizing sustainable development, African countries can improve their economic growth and quality of life for their citizens while also preserving the environment for future generations. It is important for governments in African countries to take action now to address these issues and ensure a sustainable future.

Foreign investment in Africa plays a pivotal role in paving the way toward a net zero economy. The continent's vast potential for renewable energy sources, such as solar and wind power, coupled with its abundant natural resources, offers lucrative opportunities for international investors. By harnessing these resources and implementing sustainable practices, Africa can significantly reduce its carbon footprint while simultaneously driving economic growth. Foreign investment brings not only capital but also technological expertise and knowledge transfer, enabling African nations to develop their renewable energy sectors more efficiently. This influx of investments also promotes job creation and enhances local capacity building, empowering African communities to actively participate in the transition toward a greener future. Additionally, foreign investors can collaborate with local governments and organizations to establish robust regulatory frameworks that encourage responsible environmental practices and ensure long-term sustainability. By embracing foreign investment with a focus on clean energy solutions and sustainable development, Africa can unlock its full potential as a key player in the global net zero economy while fostering economic prosperity for its people.

6.2 BENEFITS OF FOREIGN INVESTMENT IN AFRICA FOR A NET ZERO ECONOMY

6.2.1 Job Creation

Foreign investment in Africa can create job opportunities in various sectors such as renewable energy, sustainable agriculture, and eco-tourism, as well as bring in new technologies and expertise. It is important to ensure that foreign investment aligns with the local needs and priorities of African countries and does not lead to exploitation or environmental degradation. African governments should prioritize building the capacity of local businesses and entrepreneurs to ensure that they can compete with foreign investors and benefit from economic growth. This approach can also foster innovation and diversity in the economy (Ewelukwa Ofodile, 2014), make it more resilient to external shocks (Fosu, 2013), and reduce poverty and inequality by empowering marginalized groups such as women and youth (Gazzini, 2009). Investing in local businesses and promoting entrepreneurship can have a positive impact on both the social and economic aspects of African communities.

Supporting local businesses is a key step toward achieving economic independence and stability in Africa. It can foster a sense of ownership and pride among local people, reduce dependence on foreign aid, and create job opportunities for the youth (Grekou and Owoundi, 2020). It can also promote cultural preservation and diversity by allowing communities to showcase their unique products and services, which can attract tourism and generate additional revenue. This approach can also contribute to sustainable development by reducing the reliance on imported goods, promoting environmentally friendly practices such as sourcing materials locally, and reducing transportation emissions. It can create a sense of community pride and strengthen social connections and help preserve the unique character and culture of our communities in the face of increasing globalization and homogenization of culture (Gunter, 2006).

Small businesses often offer unique products and services that cannot be found elsewhere, adding to the diversity and vibrancy of African communities. Supporting small businesses not only contributes to the local economy but also helps to preserve cultural heritage and traditions. Small businesses also tend to have a smaller environmental footprint than larger corporations, as they source their materials locally and produce goods in smaller quantities. Small businesses are more likely to implement sustainable practices and use eco-friendly materials due to their close ties to the local community and environment (Hepburn, 2015), making them an ideal choice for consumers who prioritize environmental responsibility in their purchasing decisions. Small businesses that prioritize sustainability can attract a growing customer base and differentiate themselves from larger, less environmentally conscious competitors. Sustainable practices can also lead to cost savings for small businesses in the long run by reducing waste and energy consumption.

Small businesses in Africa can leverage sustainable practices to differentiate themselves from competitors and appeal to the growing number of environmentally conscious consumers. This can lead to increased revenue and a competitive advantage in the African market, as well as cost savings and improved customer loyalty. Sustainable practices can also reduce operational costs for small businesses in Africa, leading to higher profits and a more competitive edge in the market. Additionally, consumers are becoming more environmentally conscious and prefer to support eco-friendly businesses, leading to increased customer loyalty and a larger customer base. Sustainable practices can help small businesses in Africa save money and improve their financial stability and profitability.

They can also attract environmentally conscious customers and investors who prioritize ethical and sustainable business practices (Khumon, 2016), leading to increased sales and partnerships. Implementing sustainable practices can also improve the reputation of small businesses, making them more attractive to investors and potential business partners. Additionally, they can reduce operational costs by using energy-efficient technologies and reducing waste, resulting in increased profitability and competitiveness in the market.

6.2.2 Improved Access to Clean Energy

One benefit of foreign investment in Africa for a net zero economy is improved access to clean energy. Foreign investment in Africa can help accelerate the development and adoption of renewable energy technologies, which are crucial for achieving a net zero economy. This can lead to a reduced reliance on fossil fuels and contribute to mitigating climate change. Foreign investment can bring in the necessary capital and technology needed to develop renewable energy projects, such as solar and wind power, which can help African countries transition to a net zero economy. These investments can create job opportunities and boost economic growth in the region.

It is important for African countries to ensure that these investments align with their national development goals and do not lead to exploitation or environmental degradation. They should also prioritize building local capacity and knowledge transfer to ensure the long-term sustainability of these projects. This is because foreign investors bring in expertise and technology that can help Africa transition to renewable energy sources, reducing the continent's reliance on fossil fuels. Additionally, foreign investment can also create jobs and stimulate economic growth in the clean energy sector. This can lead to a more sustainable and environmentally friendly future for Africa while also improving the standard of living for its people. Furthermore, foreign investment can bring in new technologies and expertise that can help African countries develop their own clean energy industries. This can lead to greater energy independence and reduce reliance on fossil fuels, ultimately benefiting both the economy and the environment.

It is important to ensure that foreign investment is regulated and monitored to prevent exploitation and ensure that the benefits are shared equitably among all stakeholders, including local communities and workers. Additionally, African governments should prioritize policies that promote sustainable development and renewable energy sources to ensure long-term benefits for their citizens. This can be achieved through transparent and accountable governance, as well as collaboration between governments, private sector actors, and civil society organizations. Ultimately, the goal should be to create a conducive environment for foreign investment that supports inclusive economic growth and social development in Africa. This requires addressing issues such as corruption, weak institutions, and inadequate infrastructure. Additionally, promoting education and skills development can help ensure that African countries have the human capital necessary to attract and benefit from foreign investment.

Foreign investment can also provide opportunities for technology transfer and innovation, which can further enhance the development of African economies. It is important to ensure that foreign investment is sustainable and benefits local communities in the long term. To achieve this, governments should establish policies and regulations that encourage responsible investment practices and prioritize the interests of their citizens. Additionally, partnerships between foreign investors and local businesses can also facilitate knowledge sharing and promote inclusive economic growth. Such partnerships can also help in creating job opportunities and enhancing the skills of the local workforce, leading to sustainable development.

It is crucial for governments to ensure that these partnerships are transparent and equitable, with a fair distribution of benefits. This can be achieved through effective regulations and policies that promote responsible business practices and protect the rights of local communities. Additionally, regular monitoring and evaluation of these partnerships can help identify areas for improvement and ensure their long-term success. It is crucial for governments and businesses to work together toward SDGs, such as reducing poverty and protecting the environment. By doing so, we can create a more equitable and prosperous future for all. Collaboration between governments and businesses can lead to innovative solutions that address social and environmental challenges while also promoting economic growth. This approach can help create a more sustainable and resilient global economy.

6.2.3 Improved Infrastructure

Improved infrastructure is one advantage of foreign investment in Africa for a net zero economy. With increased foreign investment, African countries can build more efficient and sustainable infrastructure that reduces carbon emissions and supports a net zero economy. This can lead to increased economic growth and job creation, as well as better access to essential services such as electricity, water, and transportation. Improved infrastructure can attract further foreign investment and support the development of local industries. The construction and maintenance of infrastructure can also have negative impacts on the environment and local communities, such as displacement of people and destruction of natural habitats. Therefore, it is important to balance the benefits with the potential costs and ensure that infrastructure development is sustainable and socially responsible.

With foreign investment in Africa, there is an opportunity to build new and modern infrastructure that is sustainable and environmentally friendly. This can include renewable energy projects, efficient transportation systems, and waste management facilities. Furthermore, such investments can create job opportunities and boost economic growth in the region, ultimately leading to a better quality of life for the local population. It is important for these investments to be done responsibly and with consideration for the social and environmental impact on the communities they will affect. Investors should prioritize engaging with local stakeholders and conducting thorough impact assessments before proceeding with any investment. This will ensure that the benefits of the investment are maximized while minimizing any negative consequences.

Foreign investment in Africa can lead to the development of new transportation networks, power grids, and communication systems that are essential for a net zero economy. This improved infrastructure can help to reduce carbon emissions and promote sustainable economic growth in the region. In addition, foreign investment can also bring in new technologies and expertise that can help African countries transition to renewable energy sources and reduce their dependence on fossil fuels, further contributing to the goal of achieving a net zero economy. It is important for these investments to be made in a responsible and sustainable manner, taking into account the social and environmental impacts of such projects. This requires strong governance and regulatory frameworks to ensure that foreign investment aligns with the national development goals of African countries. It is crucial for local communities to be involved in decision-making processes and for their rights to be respected throughout the investment process.

This improved infrastructure can lead to increased efficiency and productivity, which can help African countries transition to a net zero economy. Additionally, foreign investment can bring in new technologies and expertise that can support the development of renewable energy sources and other sustainable practices. The shift toward a net zero economy can also create new job opportunities and improve the standard of living for communities in Africa. This can be achieved by investing in green industries such as solar power, wind energy, and sustainable agriculture.

Foreign investment in Africa can help improve infrastructure by providing funding for the construction of roads, bridges, and other essential facilities. This can lead to increased economic growth and improved living standards for local communities, which is crucial for achieving a net zero economy. It is important to ensure that foreign investment is sustainable and does not harm the environment or exploit local resources. It is also necessary to involve local communities in decision-making processes to ensure that their needs and concerns are taken into account. This can be achieved through implementing transparent and accountable policies, as well as conducting thorough environmental and social impact assessments before approving any foreign investment projects. Additionally, providing adequate compensation and benefits to local communities can help mitigate any negative impacts of foreign investment.

Foreign investment in Africa can lead to the development of modern transportation systems, renewable energy infrastructure, and communication networks that are essential for a net zero economy. These improvements can help African countries to reduce their carbon footprint and transition toward sustainable economic growth. Foreign investment can also create job opportunities for local communities and improve their standard of living. It is important to ensure that these investments are sustainable and benefit both the investors and the local population in the long run. To achieve this, it is necessary to establish regulations and policies that promote responsible investment practices and protect the rights of local communities. Additionally, regular monitoring and evaluation of foreign investment projects can help identify any negative impacts and ensure that appropriate measures are taken to address them.

6.2.4 Increased Access to Resources

One benefit of foreign investment in Africa for a net zero economy is increased access to resources. This increased access to resources can lead to the development of renewable energy projects, such as solar and wind farms, which can help African countries transition to a net zero economy. This can also create job opportunities and stimulate economic growth. Foreign investment can also diversify their energy mix by investing in renewable energy sources such as solar, wind, and hydropower, which can reduce their dependence on fossil fuels and contribute to the global efforts to combat climate change. It is important to ensure that local communities are not left behind in the transition to renewable energy and that they have access to the benefits of these projects. Additionally, African governments must prioritize building local capacity and expertise to ensure long-term sustainability of these renewable energy projects.

Engaging with local stakeholders can also help to address any concerns or challenges that may arise during the implementation of the projects. Foreign investment in Africa can help accelerate the development of renewable energy and other sustainable solutions, create job opportunities, and boost economic growth. It is important to ensure that foreign investment is done responsibly and sustainably (Klimova, 2022), taking into account the social and environmental impacts on local communities. This can be achieved through strong regulatory frameworks and partnerships between investors (Leiter, 2021), governments (Li, 2014), and local stakeholders (Nel and Humphrys, 1999). Responsible investment practices can lead to increased economic growth, job creation, and improved living standards for local communities.

By prioritizing environmental, social, and governance (ESG) factors in their investment decisions, foreign investors can help to address pressing global challenges such as climate change and social inequality (Nkuna, 2015). Additionally, responsible investment practices can encourage companies to improve their sustainability performance, which can lead to long-term financial benefits for both investors and the companies they invest in (Owen and Lane, 2014). It is important for African countries to prioritize investment in renewable energy sources and sustainable infrastructure to reduce their carbon footprint and contribute to global efforts to combat climate change. With the right policies and investments, other African nations can follow suit and achieve similar success.

6.2.5 Increased Economic Growth

Increased economic growth is a benefit of foreign investment in Africa for a net zero economy. Foreign investment in Africa not only boosts economic growth but also creates job opportunities for the local population, leading to poverty reduction and improved living standards. Additionally, it can also promote technology transfer and knowledge sharing (Pamba, 2023), which can contribute to the development of sustainable industries (Peat, 2018) and help achieve a net zero economy. Foreign investment in Africa has the potential to boost economic growth, create job opportunities, and promote technology transfer and knowledge sharing. Foreign investment should be carefully managed to ensure it aligns with the country's development goals and does not result in exploitation or harm to local communities and the environment (Pedler, 1955). Governments should incentivize companies to adopt sustainable practices by offering tax breaks or subsidies, while individuals can make conscious choices to reduce their carbon footprint and support eco-friendly products.

All stakeholders should work together toward achieving a common goal of preserving our planet for future generations by implementing policies and regulations that promote environmentally friendly practices, investing in renewable energy sources, and raising awareness among individuals and businesses about the importance of sustainable living. By taking these steps, we can ensure a healthier planet for ourselves and for future generations.

Foreign investment in Africa can lead to job creation and increased economic growth, but it must be accompanied by strong regulations (Qumba, 2018) and oversight to prevent exploitation and ensure benefits are shared equitably (Reinisch, 2014). African governments should prioritize investing in education and training programs to build a skilled workforce capable of managing and sustaining these industries. To achieve this balance, African countries should require foreign investors to engage in local partnerships, hire and train local workers, and invest in community development projects. Additionally, they should establish strong regulatory frameworks that ensure environmental protection and social responsibility (Rogerson, 2010). These policies and frameworks can help ensure that foreign investment contributes to sustainable economic growth and benefits local communities (Sarkanne, 2021), rather than just extracting resources and profits without regard for the long-term consequences.

Foreign investment in Africa can lead to the creation of new industries, job opportunities, and an increase in exports, which can help African countries transition toward a net zero economy. It can also lead to exploitation of natural resources and labor, exacerbating inequality and environmental degradation. It is important for African governments to carefully regulate foreign investment and ensure that it aligns with their SDGs (Sauders, 2021). This can be achieved by negotiating favorable terms and conditions with foreign investors, requiring them to use environmentally friendly technologies and pay fair wages, and prioritizing domestic investment and entrepreneurship to promote economic growth and reduce reliance on foreign investment (Schill, 2009). This approach can also help African countries maintain control over their resources and reduce the risk of exploitation by foreign investors, as well as lead to the development of local industries and the creation of job opportunities for the local population.

6.2.6 Strengthened Local Markets

An advantage of foreign investment in Africa's net zero economy is that it strengthens local markets. Foreign investment can bring in new technologies, skills, and knowledge that can help boost the productivity of local businesses. This can lead to increased competitiveness and growth, which in turn can create more job opportunities and improve the standard of living for local communities. Foreign investment can also lead to the exploitation of local resources and labor, as well as the displacement of small businesses. It is important for governments to carefully consider the potential benefits and drawbacks before welcoming foreign investment into their country. Foreign investment in Africa's net zero economy can lead to increased competitiveness and growth but can also lead to the exploitation of local resources and labor. Governments should consider the potential benefits and drawbacks before welcoming foreign investment and establish regulations and oversight mechanisms to mitigate the negative impacts. Foreign investors should prioritize building relationships with the local community and actively engaging in responsible business practices to ensure mutual benefits (Thulasidhass, 2015). Investing in the development of local talent and providing opportunities for skill-building can lead to a more skilled and productive workforce (Tit, 2015), which can attract further investment and drive economic growth (Vaksha and Singh, 2020). Additionally, foreign investors should take a holistic approach to their investments, considering the social, environmental, and economic impacts of their actions on the local community.

This approach can also help to build trust and positive relationships with local stakeholders, which can be crucial for the long-term success and sustainability of the investment. Foreign investment in Africa's net zero economy can bring new technologies, expertise, and capital that can help local businesses grow and become more competitive (Wasike, 2022). It is important to ensure that foreign investment aligns with Africa's SDGs and respects the environment and local communities (Wolfson and Erisken, 2000). This can be achieved through transparent and responsible investment practices as well as collaboration between foreign investors and local stakeholders. By prioritizing sustainable development and engaging with local communities, foreign investors can contribute to Africa's growth while also promoting long-term economic and environmental stability.

Foreign investors should recognize the cultural and historical context of each African country they invest in and work collaboratively with local stakeholders to ensure their investments align with the needs and aspirations of the community. They should prioritize social responsibility and ethical business practices in their operations to contribute to the development of the continent. Foreign investment can bring in new technologies and expertise, which can lead to job creation, increased incomes, and greater economic growth. It is important to ensure that foreign investment does not lead to the exploitation of local resources or labor. Governments must establish policies and regulations to protect the interests of their citizens while also attracting foreign investment.

Transparency and accountability in the investment process should be prioritized to prevent corruption and ensure sustainable development. Involving all stakeholders in the decision-making process can help ensure that the policies are inclusive and beneficial for all parties involved.

6.2.7 Improved Healthcare

The net zero economy of Africa benefits from foreign investment by improving healthcare. Foreign investment not only improves healthcare in Africa, but it also helps create job opportunities and boost economic growth, leading to a more sustainable and prosperous future for the continent. The net zero economy of Africa benefits from foreign investment, but it is important to ensure that these investments are made in a responsible and ethical manner that benefits local communities and protects the environment. Foreign investors should prioritize sustainable investments that align with the development goals of the host country and ensure that their investments do not exploit local resources or labor. Governments in Africa should establish policies and regulations that promote responsible foreign investment and protect the interests of their citizens. Capacity-building programs can be implemented to enhance the skills and knowledge of local communities, empowering them to actively participate in the decision-making process regarding foreign investment projects in medical services. Ultimately, sustainable foreign investment can contribute to the economic growth and development of African countries while also benefiting local communities and preserving the environment.

6.2.8 Improved Education

The net zero economy of Africa benefits from foreign investment by improving education. Foreign investment in education can help to develop the skills and knowledge needed to build a sustainable net zero economy in Africa, creating job opportunities and promoting economic growth. It can lead to the adoption of innovative technologies and practices that support the transition to a low-carbon future. This type of investment can also contribute to improving the quality and accessibility of education in Africa, which is crucial for achieving long-term SDGs and reducing poverty in the region. It can also help to address the existing skills gap and enhance the capacity of local institutions to deliver high-quality education and training programs. Furthermore, investing in education can lead to the development of a skilled workforce that can drive economic growth and innovation, creating more job opportunities and reducing unemployment rates in Africa. This, in turn, can contribute to the overall development of the continent and improve the standard of living for its people.

Foreign investment not only improves education in Africa but also helps to develop sustainable infrastructure and technology, which are essential for achieving a net zero economy. These investments can create job opportunities and promote economic growth in the region. It is important to ensure that these investments are made in a responsible and ethical manner, taking into account the social and environmental impact on local communities. This can be achieved through transparent and accountable governance structures and partnerships between foreign investors and local stakeholders. Incorporating sustainable practices and supporting local businesses can also contribute to the long-term success and stability of the investment. Ultimately, responsible investing can lead to positive outcomes for both investors and the communities in which they operate.

6.3 CHALLENGES OF FOREIGN INVESTMENT IN AFRICA FOR A NET ZERO ECONOMY

6.3.1 Political Instability

One of the factors affecting foreign investment in Africa for a net zero economy is political instability. This is because political instability can lead to a lack of trust in the government's ability to create a stable business environment, which can discourage foreign investors from committing to long-term investments in the region. Political instability can increase the risk of violence and conflict, which can disrupt business operations and cause financial losses for investors. As a result, countries such as African nations experiencing political instability often struggle to attract the necessary foreign investment needed for economic growth and development. This can perpetuate a cycle of poverty and instability, further exacerbating the underlying issues seen in most African countries.

The lack of foreign investment can lead to a brain drain as skilled workers leave in search of better opportunities, leaving behind a shortage of human capital necessary for sustainable economic growth. This can also result in a weakened education system as many teachers and professors leave for better-paying jobs abroad. Ultimately, this can hinder the development of future generations and perpetuate the cycle of poverty. Furthermore, the brain drain phenomenon can also lead to a loss of valuable skills and knowledge in the country, making it harder for local businesses and industries to compete on a global scale. This can further exacerbate economic inequality and hinder overall progress. In addition, the brain drain can also result in a shortage of skilled professionals in key sectors such as healthcare and education, which can have serious consequences for the well-being and development of the local population. Furthermore, the absence of economic diversification and overreliance on a few key exports can make African economies vulnerable to fluctuations in global commodity prices, leading to economic downturns and increased poverty.

Foreign investment is crucial for the development of African countries, but political instability and corruption hinder this process. To break this cycle, it is necessary to address the root causes of instability and create a conducive environment for foreign investment. This can be achieved through the implementation of good governance practices, such as transparency and accountability, and the establishment of strong institutions that can effectively regulate and monitor foreign investment. African countries can also leverage their natural resources and invest in infrastructure development to attract more foreign investment. It is important for African countries to strike a balance between attracting foreign investment and protecting their own interests, such as ensuring that local communities benefit from the investment. African governments should prioritize building local capacity and skills to maximize the benefits of foreign investment in the long run.

Foreign investment can bring much-needed capital and expertise to African countries, but it should not come at the expense of local communities or the environment. Therefore, African governments should establish clear regulations and standards to ensure that foreign investors operate in a socially and environmentally responsible manner. This can help prevent exploitation and ensure that the benefits of foreign investment are shared fairly. African governments should also prioritize building local capacity and promoting domestic investment to reduce reliance on foreign investors. This will not only promote economic growth but also provide a more sustainable and stable source of revenue for the country. It is important for African countries to strike a balance between foreign investment and domestic investment to achieve long-term development goals. Foreign investment can bring in new technology, expertise, and access to global markets, but it can also create dependency and vulnerability to external shocks. Therefore, African countries should prioritize domestic investment in sectors such as education, healthcare, and infrastructure to build a strong foundation for sustainable development. This will enable them to attract more foreign investment in the long run, while also ensuring that their economies are not solely reliant on external funding.

Investing in these sectors will help to improve the standard of living for citizens and create a skilled workforce, which can further drive economic growth. It will also promote innovation and technological advancements, which can lead to the creation of new industries and job opportunities. This, in turn, can boost the overall economic development of the country. For instance, the development of renewable energy sources can not only reduce carbon emissions but also create new job opportunities in the field of green technology. The adoption of artificial intelligence and automation can increase productivity and efficiency in various industries, leading to economic growth.

The implementation of sustainable practices in agriculture and forestry can help preserve natural resources and promote biodiversity while also providing economic benefits to local communities. Furthermore, investing in education and training programs can equip individuals with the skills needed to succeed in emerging industries and contribute to a more innovative and competitive economy. In addition, promoting sustainable tourism can also have a positive impact on both the environment and local economies by encouraging responsible travel practices and supporting conservation efforts. Finally, it is important for governments and businesses to work together to prioritize sustainability in their policies and operations in order to ensure a more sustainable future for all. This can be achieved through implementing eco-friendly practices such as reducing waste and carbon emissions, as well as supporting local communities through fair trade and employment opportunities.

By doing so, we can create a more equitable and sustainable tourism industry that benefits both people and the planet. Sustainable tourism not only benefits the environment but also enhances the overall tourist experience by providing unique and authentic cultural experiences. This can lead to increased visitor satisfaction and repeat business, as well as positive word-of-mouth recommendations. Sustainable tourism can also create job opportunities for local communities and contribute to their economic development while preserving their cultural heritage and traditions. This way, it promotes a more responsible and respectful approach to travel that benefits both the visitors and the destinations they visit. Sustainable tourism can also help reduce the negative impacts of tourism on the environment by promoting conservation and reducing waste. By encouraging responsible travel practices, it can help to protect natural resources and ecosystems for future generations to enjoy. Sustainable tourism can also benefit local communities by providing economic opportunities and supporting cultural preservation. By involving and empowering local people in the tourism industry, it can promote a more equitable distribution of benefits and help to preserve traditional ways of life.

Furthermore, sustainable tourism can also contribute to the conservation of natural resources and biodiversity by promoting responsible practices such as reducing waste, conserving water and energy, and protecting wildlife habitats. This can help to ensure that future generations can continue to enjoy the beauty and diversity of our planet. In addition, sustainable tourism can also have positive economic impacts on local communities by creating jobs and generating income through the promotion of local products and services. This can help improve the standard of living for residents in the area and reduce poverty. Furthermore, sustainable tourism can also encourage the preservation of cultural heritage and natural resources, which can attract more visitors and boost the local economy in the long run. This can lead to a more stable and resilient community that is better equipped to handle economic challenges.

6.3.2 Lack of Regulatory Framework

The absence of a regulatory framework in Africa in a net zero economy leads to uncertainty and risk for investors, making it difficult to plan for the long term. Investors are often hesitant to invest in African countries due to the lack of legal and regulatory frameworks that provide stability and predictability. The absence of reliable infrastructure and access to finance hinders foreign investment in Africa's net zero economy. Some African countries are making efforts to address these challenges by implementing reforms and creating a more conducive environment for investment. For instance, the African Continental Free Trade Area (AfCFTA) agreement aims to create a single market for goods and services in Africa, which could attract more foreign investment. Some countries are also investing in infrastructure development, such as building roads and expanding access to electricity, which could improve the business environment and attract more investors.

The implementation of digital technologies and the adoption of e-commerce could also enhance trade and investment opportunities in Africa, as they can reduce transaction costs and increase efficiency. This could lead to greater regional integration and economic growth on the continent. In addition, the use of digital technologies and e-commerce can also provide greater access to global markets for African businesses, allowing them to reach customers beyond their local regions. This can help them diversify their customer base and increase their revenue streams. Digital technologies can also help African businesses streamline their operations and reduce costs, leading to greater efficiency and profitability. This can enable them to compete more effectively with larger, established companies in the global marketplace. For instance, digital payment systems can help businesses avoid the high costs associated with traditional banking services, while e-commerce platforms can provide access to new markets and customers. As a result, African businesses can leverage technology to overcome some of the barriers to growth and development that they face.

To attract foreign investment, African countries need to prioritize creating a stable and predictable regulatory environment that encourages long-term planning. They must invest in building reliable infrastructure and financial systems that can support the growth of a net zero economy. By doing so, African countries can demonstrate their commitment to sustainable development and attract socially responsible investors who are looking for opportunities to invest in environmentally friendly projects. This will not only help boost economic growth but also contribute to the global effort to combat climate change. Furthermore, promoting sustainable development in Africa can also lead to job creation and poverty reduction, as it can provide opportunities for small businesses and entrepreneurs to thrive. This can lead to a more inclusive and equitable society in the long run.

Sustainable development can also lead to improved health outcomes and access to education, which are essential components of human development. By investing in these areas, African countries can build a brighter future for their citizens and contribute to global efforts toward achieving the SDGs. Furthermore, sustainable development can also promote economic growth and create job opportunities, which can reduce poverty and inequality. By adopting sustainable practices, African countries can improve their economic prospects and ensure a more equitable distribution of resources. In addition, sustainable development can also help African countries mitigate the impacts of climate change and protect their natural resources, which are crucial for their long-term development. This can also contribute to building resilience against future environmental shocks and crises.

To attract more foreign investment, African governments should prioritize creating a stable and predictable business environment, investing in infrastructure development, and improving access to finance. This will help to build investor confidence and facilitate the transition toward a net zero economy in Africa. African governments can also consider offering tax incentives and regulatory reforms to attract more foreign investment in sustainable industries such as renewable energy and green technology. These measures will not only help address climate change but also promote economic growth and job creation in the region. Furthermore, African governments can collaborate with international organizations and private sector entities to develop and implement large-scale projects that promote sustainable development. This will not only enhance the region's economic potential but also help mitigate the adverse effects of climate change.

For instance, the African Development Bank has launched the Desert to Power initiative, which aims to provide electricity to 250 million people in the Sahel region through the development of solar power projects. Similarly, private sector entities like M-KOPA Solar have provided access to clean energy for over 1 million households in East Africa through innovative pay-as-you-go financing models. These initiatives not only address the issue of energy poverty but also contribute to reducing carbon emissions and mitigating the effects of climate change. Such efforts are crucial in achieving SDGs and creating a better future for all. By providing access to clean and affordable energy, these initiatives can also improve the quality of life for people in underserved communities and promote economic growth. It is important for governments and organizations to continue investing in such initiatives to ensure a more equitable and sustainable future. In addition, the adoption of renewable energy sources can also contribute to reducing greenhouse gas emissions and mitigating the effects of climate change. Therefore, supporting these initiatives is not only a matter of social responsibility but also a crucial step toward achieving global environmental goals.

6.3.3 Environmental Concerns

Environmental concerns are one of the factors influencing foreign investment in Africa for a net zero economy. Foreign investors are increasingly looking for opportunities to invest in sustainable projects in Africa that align with their environmental goals. This trend is likely to continue as more countries commit to achieving net zero emissions by 2050. In addition, sustainable investments in Africa can also contribute to the continent's economic growth and social development, creating a win-win situation for both investors and local communities. It is important for investors to conduct thorough due diligence and ensure that their investments have a positive impact on the environment and society. Sustainable investments can also help to address some of the pressing environmental and social challenges facing Africa, such as climate change, poverty, and inequality.

By supporting sustainable projects and businesses, investors can help build a more resilient and inclusive economy that benefits everyone. Furthermore, sustainable investments can attract more foreign investment and create job opportunities, which can contribute to the economic growth of the continent. This can also lead to a reduction in migration as people will have access to better living conditions and job opportunities. Investors should also consider partnering with local organizations and businesses to ensure that their investments align with the needs and priorities of the communities they are investing in, promoting long-term sustainability and mutual benefit. Incorporating sustainable practices into their investment strategies can also help mitigate risks and increase returns in the long run. This approach not only benefits the environment but also enhances the investors' reputation as socially responsible. It can attract more investors who prioritize sustainability and ethical investments, leading to a positive impact on the overall financial performance of the investment portfolio.

Furthermore, companies that adopt sustainable practices are more likely to have a competitive advantage in the long run as consumers increasingly demand environmentally conscious products and services. This trend toward sustainability is not limited to consumer-facing companies, as more and more B2B companies are also recognizing the benefits of sustainable practices and incorporating them into their operations. As a result, investing in sustainable companies can provide both financial returns and align with ethical values. This can result in increased profitability and market share for the company. In addition, sustainable practices can also lead to cost savings for companies through reduced energy and resource consumption. By implementing sustainable practices, companies can improve their bottom line while also benefiting the environment. Sustainable practices can also enhance a company's reputation and brand image, as consumers are increasingly demanding environmentally responsible products and services. This can lead to increased customer loyalty and market share for companies that prioritize sustainability.

6.3.4 Unequal Distribution of Benefits

One of the factors influencing foreign investment in Africa for a net zero economy is unequal benefit distribution. This is because investors may be hesitant to invest in countries where the benefits of their investments are not distributed fairly among the population, leading to potential social and political instability. Companies that prioritize sustainable and equitable business practices may be more attractive to investors looking to support a net zero economy in Africa. Therefore, promoting social and environmental responsibility in business practices can lead to more stable and attractive investment opportunities in Africa. This can ultimately drive economic growth and development while also benefiting local communities and the environment. Investors are becoming increasingly aware of the impact their investments have on society and the environment. By prioritizing social and environmental responsibility, businesses in Africa can attract a wider range of socially conscious investors, which can lead to long-term sustainable growth. In addition, socially responsible businesses in Africa can also benefit from increased customer loyalty and a positive brand reputation as consumers become more conscious of the impact their purchases have on the world around them. Therefore, prioritizing social and environmental responsibility can have a positive impact on both the financial and ethical aspects of a business. Socially responsible businesses can attract and retain talented employees who are passionate about making a difference in their communities and the world. This can lead to a more motivated and engaged workforce, ultimately driving innovation and growth within the company.

In addition, consumers are increasingly demanding that businesses take a stand on social and environmental issues and are more likely to support companies that align with their values. By prioritizing social responsibility, businesses can build trust and loyalty with their customers, ultimately leading to increased sales and revenue. This shift toward socially responsible business practices is not only beneficial for the company's bottom line, but it also has a positive impact on society and the environment. By taking action on issues such as climate change and human rights, businesses can make a meaningful difference in the world. Consumers are becoming increasingly aware of the impact their purchasing decisions have on the world and are more likely to support companies that align with their values. This makes socially responsible business practices not only a moral imperative but also a smart business strategy. Companies that prioritize social responsibility can build a loyal customer base and improve their reputation, which can lead to increased sales and profits. Implementing sustainable practices can also reduce costs and improve efficiency in the long run. This is because socially responsible companies are viewed as trustworthy and ethical, which can attract more customers and create a positive brand image. Sustainable practices such as reducing waste and conserving energy can lead to cost savings and improve the company's bottom line.

6.3.5 Potential for Corruption

The potential for corruption is one of the factors influencing foreign investment in Africa for a net zero economy. Corruption creates an uncertain business environment that discourages foreign investors from committing their resources to Africa's net zero economy. Corruption undermines investor confidence and increases the risk of financial loss, which can discourage foreign investment in Africa's transition to a net zero economy. To attract more investment, African countries need to prioritize anti-corruption measures and create a transparent business environment. This is particularly important as the transition to a net zero economy requires significant investments in renewable energy, sustainable infrastructure, and other green technologies. Without a strong commitment to anti-corruption measures, these investments may be at risk of being misused or diverted for personal gain. Corruption can also undermine public trust in the transition to a net zero economy, making it more difficult to secure the necessary political and social support for these investments.

Implementing effective anti-corruption measures is crucial for ensuring the success of the transition to a sustainable and equitable future. In addition, corruption can also divert resources away from important climate change mitigation and adaptation efforts, exacerbating the negative impacts of global warming on vulnerable communities and ecosystems. It is imperative that we address corruption as a key obstacle to achieving a just and sustainable future for all. Corruption undermines trust in institutions and reduces public support for climate action, making it more difficult to implement effective policies. Therefore, tackling corruption is not only a moral imperative but also a practical necessity for achieving a sustainable and resilient future. Furthermore, corruption also leads to a misallocation of resources and slows down economic growth, hindering the transition toward a low-carbon economy. Addressing corruption can therefore contribute to both environmental and economic sustainability. By promoting transparency and accountability, corruption can be effectively tackled. This can lead to increased investor confidence, improved governance, and a more conducive environment for sustainable development. Transparency and accountability can also help reduce poverty and inequality by ensuring that resources are distributed fairly and used for the benefit of all. It can lead to the creation of a level playing field for businesses, thereby promoting healthy competition and innovation.

To attract foreign investment, African governments must prioritize anti-corruption measures and enforce strict laws and regulations. Foreign investors seek stable and predictable environments to invest their resources, and corruption undermines this stability. In addition, African governments should also invest in infrastructure development and human capital to create an enabling environment for foreign investors. This will not only attract foreign investment but also promote economic growth and development on the continent. The African governments should establish policies that encourage and protect foreign investments, such as tax incentives and legal protection for investors. These measures will help build trust between foreign investors and the host country, leading to increased investment inflows. In addition, the government should also focus on developing the local workforce by providing education and training programs to improve the skills of the labor force. This will not only attract more foreign investors but also create job opportunities for the local population.

By investing in education and training programs, the government can ensure that the local workforce is equipped with the necessary skills to meet the demands of a changing job market. This will not only benefit foreign investors but also improve the overall economic growth and development of the country. Furthermore, providing access to education and training opportunities can also help reduce poverty and inequality by creating more job opportunities for marginalized communities. This can lead to a more inclusive and sustainable economy. In addition, investing in education and training can also improve the overall quality of life for individuals and communities by increasing their knowledge and skills, which can lead to better health outcomes, increased civic engagement, and a greater sense of empowerment. This can ultimately contribute to a more just and equitable society.

6.3.6 Lack of Infrastructure

Inadequate infrastructure is one of the barriers to foreign investment in Africa for a net zero economy. This challenge is particularly acute in rural areas where access to reliable electricity and internet connectivity is limited. Addressing these infrastructure gaps will be critical to unlocking the full potential of Africa for sustainable development. Without reliable electricity and internet connectivity, rural areas in Africa will continue to lag behind in terms of technological advancement and economic growth. Investments in infrastructure development should be prioritized by African governments and international organizations to ensure that all citizens have access to the necessary resources for technological innovation. This will not only improve the quality of life for rural Africans but also create opportunities for businesses and entrepreneurs to thrive in these areas.

It will help bridge the digital divide between urban and rural areas, leading to more inclusive and equitable economic growth. This will not only improve economic opportunities but also enhance social and educational outcomes. Investments in infrastructure development will not only benefit the tech sector but also have a positive impact on other industries such as agriculture, healthcare, and education, ultimately contributing to the overall socio-economic development of Africa. The improved infrastructure will attract foreign investment and create more job opportunities, leading to a reduction in poverty and an increase in the standard of living for African citizens. This will also promote regional integration and cooperation, fostering a more stable and prosperous continent. In addition, the development of infrastructure will facilitate the movement of goods and services, reducing transportation costs and improving access to markets. This will enhance trade within Africa and with other regions, contributing to economic growth and development.

The improvement of infrastructure will also attract foreign investment, as it creates a more conducive environment for businesses to operate and expand. This will lead to job creation and further boost economic activity in the region. In addition, the increased connectivity and efficiency in transportation and communication will also facilitate the movement of people, goods, and services, promoting cultural exchange and knowledge sharing among African countries. Ultimately, this will strengthen regional integration and cooperation toward achieving SDGs. The enhanced collaboration and partnerships between African nations will create a more stable and secure environment, attracting foreign investment and boosting economic growth in the region. This will further contribute to poverty reduction and improved living standards for the people of Africa.

The increased cooperation between African nations will also lead to the sharing of knowledge and expertise, which can help address common challenges such as climate change, health crises, and security threats. This will promote regional integration and strengthen Africa's position in the global community. Collaboration among African countries can foster economic growth and development through increased trade and investment. This will create more job opportunities and improve the standard of living for people across the continent. In addition, the pooling of resources and expertise can lead to the development of infrastructure, such as transportation networks and energy systems, which are critical for economic growth and development. This will further enhance the continent's competitiveness in the global market. Collaborations between countries can also lead to the sharing of best practices and knowledge, which can help address common challenges such as poverty reduction, disease control, and environmental sustainability. These partnerships can ultimately contribute to a more stable and prosperous Africa. Through the exchange of ideas and experiences, countries can learn from each other's successes and failures, leading to more effective solutions to complex problems.

This can foster a sense of unity and cooperation among African nations, paving the way for long-term progress and development. In addition, this can also create opportunities for collaboration and partnerships, allowing countries to pool their resources and expertise to tackle common challenges such as poverty, climate change, and economic inequality. Ultimately, this can lead to a more prosperous and stable Africa for all its citizens. By working together, African countries can also leverage their collective bargaining power in international trade negotiations and secure better deals for their economies. This can help boost economic growth and create more job opportunities for people across the continent. Furthermore, cooperation among African countries can lead to the development of regional value chains, which can increase intra-African trade and reduce reliance on external markets. This can contribute to a more sustainable and resilient economic future for the continent.

The development of regional value chains can also create job opportunities and enhance technological advancements within Africa. This can lead to a more diversified and competitive economy, ultimately improving the standard of living for African citizens. Furthermore, regional value chains can also reduce Africa's dependence on foreign imports and increase intra-African trade, promoting economic growth and stability. This can help address some of the challenges faced by African countries, such as poverty, inequality, and unemployment. In addition, regional value chains can also foster innovation and technological advancement in Africa as local firms collaborate and learn from each other to improve their competitiveness and productivity. This can lead to the emergence of new industries and the creation of higher-skilled jobs, contributing to long-term sustainable development. Regional value chains can help to reduce Africa's dependence on commodity exports and diversify its economy, thereby reducing its vulnerability to external shocks. This can also contribute to poverty reduction and inclusive growth by providing opportunities for small and medium-sized enterprises to participate in regional trade and investment.

Regional value chains can promote technology transfer and knowledge sharing among countries in the region, leading to increased productivity and competitiveness. This can ultimately lead to higher incomes and better living standards for people in Africa. Furthermore, regional value chains can also attract foreign investment and enhance the bargaining power of African countries in global trade negotiations. This can help to diversify their economies and reduce their dependence on a few primary commodities. In addition, regional value chains can lead to the creation of more jobs and the transfer of technology and skills, which can contribute to the overall development of African countries. It is important for these value chains to be inclusive and benefit all stakeholders involved, especially small and medium-sized enterprises. To ensure inclusivity, governments and policymakers need to create an enabling environment that supports the growth of small and medium-sized enterprises, such as by providing access to finance, training, and technology. Partnerships between the private sector, civil society organizations, and governments can also help to promote inclusive regional value chains. These efforts can lead to the creation of job opportunities, increased income, and economic growth, particularly in marginalized communities.

Ultimately, fostering inclusive economic development can contribute to reducing poverty and promoting sustainable development. African governments and organizations can play a crucial role in achieving inclusive economic development by implementing policies and programs that prioritize marginalized communities. This can include providing access to education, training, and financial resources to help individuals and businesses thrive. Investing in infrastructure and technology can also create opportunities for economic growth and development in underserved areas. By taking these steps, African nations can work toward reducing poverty and promoting sustainable economic progress. Promoting entrepreneurship and small business development can also contribute to job creation and economic empowerment in these regions. It is important for governments to create policies and programs that support these initiatives. These policies can include providing access to funding and resources, reducing bureaucratic barriers, and offering training and mentorship programs. By doing so, governments can help foster a culture of entrepreneurship and innovation that can drive economic growth in underdeveloped regions.

These policies can also encourage the creation of new businesses and industries, which can create job opportunities and improve the standard of living for individuals in these regions. This can lead to a more diversified economy and reduce dependence on a single industry or sector. In addition, the implementation of these policies can also attract foreign investment and increase international trade, which can further boost economic growth and development. It is important to ensure that these policies are sustainable and do not have negative impacts on the environment or local communities. Governments should also consider the potential long-term consequences of these policies and prioritize the well-being of their citizens over short-term economic gains. This can be achieved through effective monitoring and evaluation mechanisms that ensure the policies are achieving their intended goals without causing harm. Furthermore, governments should involve stakeholders and experts in the decision-making process to ensure that policies are evidence-based and take into account diverse perspectives. This can lead to more sustainable and equitable outcomes for all members of society.

Transparency and accountability in the decision-making process can help build trust between governments and their constituents. This can foster a sense of ownership and participation in the policymaking process, ultimately leading to more effective and widely accepted policies. Involving diverse stakeholders in the decision-making process can bring different perspectives and ideas to the table, leading to more comprehensive and innovative solutions. This can also increase the legitimacy of policies by ensuring that they reflect the needs and interests of a broad range of people. In addition, when stakeholders are involved in the decision-making process, they are more likely to support and implement the policies that are created, leading to greater success in achieving desired outcomes. This can also foster a sense of ownership and accountability among stakeholders, which can lead to long-term commitment and sustainability. Furthermore, involving stakeholders can bring diverse perspectives and ideas to the table, leading to more innovative and effective solutions. This can also help build trust and relationships between stakeholders, which can be beneficial for future collaborations. In addition, involving stakeholders can increase the likelihood of successful implementation and sustainability of solutions, as they are more likely to support and take ownership of initiatives that they have had a hand in developing. This can lead to long-term benefits for all parties involved. Furthermore, engaging stakeholders can bring diverse perspectives and ideas to the table, leading to more innovative and effective solutions that address the root causes of problems. This collaborative approach can also foster trust and positive relationships between stakeholders, creating a foundation for future cooperation and problem-solving.

6.4 SOLUTIONS TO ADDRESS THE CHALLENGES OF FOREIGN INVESTMENT IN AFRICA FOR A NET ZERO ECONOMY

6.4.1 Strengthening of Political Institutions

One way to deal with the difficulties of foreign investment in Africa for a net zero economy is to strengthen political institutions. By strengthening political institutions, African countries can create a stable and predictable investment environment that will attract foreign investors. This will also help to ensure that investments are made in a sustainable manner, contributing to the achievement of a net zero economy. In addition, promoting transparency and accountability in the governance system will also build trust between investors and governments, which is crucial for long-term investments and economic growth. Ultimately, this will lead to job creation, poverty reduction, and a better standard of living for African citizens. Furthermore, implementing effective policies to combat corruption and bribery will also attract more foreign investment and improve the business environment in Africa. This will create more opportunities for local entrepreneurs and increase competition, leading to innovation and economic diversification.

Promoting transparency and accountability in government institutions will enhance public trust and confidence in the system, which can lead to increased civic engagement and participation. This can also contribute to the overall development of democratic institutions in Africa. Providing citizens with access to information and opportunities to participate in decision-making processes can help ensure that their needs and priorities are taken into account by policymakers. This can ultimately lead to more effective and responsive governance, and improved service delivery for citizens. In addition, transparency and citizen engagement can also help to build trust between the government and its citizens, fostering a sense of ownership and accountability among all stakeholders. This can create a more collaborative and inclusive approach to governance, leading to better outcomes for everyone involved.

Transparency can be achieved through open data policies and public reporting of government activities, while citizen engagement can involve participatory budgeting, town hall meetings, and other forms of direct democracy. By implementing these practices, governments can ensure that their decisions are informed by the needs and perspectives of the people they serve. This can lead to increased trust and accountability between the government and its citizens, ultimately resulting in more effective and equitable policies that better serve the public interest. It also promotes a culture of collaboration and cooperation between government officials and citizens, fostering a stronger sense of community ownership over decision-making processes. Citizen participation can help identify and address the needs and concerns of marginalized groups that may otherwise be overlooked by the government. This can lead to more inclusive policies that promote social justice and equality.

Citizen participation can also increase transparency and accountability in government decision-making processes, as it allows for greater scrutiny and feedback from the public. This can ultimately result in more effective and responsive governance. Citizen participation can also lead to a greater sense of ownership and responsibility among the public toward their community and government. This can foster a culture of active citizenship and civic engagement, which is essential for a healthy democracy. When citizens are actively involved in decision-making processes, they are more likely to hold their elected officials accountable and demand transparency. This can ultimately lead to more effective and responsive governance. Furthermore, citizen participation can also help identify and address issues that may have been overlooked by officials, leading to more inclusive policies that better reflect the needs and values of the community. Overall, citizen engagement is crucial for promoting a healthy democracy and improving the quality of governance.

6.4.2 Strengthen Regulations

Strengthening regulations is one strategy to address the challenges of foreign investment in Africa for a net zero economy. Other strategies include promoting domestic investment, fostering technology transfer and capacity building, and encouraging public–private partnerships to ensure that foreign investment aligns with the goals of a net zero economy. Creating a favorable business environment through stable policies and transparent regulations can also attract responsible foreign investors. These measures can help countries achieve their climate goals while also boosting their economies and creating jobs. It is important for governments to prioritize sustainable investment and work with international partners to ensure a smooth transition toward a net zero future. Investing in renewable energy, improving energy efficiency, and transitioning to low-carbon transportation are some of the key steps that countries can take to achieve their climate goals.

By doing so, they can also create new job opportunities and stimulate economic growth in these sectors. Furthermore, investing in these areas can also lead to a reduction in air pollution and health-related costs, as well as increased energy security and resilience to climate impacts. It is important for countries to prioritize these actions in order to mitigate the effects of climate change and ensure a sustainable future for all. Governments can encourage private sector investment in these areas through policy incentives and regulations. Public–private partnerships can be established to facilitate the implementation of sustainable infrastructure projects. These measures can help address the funding gap for sustainable infrastructure projects, which often require significant upfront investment. By working together, governments and the private sector can create more sustainable and resilient infrastructure that benefits both society and the economy. In addition, promoting green bonds and other sustainable finance instruments can also attract more investors to support sustainable infrastructure projects.

This can help accelerate the transition toward a low-carbon and more sustainable future. Sustainable finance instruments can also provide long-term benefits for investors, as they often generate stable and predictable returns. This can incentivize more investors to shift their investments toward sustainable projects, creating a positive cycle of investment and sustainability. In addition, sustainable finance can also contribute to the achievement of global environmental and social goals, such as the United Nations SDGs. By directing investments toward sustainable projects, we can work toward a more equitable and environmentally responsible future for all. Furthermore, sustainable finance can also help mitigate climate change by reducing greenhouse gas emissions and promoting renewable energy sources. It can also encourage companies to adopt more socially responsible practices, such as fair labor standards and community engagement.

In addition, sustainable finance can improve the long-term financial performance of companies by promoting efficient resource use and reducing waste. This can lead to cost savings and increased profitability, benefiting both the company and the environment. Furthermore, sustainable finance can also attract more socially responsible investors who prioritize environmental and social impact alongside financial returns, increasing the company's access to capital and enhancing its reputation. As a result, sustainable finance can be a win-win solution for both businesses and the planet. In addition, incorporating sustainable practices into a company's operations can lead to cost savings in the long run, as they reduce waste and improve efficiency. This can further enhance the financial benefits of sustainable finance for businesses.

6.4.3 Sustainable Investment Practices

One method for overcoming the difficulties of foreign investment in Africa for a net zero economy is to use sustainable investment practices. Sustainable investment practices can help ensure that investments are made in a way that benefits both the environment and local communities, while also providing long-term economic benefits. This approach can help attract more foreign investment to Africa and promote a net zero economy. In addition, sustainable investment practices can also contribute to the achievement of the United Nations SDGs and support the transition to a more resilient and inclusive economy in Africa. This can lead to improved social and environmental outcomes, as well as increased investor confidence and financial returns. Furthermore, sustainable investment practices can also help to mitigate the risks associated with climate change and other environmental factors, such as water scarcity and deforestation. By investing in companies that prioritize sustainability, investors can play a role in promoting positive change and creating a more sustainable future for Africa.

Sustainable investment practices can also have a positive impact on the social and economic development of African communities, by supporting businesses that prioritize fair labor practices and community engagement. In this way, sustainable investing can contribute to the overall well-being and resilience of African societies. Furthermore, sustainable investment practices can also help to address environmental challenges in Africa, such as deforestation and water scarcity, which can have a significant impact on local communities. By investing in companies that prioritize sustainable resource management, investors can help mitigate these challenges and promote long-term environmental sustainability. This can also lead to improved social and economic outcomes for local communities, as sustainable investments often prioritize fair labor practices and community engagement. Overall, sustainable investing has the potential to create positive impact across multiple sectors in Africa.

Sustainable investing can also help address environmental challenges such as climate change and biodiversity loss, which are particularly pressing issues in many African countries. By directing capital toward sustainable solutions, investors can contribute to building a more resilient and sustainable future for the continent. Furthermore, sustainable investing can also have positive social impacts by promoting good governance, human rights, and labor standards. This can lead to improved living conditions and greater economic opportunities for communities in Africa. In addition, sustainable investing can help reduce poverty and inequality by supporting businesses that prioritize fair wages and safe working conditions for their employees. By doing so, investors can contribute to the overall well-being of African societies while also generating financial returns. Furthermore, sustainable investing can also promote environmental protection by supporting companies that prioritize renewable energy and responsible resource management. This can lead to a healthier planet for future generations while also creating economic opportunities in the green economy.

Investing in sustainable companies can also have a positive impact on social issues such as labor practices and community development, as these companies often prioritize fair treatment of workers and engagement with local communities. By investing in these companies, individuals and organizations can contribute to a more just and equitable society. Sustainable companies tend to have a long-term perspective on their business practices, which can lead to greater stability and resilience in the face of economic and environmental challenges. This makes them attractive investments for those looking for both financial returns and positive social impact. Investing in sustainable companies not only benefits the investors financially, but it also contributes to a more sustainable future for everyone. By supporting these companies, investors can help drive positive change toward a more environmentally conscious and socially responsible economy. Sustainable companies are often more resilient to economic and environmental shocks, making them a safer long-term investment option. By investing in these companies, investors can also align their values with their financial goals. This is because sustainable companies tend to have better risk management practices and are more adaptable to changing market conditions. As a result, they are better positioned to weather economic downturns and environmental crises.

6.4.4 Increase Transparency

A strategy to address the difficulties of foreign investment in Africa for a net zero economy is to increase transparency. By improving transparency, foreign investors will have a better understanding of the risks and opportunities associated with investing in Africa's net zero economy. This can help build trust and encourage more foreign investment in the region. In turn, this increased investment can drive economic growth and create job opportunities for the local population, ultimately contributing to the development of a sustainable and thriving African economy. It is crucial for African governments to prioritize creating a stable and favorable business environment to attract foreign investors. This includes implementing policies that protect investments, reducing corruption, and improving infrastructure. By attracting foreign investors, African governments can create job opportunities, increase economic growth, and ultimately improve the standard of living for their citizens. This requires a long-term commitment to creating an environment that fosters sustainable business practices and encourages investment in key sectors such as agriculture, energy, and technology.

To overcome the challenges of foreign investment in Africa for a net zero economy, greater transparency is one approach. Furthermore, increased foreign investment can lead to the transfer of technology and knowledge, which can further enhance Africa's transition to a net zero economy. This can also create job opportunities and boost economic growth in the region. Foreign investment can also help African countries diversify their economies, reducing their dependence on fossil fuels and other non-renewable resources. This can contribute to the long-term sustainability of the region's economy and environment. Foreign investment can also bring in new technologies and expertise that can help African countries develop their renewable energy sector and increase their energy efficiency. This can lead to a reduction in greenhouse gas emissions and a more sustainable future for the continent.

Increasing transparency is one way to combat the challenges of foreign investment in Africa for a net zero economy. This can be achieved by implementing clear regulations and policies that promote accountability and by creating platforms for investors to access information on investment opportunities. Involving local communities in the decision-making process can help build trust and foster a conducive environment for foreign investment. By doing so, foreign investors will be more confident in their investments and will be more likely to invest in the country, leading to economic growth and development. It is important for governments to prioritize transparency and community involvement in order to attract foreign investment. Furthermore, promoting stability and predictability in policies and regulations can also encourage foreign investors to invest in the country. This can be achieved through establishing clear and consistent rules for doing business and minimizing corruption. In addition, providing incentives such as tax breaks or subsidies for foreign investors can also attract more investment and stimulate economic growth. It is important to ensure that these incentives are not excessive and do not create an unfair advantage for foreign companies over domestic ones.

In order to address the challenges of foreign investment in Africa for a net zero economy, greater transparency is one solution. By increasing transparency, investors will have access to reliable information on the investment climate in Africa, which will help them make informed decisions. Transparency can help to build trust between investors and local communities, leading to more sustainable investments. This can also lead to improved governance and accountability, as companies are more likely to adhere to ethical and environmental standards when their practices are open to scrutiny. Ultimately, increased transparency can contribute to the long-term economic growth and development of African countries. Furthermore, it can attract foreign investment and promote a positive image of the country, which can lead to more opportunities for trade and partnerships with other nations. Transparency alone is not enough and must be accompanied by strong regulatory frameworks and enforcement mechanisms to ensure compliance. This will help prevent corruption and ensure that businesses operate ethically, which can ultimately benefit the economy and society as a whole. It is important for governments to strike a balance between promoting transparency and implementing effective regulations to create a fair and sustainable business environment.

Increasing transparency is one way to combat the challenges faced by foreign investment in Africa in a net zero economy. By increasing transparency, foreign investors will have a better understanding of the policies and regulations governing their investments in Africa. This will help to build trust and confidence, leading to more sustainable and long-term investments that support the transition to a net zero economy. Furthermore, this will also attract more responsible investors who prioritize ESG factors in their investment decisions, thereby promoting sustainable development in Africa. Ultimately, increased transparency can contribute to the achievement of the United Nations SDGs in the region. By implementing ESG factors in their investment decisions, responsible investors can positively impact local communities and the environment. This can lead to long-term benefits for both the investors and the African region as a whole. In addition, investing in companies with strong ESG practices can also lead to better financial performance and better risk management, making it a win-win situation for both investors and the African region. It is crucial for investors to consider the social and environmental impact of their investments in order to create a sustainable future for all.

Increasing transparency is one method for addressing the challenges of foreign investment in Africa for a net zero economy. This can be achieved by implementing policies that require companies to disclose their environmental and social impact, as well as ensuring that government officials are held accountable for any corrupt practices that may hinder foreign investment. Promoting sustainable business practices and investing in renewable energy can also attract more foreign investors to Africa. These measures can not only attract foreign investment but also contribute to the long-term economic growth and development of the African continent. It is crucial for African governments to prioritize sustainable and ethical business practices to ensure a prosperous future for their citizens. By promoting transparency, accountability, and good governance, African governments can create a conducive environment for businesses to thrive. Investing in education and infrastructure can help build a skilled workforce and improve the overall business climate on the continent. This will attract foreign investment and increase economic growth, leading to job creation and poverty reduction. It is crucial for African governments to prioritize these efforts in order to achieve sustainable development and improve the standard of living for their citizens.

6.4.5 Invest in Local Communities

Investing in local communities is one way to overcome obstacles to foreign investment in Africa and achieve a net zero economy. This approach can build trust and relationships with the people and governments of African countries, which can lead to greater business opportunities and partnerships. It can also help address some of the social and environmental challenges faced by African countries, such as poverty and climate change, and create jobs and improve the standard of living for residents. Investing in education and healthcare can also have a positive impact on the workforce, leading to increased productivity and economic growth. A healthy and educated workforce can lead to a reduction in healthcare costs and an increase in innovation, which can further drive economic growth and improve the overall quality of life.

Local investments and sustainable infrastructure can help Africa achieve a net zero economy, reducing its dependence on foreign aid and improving its economic self-sufficiency. This approach can also create job opportunities for the local population and encourage entrepreneurship, which can further boost economic growth. It is important for African governments to prioritize such investments and work toward building a self-sufficient economy. Intra-African trade can also create employment opportunities and reduce poverty, especially in rural areas. By increasing trade within the continent, African countries can harness their full economic potential and achieve sustainable development.

Foreign investors can create jobs, improve infrastructure, and support sustainable development by investing in local communities. Governments must create a conducive environment for foreign investment by implementing policies that promote transparency, accountability, and the rule of law. By engaging with the local communities, investors can build trust and establish long-term relationships, which can be beneficial for future projects. Involving the communities in the decision-making process and considering their needs and concerns can lead to better project outcomes and reduce potential conflicts. It is important for investors to recognize the value of community engagement as a crucial aspect of responsible and sustainable investing.

Foreign investors can establish a positive relationship with the local population and gain their trust, leading to a more stable investment environment and a net zero economy in Africa. By investing in local communities, foreign investors can also establish long-term relationships with the community and gain a better understanding of the local culture and customs. This can lead to a mutually beneficial partnership between the investors and the community. Investing in local communities can also create job opportunities and stimulate economic growth, which can have a positive impact on the overall development of the region. Supporting local businesses and industries can also lead to a reduction in carbon emissions and environmental impact as goods and services are produced and consumed within the region, reducing the need for long-distance transportation. This can contribute to a more sustainable future for both the community and the planet.

6.4.6 Promote Sustainability

One strategy for removing barriers to foreign investment in Africa and achieving a net zero economy is to promote sustainability. This can be done by encouraging investments in renewable energy, sustainable agriculture, and eco-tourism. Governments can implement policies that incentivize companies to adopt sustainable practices and reduce their carbon footprint. Investing in renewable energy and promoting sustainable practices can lead to the creation of new jobs in various sectors such as engineering, construction, and technology, while also reducing carbon emissions and preserving natural resources for future generations. Investing in renewable energy can reduce the dependence on fossil fuels and enhance energy security, which can benefit both developed and developing countries.

Governments should also provide incentives and support for businesses to transition toward sustainable practices and technologies, such as tax breaks or subsidies for companies that invest in renewable energy sources or implement eco-friendly production methods, and establish regulations and standards to ensure that businesses operate in an environmentally responsible manner. Promoting sustainability is one way to get rid of obstacles to foreign investment in Africa and create a net zero economy. Policies that encourage renewable energy sources, reduce carbon emissions, and promote sustainable practices in agriculture and industry can help drive investment in sustainable projects. Partnerships between governments, private sector entities, and civil society organizations can also facilitate knowledge sharing and technology transfer, enabling African countries to leapfrog traditional development pathways and adopt sustainable solutions that are tailored to their specific needs. These efforts can help Africa achieve its development goals while also addressing the urgent challenge of climate change.

Partnerships can promote the growth of local businesses and create job opportunities, leading to economic empowerment and poverty reduction in African communities. Knowledge transfer and technology sharing can enhance the capacity of African communities to address pressing issues such as climate change and food insecurity. These partnerships can promote cultural exchange and mutual understanding between African communities and their international counterparts, fostering a sense of global interconnectedness and cooperation.

In order to achieve a net zero economy and remove obstacles to foreign investment in Africa, one strategy is to promote sustainability. This can be done by encouraging the adoption of renewable energy sources, promoting green infrastructure, and implementing policies that prioritize environmental protection. Collaboration between governments and businesses can also lead to innovative solutions and technologies that can help reduce carbon emissions, conserve natural resources, and promote sustainable development. For example, the partnership between the US government and Tesla has resulted in the development of electric cars that have significantly reduced carbon emissions and created new job opportunities for people. This shift toward sustainable transportation has also led to the development of new technologies such as autonomous driving and improved battery life, which have further increased the appeal of electric cars to consumers.

As more people make the switch to electric vehicles, it is likely that we will see a significant impact on the environment and a move toward a more sustainable future.

The promotion of sustainability is one method for lowering obstacles to foreign investment in Africa and achieving a net zero economy. This can be done by implementing policies that encourage the use of renewable energy sources, promoting green technologies and practices, and ensuring natural resources are used in a responsible and sustainable manner. Partnerships between governments, businesses, and communities can help to create a more sustainable future for Africa. Collaborative efforts can bring diverse perspectives, resources, and expertise together to address complex problems more effectively and create sustainable solutions that benefit all members of society.

Collaboration also promotes a sense of ownership and shared responsibility, which can lead to increased commitment and accountability toward achieving common goals. This can result in long-lasting positive impacts on communities and the environment and foster learning and skill-building among collaborators.

REFERENCES

AFRICA–EU: Investment promises. (2022, March). *Africa Research Bulletin: Economic, Financial and Technical Series*, 59(1). https://doi.org/10.1111/j.1467-6346.2022.10412.x

Africanacity of international investment law: A reflection on investment agreements in Africa. (2021, September). *Journal of Law, Policy and Globalization*. https://doi.org/10.7176/jlpg/113-01

Amendolagine, V., & Prota, F. (2021, May). Bilateral investment treaties and backwards linkages in sub-Saharan Africa. *International Economics*, *165*, 172–185. https://doi.org/10.1016/j.inteco.2021.01.001

Asiedu, E. (2004, January). Policy reform and foreign direct investment in Africa: Absolute progress but relative decline. *Development Policy Review*, *22*(1), 41–48. https://doi.org/10.1111/j.1467-8659.2004.00237.x

Baltag, C. (2009, January 1). International investment arbitration: Substantive principles, principles of international investment law and the fair and equitable treatment standard in international foreign investment law. *British Yearbook of International Law*, *79*(1), 378–388. https://doi.org/10.1093/bybil/79.1.378

Capital investment in Africa: Its course and effects. (1939, March 1). *International Affairs*. https://doi.org/10.2307/3019884

Cheng, W. (2023, March 13). The green investment principles: From a nodal governance perspective. *International Environmental Agreements: Politics, Law and Economics*. https://doi.org/10.1007/s10784-023-09595-w

Cheung, Y. W., de Haan, J., Qian, X., & Yu, S. (2012, April 16). China's outwards direct investment in Africa. *Review of International Economics*, *20*(2), 201–220. https://doi.org/10.1111/j.1467-9396.2012.01017.x

Clark, W. (1997). The global information economy and its effect on local economic development. *New Directions for Higher Education*, *1997*(97), 51–61. https://doi.org/10.1002/he.9705

Ewelukwa Ofodile, U. (2014, January 1). Managing foreign investment in agricultural land in Africa: The role of bilateral investment treaties and international investment contracts. *Law and Development Review*, *7*(2). https://doi.org/10.1515/ldr-2014-0010

Fosu, A. K. (2013, November). Impact of the global financial and economic crisis on development: Whither Africa? *Journal of International Development*, *25*(8), 1085–1104. https://doi.org/10.1002/jid.2971

Gazzini, T. (2009). General principles of law in the field of foreign investment. *The Journal of World Investment & Trade*, 103–119. https://doi.org/10.1163/221190009x00240

Grekou, C., & Owoundi, F. (2020, December). Understanding how foreign direct investment inflows impact urbanization in Africa. *International Economics*, *164*, 48–68. https://doi.org/10.1016/j.inteco.2020.07.002

Gunter, A. (2006, June 30). Integrated development plans and local economic development: The case of Mpumalanga Province, South Africa. *Africa Insight*, *35*(4). https://doi.org/10.4314/ai.v35i4.22452

Hepburn, J. (2015, September 4). The UNIDROIT principles of international commercial contracts and investment treaty arbitration: A limited relationship. *International and Comparative Law Quarterly*, *64*(4), 905–933. https://doi.org/10.1017/s0020589315000354

Khumon, P. (2016). Development of rule-based investment protection provisions in Mega-FTAs: A manifestation or a Negation from customary international law principles? *SSRN Electronic Journal*. https://doi.org/10.2139/ssrn.2801917

Klimova, E. (2022). Opportunities and directions for the development of global e-commerce in the context of digitalization. *Economic Development Research Journal*, *4*, 45–49. https://doi.org/10.54092/25420208_2022_4_45

Larmer, M. (2010). A new scramble for Africa? Imperialism, investment and development. *Review of African Political Economy*, *37*(123). https://doi.org/10.1080/03056241003630354

Leiter, A. (2021, November 26). Protecting concessionary rights: General principles and the making of international investment law. *Leiden Journal of International Law*, *35*(1), 55–69. https://doi.org/10.1017/s0922156521000601

Li, T. M. (2014, July 14). What is land? Assembling a resource for global investment. *Transactions of the Institute of British Geographers*, *39*(4), 589–602. https://doi.org/10.1111/tran.12065

Nel, E. L., & Humphrys, G. (1999, June). Local economic development: Policy and practice in South Africa. *Development Southern Africa*, *16*(2), 277–289. https://doi.org/10.1080/03768359908440077

Nkuna, O. (2015, December 24). Intra-SADC foreign direct investment: A gravity approach to South Africa outwards investment. *International Journal of Economics and Finance*, *8*(1), 79. https://doi.org/10.5539/ijef.v8n1p79

Nwokolo, S. C., Obiwulu, A. U., & Ogbulezie, J. C. (2023). Machine learning and analytical model hybridization to assess the impact of climate. *Physics and Chemistry of the Earth*, *130*(February), 103389. https://doi.org/10.1016/j.pce.2023.103389

Nwokolo, S. C., & Ogbulezie, J. C. (2018a). A qualitative review of empirical models for estimating diffuse solar radiation from experimental data in Africa. *Renewable and Sustainable Energy Reviews*. https://doi.org/10.1016/j.rser.2018.04.118

Nwokolo, S. C., & Ogbulezie, J. C. (2018b). A quantitative review and classification of empirical models for predicting global solar radiation in West Africa. *Beni-Suef University Journal of Basic and Applied Sciences*, *7*(4), 367–396. https://doi.org/10.1016/j.bjbas.2017.05.001

Nwokolo, S. C., Ogbulezie, J. C., & Ushie, O. J. (2023). A multimodel ensemble-based CMIP6 assessment of future solar radiation and PV potential under various climate warming scenarios. *Optik*, *285*(May), 170956. https://doi.org/10.1016/j.ijleo.2023.170956

Nwokolo, S., & Otse, C. (2019). Impact of sunshine duration and clearness index on diffuse solar radiation estimation in mountainous climate. *Trends in Renewable Energy*, *5*(3), 307–332. https://doi.org/10.17737/tre.2019.5.3.00107

Obiwulu, A. U., Chendo, M. A. C., Erusiafe, N., & Nwokolo, S. C. (2020). Implicit meteorological parameter-based empirical models for estimating back temperature solar modules under varying tilt-angles in Lagos, Nigeria. *Renewable Energy*, 145. https://doi.org/10.1016/j.renene.2019.05.136

Obiwulu, A. U., Erusiafe, N., Olopade, M. A., & Nwokolo, S. C. (2020). Modelling and optimization of back temperature models of mono-crystalline silicon modules with special focus on the effect of meteorological and geographical parameters on PV performance. *Renewable Energy, 154*. https://doi.org/10.1016/j.renene.2020.02.103

Obiwulu, A. U., Erusiafe, N., Olopade, M. A., & Nwokolo, S. C. (2022). Modelling and estimation of the optimal tilt angle, maximum incident solar radiation, and global radiation index of the photovoltaic system. *Heliyon*, *8*(6), e09598. https://doi.org/10.1016/j.heliyon.2022.e09598

Owens, T. L., & Lane, J. E. (2014, December). Cross-border higher education: Global and local tensions within competition and economic development. *New Directions for Higher Education*, *2014*(168), 69–82. https://doi.org/10.1002/he.20114

Pamba, D. (2023, January 5). Does foreign direct investment crowd in or crowd out domestic investment in South Africa? An ARDL-ECM approach. *International Journal of Management Research and Economics*, *3*(1). https://doi.org/10.51483/ijmre.3.1.2023.18-35

Peat, D. (2018, August 14). International investment law and the public law analogy: The fallacies of the general principles method. *Journal of International Dispute Settlement*, *9*(4), 654–678. https://doi.org/10.1093/jnlids/idy028

Pedler, F. J. (1955, October). Foreign investment in West Africa. *International Affairs*, *31*(4), 459–468. https://doi.org/10.2307/2604821

Qumba, M. F. (2018, July 3). Safeguarding foreign direct investment in South Africa: Does the protection of investment act live up to its name? *South African Journal of International Affairs*, *25*(3), 341–364. https://doi.org/10.1080/10220461.2018.1523024

Reinisch, A. (2014, November 29). The relevance of the UNIDROIT principles of international commercial contracts in international investment arbitration. *Uniform Law Review – Revue De Droit Uniforme*, *19*(4), 609–622. https://doi.org/10.1093/ulr/unu035

Rogerson, C. M., & Rogerson, J. M. (2010, October). Local economic development in Africa: Global context and research directions. *Development Southern Africa*, *27*(4), 465–480. https://doi.org/10.1080/0376835x.2010.508577

Särkänne, K. (2021, September 1). A guide to general principles of law in international investment arbitration. *ICSID Review – Foreign Investment Law Journal*, *36*(3), 657–660. https://doi.org/10.1093/icsidreview/siab040

Saunders, I. (2021, December 15). Chasing the rabbit: The resurgence of general principles of law in international investment law and beyond. *The Journal of World Investment & Trade*, *23*(3), 457–475. https://doi.org/10.1163/22119000-12340233

Schill, S. W. (2009, April 1). Principles of international investment law. *European Journal of International Law*, *20*(2), 471–473. https://doi.org/10.1093/ejil/chp019

Thulasidhass, P. (2015, June 1). Most-favoured-nation treatment in international investment law: Ascertaining the limits through interpretative principles. *Amsterdam Law Forum*, *7*(1), 3. https://doi.org/10.37974/alf.272

Titi, C. (2015, August 1). International investment law and the European Union: Towards a new generation of international investment agreements. *European Journal of International Law*, *26*(3), 639–661. https://doi.org/10.1093/ejil/chv040

Vaksha, A. K., & Singh, S. P. (2020). Revisiting Vattel's law of nations to discern the classical principles of the international investment law. *Beijing Law Review*, *11*(01), 277–291. https://doi.org/10.4236/blr.2020.111019

Wasike, S. (2022, May 31). Foreign direct investment and economic growth in Africa. *International Journal of Current Science Research and Review*, *05*(05). https://doi.org/10.47191/ijcsrr/v5-i5-50

Wolfson, J., & Frisken, F. (2000, December). Local response to the global challenge: Comparing local economic development policies in a regional context. *Journal of Urban Affairs*, *22*(4), 361–384. https://doi.org/10.1111/0735-2166.00062

7 Current State of Energy and Emissions in Africa

7.1 BACKGROUND INFORMATION ON AFRICA'S ENERGY AND EMISSIONS

Africa is a continent with diverse energy resources, including fossil fuels, hydropower, and renewable energy sources. Despite this abundance of resources, access to electricity remains limited in many areas, particularly in rural communities. As a result, Africa has one of the lowest per capita energy consumption rates in the world. This is changing as many African countries are investing in renewable energy technologies such as solar and wind power. In addition to improving access to electricity, these efforts are also helping to reduce carbon emissions on the continent. While Africa currently accounts for a relatively small percentage of global greenhouse gas emissions, it is expected that emissions will increase as the continent continues to develop economically. Therefore, it is important for African countries to continue investing in clean energy technologies and adopting sustainable development practices to mitigate the impact of climate change on their populations and ecosystems.

Investing in clean energy technologies and adopting sustainable development practices are crucial steps for African countries to mitigate the impact of climate change on their populations and ecosystems (Okorie and Lin, 2022). As Africa continues to develop economically, it is expected that emissions will increase (Ouedraogo, 2017), making it all the more important to prioritize these efforts. Not only do these initiatives reduce carbon emissions, but they also have the potential to create jobs and stimulate economic growth (Inal et al., 2022). By embracing renewable energy sources such as solar and wind power, African countries can reduce their dependence on fossil fuels and increase energy security (Tyler and Hochstetler, 2021). It is clear that investing in clean energy and sustainable development practices is a win–win for both the environment and the economy in Africa.

According to recent studies, Africa is responsible for only a small fraction of global greenhouse gas emissions, accounting for less than 4% of the world's total (IEA, 2023). This figure is expected to rise as the continent undergoes rapid economic development and urbanization. At the same time, energy consumption in Africa remains low compared to other regions, with many people lacking access to modern forms of energy such as electricity and clean cooking fuels. This has significant

DOI: 10.1201/9781003483175-8

implications for efforts to combat climate change (Oladunni et al., 2022) and promote sustainable development in Africa (Jahanger et al., 2022). A complex picture can be seen in the statistical data on emissions and energy consumption in Africa (Yetano Roche, 2023). On one hand, there are countries that have made significant progress in reducing their carbon footprint and transitioning to renewable energy sources. For example, Morocco has invested heavily in wind and solar power, while Ethiopia has implemented a national reforestation program to offset its emissions. These success stories are overshadowed by the fact that many African countries still rely heavily on fossil fuels and have some of the highest rates of deforestation in the world. Access to electricity remains a major challenge for millions of people on the continent.

On one hand, Africa is responsible for only a small fraction of global greenhouse gas emissions. The continent is also home to some of the fastest-growing economies in the world (Bamisile et al., 2021), which means that energy consumption is increasing rapidly (Ayompe et al., 2020). This presents a challenge for policymakers who must balance economic growth with environmental sustainability. Many African countries are still heavily reliant on fossil fuels for energy production, which contributes to air pollution and other environmental problems. On a positive note, there are many renewable energy resources available in Africa, including solar and wind power. With the right investments and policies in place, these resources could be harnessed to power sustainable development across the continent. It is clear that addressing emissions and energy consumption in Africa will require a multifaceted approach that takes into account economic development (Oluochi et al., 2021), environmental concerns (Calvin et al., 2016), and the unique challenges facing each country on the continent (Ameyaw and Yao, 2018). Despite these challenges, there is reason for optimism as more and more African governments are recognizing the importance of sustainable development and taking steps to address climate change. International support and investment will be crucial in helping Africa achieve its goals for a cleaner, greener future.

To address these challenges, policymakers and stakeholders must work together to promote renewable energy sources, improve energy efficiency, and increase access to modern energy services across the continent. This will require significant investments in infrastructure, technology transfer, and capacity building, as well as strong political leadership and public engagement (Qudrat-Ullah and Nevo, 2021). The goal must be to create a more sustainable and equitable future for all Africans while minimizing the environmental impact of economic growth and development (Jinapor et al., 2023). To achieve this vision, African governments must prioritize the development of renewable energy sources such as solar, wind, and hydropower. This will require investments in infrastructure to support the construction of new power plants (Obiwulu, Chendo, et al., 2020) and transmission lines (Obiwulu et al., 2022). Governments must work to improve energy efficiency by promoting the use of energy-efficient appliances and buildings (Obiwulu, Erusiafe, et al., 2020). This can be achieved through public education campaigns and financial incentives for businesses and households that adopt sustainable practices.

To increase access to modern energy services, governments must focus on expanding electricity grids (Nwokolo, Meyer, et al., 2024) and improving access to clean cooking fuels such as liquefied petroleum gas (LPG) and biogas (Nwokolo,

Echeng, et al., 2024). This will require significant investments in rural electrification programs (Nwokolo and Otse, 2019) and the development of new distribution networks (Ituen et al., 2012). It is crucial for governments to prioritize the development of renewable energy sources such as solar, wind, and hydropower (Nwokolo, 2017). These sources of energy are not only more sustainable but also have the potential to create new job opportunities and stimulate economic growth. Policies should be put in place to encourage the adoption of energy-efficient technologies in both residential and commercial settings (Nwokolo and Ogbulezie, 2018a). This can include incentives for the installation of energy-efficient appliances and building materials (Nwokolo and Ogbulezie, 2018b), as well as regulations that require new buildings to meet certain energy efficiency standards. By taking these steps, governments can work toward a more sustainable future while also improving access to modern energy services for all.

Finally, strong political leadership is essential to drive these efforts forward. Governments must create policies that support renewable energy development and promote sustainable economic growth. They must also engage with the public to build support for these initiatives and ensure that all stakeholders are involved in decision-making processes. In addition to these actions, governments can also invest in research and development of new technologies that can further reduce carbon emissions and increase energy efficiency. This can include advancements in battery storage, smart grid systems, and renewable energy sources such as solar and wind power. By prioritizing these innovations, governments can accelerate the transition to a low-carbon economy and create new opportunities for job growth and economic development.

It is important for governments to collaborate with other countries and international organizations to address global climate challenges (Nwokolo, Proutsos, et al., 2023). This can involve sharing best practices, coordinating efforts on climate mitigation and adaptation (Nwokolo, Proutsos, et al., 2023), and providing financial support to developing countries to help them transition toward sustainable energy systems. In addition to government action, individuals and businesses can also play a crucial role in reducing carbon emissions and promoting sustainability. This can involve adopting energy-efficient practices, investing in renewable energy sources, and supporting environmentally responsible companies. Education and awareness campaigns can help raise public consciousness about the importance of climate action and encourage widespread participation in efforts to combat global warming. Addressing climate change requires a comprehensive approach that involves all sectors of society working together toward a common goal of creating a more sustainable future for generations to come. By taking bold action now, we can pave the way for a cleaner, healthier planet and a more prosperous future for all.

Achieving a more sustainable future requires a collective effort from all sectors of society. Governments must work closely with businesses, civil society organizations, and individuals to build a shared vision for a more sustainable future. By working together toward this common goal, we can create a world that is cleaner, healthier, and more prosperous for all. To achieve this, we need to prioritize renewable energy sources and invest in innovative technologies that reduce our carbon footprint. We must also promote sustainable agriculture practices and protect our oceans and forests from further destruction. We need to shift toward a circular economy model that prioritizes recycling and reducing waste. Education and awareness campaigns can also play a crucial role in encouraging individuals to adopt more sustainable lifestyles. By making these changes now, we can ensure that future generations inherit a planet that is thriving, healthy, and sustainable. It is up to all of us to take bold action today for a better tomorrow.

In conclusion, achieving a more sustainable future is a complex and challenging task that requires a multifaceted approach. It is not enough for governments to simply set targets and regulations; businesses must also play their part by adopting sustainable practices and investing in renewable energy sources. Civil society organizations have an important role to play in advocating for change and holding decision-makers accountable. At the same time, individuals can make a difference by making small changes in their daily lives, such as reducing their energy consumption and waste. Achieving a more sustainable future will require a shift in mindset and values, toward a greater appreciation of the interconnectedness of all living things and the importance of preserving our planet for future generations. By working together toward this common goal, we can create a world that is not only more sustainable but also more just, equitable, and resilient.

The lowest access rate in the world is in sub-Saharan Africa, where less than 20% of households have access to electricity (IEA, 2023). This lack of access to electricity has significant implications for the daily lives of people in the region. Without electricity, many households are unable to power basic appliances such as refrigerators or lights, making it difficult to store food or study after dark. A lack of access to electricity limits economic opportunities for individuals and communities. Businesses are unable to operate at full capacity without reliable power sources, and many people are forced to rely on expensive and environmentally harmful alternatives such as kerosene lamps. The situation is further complicated by the fact that many countries in sub-Saharan Africa are experiencing rapid population growth, which puts additional strain on already overburdened energy systems. Addressing this issue will require significant investment in infrastructure and innovative solutions that can provide affordable and sustainable energy sources for all.

This lack of access to electricity has significant consequences for the region's economic development, as well as the health and well-being of its residents. Without electricity, it is difficult for businesses to operate effectively, limiting job opportunities and economic growth. Lack of access to electricity means that many households rely on dangerous and unhealthy sources of light and heat, such as kerosene lamps or open fires. This can lead to respiratory problems and other health issues. To address this issue, governments and international organizations are working to increase access to electricity in sub-Saharan Africa through a variety of initiatives, including investing in renewable energy sources like solar power and wind turbines. While progress has been made in recent years, there is still a long way to go before all households in the region have access to reliable electricity.

Africa's energy and emissions have been a topic of discussion for many years. The continent has abundant natural resources, including oil, gas, and coal, but these resources have not been fully utilized to meet the energy needs of the population. The lack of access to modern energy services is a major challenge for Africa's development. This is why the background information on Africa's energy and emissions is significant. It provides insights into the current state of energy production and consumption in Africa, as well as the challenges faced by the continent in meeting its energy needs.

One of the main challenges facing Africa is its high dependence on traditional biomass for cooking and heating. This has significant health implications for women and children who are exposed to indoor air pollution from burning wood or charcoal. Africa's electricity generation capacity is low compared to other regions of the world. This limits economic growth and development opportunities for African countries. The lack of access to modern energy services hinders progress in sectors such as education and healthcare. It also affects agriculture and industrial productivity, which are crucial for creating jobs and reducing poverty. To address these challenges, African governments and international partners are investing in renewable energy sources such as solar, wind, and hydropower. This has the potential to not only increase energy access but also reduce greenhouse gas emissions. There are still barriers to scaling up renewable energy deployment, including financing constraints and inadequate regulatory frameworks. Therefore, it is important for stakeholders to work together to overcome these obstacles and accelerate the transition toward a sustainable energy future for Africa.

Another significant issue is Africa's contribution to global greenhouse gas emissions. Although Africa's emissions are relatively low compared to other regions, the continent is still vulnerable to the impacts of climate change. The effects of climate change are already being felt in Africa, with increased frequency and severity of droughts, floods, and other extreme weather events. These events have a significant impact on food security and livelihoods, particularly for rural communities who rely on agriculture. Climate change is exacerbating existing challenges in Africa such as poverty, disease, and conflict. The lack of access to basic resources such as water and energy is hindering development efforts in many parts of the continent. It is therefore crucial that African countries are supported in their efforts to transition to low-carbon economies and adapt to the impacts of climate change. This requires a global effort to provide funding, technology transfer, and capacity building for African nations. It is important that developed countries take responsibility for their historical emissions and provide support for developing countries to mitigate and adapt to climate change. Only through a collaborative effort can we ensure a sustainable future for Africa and the world at large.

To address these challenges, there is a need for investment in renewable energy sources such as solar, wind, and hydropower. This will not only reduce indoor air pollution but also increase access to electricity for households and businesses. There is a need for policies that promote sustainable land-use practices and reduce deforestation rates. This will help to mitigate greenhouse gas emissions while also protecting biodiversity and ecosystem services. It is crucial to promote sustainable transportation systems that reduce reliance on fossil fuels. This can be achieved through the adoption of electric vehicles and the development of public transportation infrastructure. There is a need for increased investment in research and development of clean energy technologies to improve their efficiency and affordability. Education and awareness campaigns can also play a significant role in promoting sustainable lifestyles and encouraging individuals to adopt eco-friendly practices. Finally, international cooperation and collaboration are essential to address global environmental challenges such as climate change. By working together, we can create a more sustainable future for ourselves and future generations.

Finally, it is important to recognize that addressing these challenges requires a collaborative effort between governments, civil society organizations, and the private sector. There is a need for increased funding for research and development of innovative technologies that can help to address these challenges in a sustainable way. Governments must also implement policies and regulations that incentivize businesses to adopt sustainable practices, such as reducing carbon emissions and minimizing waste. Civil society organizations can play a crucial role in raising awareness and advocating for change, while the private sector can drive innovation and investment in sustainable solutions. It is also important to prioritize education and public engagement to empower individuals to make more sustainable choices in their daily lives. By taking a holistic approach that addresses the interconnectedness of environmental, social, and economic issues, we can create a more resilient and equitable world for all.

7.2 OVERVIEW OF ENERGY IN AFRICA

Energy consumption patterns in Africa are unique and complex, with significant variations across different regions and countries. In general, energy consumption in Africa is dominated by traditional biomass fuels such as wood, charcoal, and animal dung, which are used for cooking and heating (Jinapor et al., 2023). There is also a growing demand for modern energy sources such as electricity and fossil fuels, driven by urbanization, industrialization, and economic growth (Udeagha and Ngepah, 2022). This demand is expected to increase rapidly in the coming years as Africa's population continues to grow and its economies develop (Udeagha and Ngepah, 2023). To meet this demand sustainably, African countries will need to invest in renewable energy sources such as solar, wind, hydroelectricity, and geothermal power. They will also need to improve energy efficiency through measures such as building insulation, efficient lighting systems, and better transportation infrastructure.

Finally, they will need to address the issue of energy poverty by providing access to affordable and reliable energy services for all their citizens. Achieving these goals will require strong political leadership, effective policies and regulations, and innovative financing mechanisms to support the transition to a sustainable energy system (Udeagha and Breitenbach, 2023). Governments must work closely with industry and civil society to ensure that the benefits of this transition are shared equitably, and that no one is left behind. This will require a concerted effort to educate the public about the importance of energy conservation and renewable energy sources, as well as to invest in research and development of new technologies. Governments must work together at the international level to address global challenges such as climate change, which requires a coordinated response from all nations. By taking these steps, governments can create a more sustainable future for their citizens and for generations to come.

Africa is a continent that is rich in natural resources, including various forms of energy. The energy mix of Africa includes traditional sources such as biomass, which is used for cooking and heating, as well as hydropower, which is generated from the continent's many rivers and waterfalls. In recent years, Africa has also begun to tap into its vast reserves of fossil fuels, including oil and natural gas. There has been a growing interest in renewable energy sources such as solar and wind power. These forms of energy have the potential to provide reliable and sustainable sources of electricity for communities across the continent. There are still significant challenges to overcome in terms of infrastructure development and investment in order to fully realize Africa's potential as an energy powerhouse. With careful planning and strategic investments, it is possible for Africa to achieve a diverse and sustainable energy mix that meets the needs of its growing population while also protecting the environment for future generations. One of the key challenges facing Africa's energy sector is the lack of infrastructure. Many communities across the continent still lack access to reliable electricity, which limits their economic potential and quality of life. To address this issue, governments and private sector actors must invest in the development of transmission and distribution networks, as well as renewable energy technologies like solar and wind power. There is a need for greater investment in research and development to improve the efficiency and affordability of these technologies.

Another challenge facing Africa's energy sector is financing. Many renewable energy projects require significant upfront investments, which can be difficult to secure without access to capital markets or other sources of funding. To overcome this challenge, governments can work with international financial institutions like the World Bank or African Development Bank to secure loans or grants for energy projects. Private sector actors can also play a role by investing in renewable energy projects or partnering with governments on public–private partnerships. In addition to securing funding, governments can also implement policies and regulations that incentivize the development of renewable energy projects. For example, feed-in tariffs can be implemented to guarantee a fixed price for renewable energy generated by private sector actors. Net metering policies can also be put in place to allow individuals and businesses to sell excess energy back to the grid. Governments can invest in research and development to improve the efficiency and affordability of renewable energy technologies. This can lead to a reduction in costs over time, making renewable energy more accessible for all. Finally, education and awareness campaigns can be launched to promote the benefits of renewable energy and encourage individuals and businesses to make the switch. By taking a multifaceted approach, governments can overcome the challenges associated with financing and promoting renewable energy projects, ultimately leading to a more sustainable future for all.

Despite these challenges, there are many reasons to be optimistic about Africa's energy future. The continent is rich in renewable energy resources, including solar, wind, hydropower, and geothermal power. With the right policies and investments in place, Africa has the potential to become a leader in renewable energy production. This could not only provide access to electricity for millions of people but also create new job opportunities and spur economic growth. Renewable energy can help mitigate the effects of climate change by reducing greenhouse gas emissions. There are still obstacles to overcome, such as inadequate infrastructure and regulatory frameworks that do not incentivize investment in renewable energy. Addressing these challenges will require collaboration between governments, private sector actors, and civil society organizations. By working together toward a common goal of sustainable energy for all, Africa can unlock its full potential and build a brighter future for generations to come.

Energy sources in Africa are diverse, with a mix of traditional and modern sources. Traditional sources such as wood, charcoal, and animal waste still play a significant role in the energy mix, particularly in rural areas. Modern sources such as fossil fuels, hydropower, and renewable energy are becoming increasingly important. Fossil fuels such as oil and gas are major contributors to the energy mix in countries such as Nigeria and Angola. Hydropower is also an important source of energy in countries like Ethiopia and Zambia, where large rivers provide ample opportunities for hydropower generation. Renewable energy sources such as solar, wind, and geothermal are gaining momentum across the continent, with countries like South Africa leading the way in solar power generation. The potential for renewable energy is enormous in Africa, given its abundant natural resources. Significant investment is needed to fully exploit this potential and ensure that all Africans have access to clean and affordable energy. In addition to hydropower and solar energy, wind and geothermal sources also hold great promise for Africa's renewable energy future. Wind turbines can be installed in areas with consistent winds, such as coastal regions, while geothermal power plants can tap into the earth's natural heat to generate electricity. The benefits of renewable energy go beyond just reducing carbon emissions; it can also create jobs and stimulate economic growth. The high upfront costs of building renewable energy infrastructure remain a barrier for many African countries. International partnerships and funding initiatives will be crucial in helping to bridge this gap and unlock Africa's vast potential for clean energy. With the right investments and policies in place, Africa could become a global player in sustainable development and a shining example of how renewable energy can power a brighter future for all.

Africa is a continent blessed with abundant renewable energy sources. The sun is one of the most significant resources, and solar energy is being harnessed through photovoltaic panels and concentrated solar power plants. Wind power is also becoming increasingly popular, with wind turbines being installed in countries such as Egypt, Morocco, and South Africa. Hydropower is another vital source of renewable energy in Africa, with dams such as the Grand Ethiopian Renaissance Dam providing electricity to millions of people. Geothermal energy is also being harnessed in countries such as Kenya and Ethiopia, where hot springs are used to generate electricity. Biomass energy from agricultural waste and forestry residues is also an essential source of renewable energy in Africa. In addition to these sources, wave and tidal power have the potential to provide significant amounts of clean energy along the continent's coastline. With its vast potential for renewable energy, Africa has a unique opportunity to transition toward a more sustainable future while providing access to electricity for its growing population.

Renewable energy is a critical component of Africa's sustainable future. The continent has abundant sources of renewable energy, including solar, wind, hydropower, and biomass. Solar power is particularly promising in Africa due to its high levels of sunshine. Wind power is also gaining traction, with several large-scale wind farms already operational in countries such as Egypt and Morocco. Hydropower from rivers and dams is another significant source of renewable energy in Africa. Biomass energy from agricultural waste and forestry residues is also an essential contributor to the continent's renewable energy mix. Wave and tidal power have the potential to provide significant amounts of clean energy along the coastline. With the right investments and policies in place, Africa can harness its vast potential for renewable energy to provide access to electricity for its growing population while transitioning toward a more sustainable future.

Renewable energy sources are becoming increasingly important in African countries as they seek to reduce their dependence on fossil fuels and mitigate the impacts of climate change. One of the most promising sources of renewable energy in Africa is solar power, which is abundant across the continent and has the potential to provide electricity to millions of people who currently lack access to it. Wind power is also gaining traction in some areas, particularly along the coastlines where there are strong and consistent winds. Hydropower is another important source of renewable energy in Africa, with many countries investing in large-scale dam projects to generate electricity. Biomass energy is widely used throughout Africa for cooking and heating, although there are concerns about its sustainability and impact on local ecosystems (Oladipupo et al., 2022). Despite these challenges, there is growing recognition that renewable energy has the potential to transform Africa's energy landscape and bring about significant social, economic, and environmental benefits for its people.

Renewable energy has been gaining momentum in Africa as many countries have started investing in large-scale dam projects to generate electricity. These projects have the potential to transform the energy landscape of the continent and bring about significant social, economic, and environmental benefits for its people. Biomass energy is also widely used throughout Africa for cooking and heating, but there are concerns about its sustainability and impact on local ecosystems. Despite these challenges, there is growing recognition that renewable energy can provide a sustainable solution to Africa's energy needs. The use of renewable energy sources such as solar, wind, and geothermal power can help reduce the continent's reliance on fossil fuels and mitigate the effects of climate change. Renewable energy can create job opportunities and stimulate economic growth in rural areas where access to electricity is limited. With continued investment in renewable energy infrastructure and technology, Africa has the potential to become a leader in sustainable development and pave the way for a brighter future for its people.

7.3 EMISSIONS IN AFRICA

Africa is a continent that is home to a diverse range of ecosystems and communities. Despite this diversity, one common thread that ties the region together is the impact of greenhouse gas emissions (Khobai and Sithole, 2022). In recent years, there has been growing concern about the effects of these emissions on the environment and human health (Tagwi, 2022). One major source of greenhouse gas emissions in Africa is the burning of fossil fuels for energy production. This includes both coal-fired power plants and oil and gas extraction operations. In addition to these sources, agriculture and deforestation also contribute significantly to greenhouse gas emissions in Africa. These activities release carbon dioxide into the atmosphere, which contributes to global warming and climate change. Reducing greenhouse gas emissions in Africa is crucial for mitigating the impacts of climate change on both the environment and human health (Amowine, 2023).

To achieve this, it is important to shift toward cleaner energy sources such as solar, wind, and hydropower. Improving agricultural practices through the use of sustainable farming techniques and reducing deforestation can also help decrease carbon dioxide emissions. Investing in public transportation systems and promoting walking and cycling can also reduce reliance on fossil fuels for transportation. It is important for governments to implement policies that incentivize these changes and encourage individuals and businesses to adopt more sustainable practices (van der Zwaan et al., 2018). By adopting a multifaceted approach to reducing carbon dioxide emissions, we can make significant progress toward mitigating the effects of climate change. One key area for improvement is the development of renewable energy sources such as solar, wind, and hydropower. By investing in these technologies and expanding their use, we can reduce our reliance on fossil fuels and decrease carbon emissions.

In addition to this, improving agricultural practices through sustainable farming techniques and reducing deforestation can also play a critical role in reducing carbon dioxide emissions. By protecting our forests and promoting sustainable agriculture, we can help preserve our planet's natural resources for future generations. Another important step is to invest in public transportation systems and promote alternative modes of transportation such as walking and cycling. By reducing our reliance on cars and other fossil fuel-powered vehicles, we can significantly decrease carbon emissions from transportation. Finally, it is crucial for governments to implement policies that incentivize these changes and encourage individuals and businesses to adopt more sustainable practices. By taking action now, we can help create a more sustainable future for Africa and the planet as a whole.

Major sources of emissions in African countries include the burning of fossil fuels for energy production, transportation, and industrial processes. Deforestation and land-use changes contribute significantly to greenhouse gas emissions. Agricultural practices such as livestock rearing and rice cultivation also release methane into the atmosphere. In many African countries, inadequate waste management systems lead to significant emissions from landfill sites. The use of inefficient cooking stoves and traditional biomass fuels for cooking and heating is another significant source of emissions. Rapid urbanization in many African countries has led to increased energy demand and emissions from buildings and infrastructure development. Addressing these sources of emissions will require a combination of policy interventions, technological innovations, and behavioral changes across sectors. In order to tackle the significant emissions from landfill sites, it is important to implement effective waste management policies that prioritize recycling and composting.

Incentivizing businesses and individuals to reduce their waste production can also have a positive impact. To address the use of inefficient cooking stoves and traditional biomass fuels, promoting the adoption of clean energy technologies such as solar cookers and biogas systems can help reduce emissions while also improving health outcomes for communities. Increasing access to modern energy services in urban areas through investments in renewable energy infrastructure can help meet growing energy demand while reducing emissions from buildings and transportation. A multi-faceted approach that involves policy interventions, technological innovations, and behavioral changes will be necessary to effectively address Africa's emissions challenges.

The principal emitting countries in Africa are Nigeria, South Africa, and Egypt. These countries are responsible for a significant portion of the continent's greenhouse gas emissions, largely due to their reliance on fossil fuels for energy production. There are also efforts underway in each of these countries to transition toward cleaner energy sources and reduce emissions. Nigeria has set a target of generating 30% of its electricity from renewable sources by 2030, while South Africa has implemented a carbon tax and is investing in renewable energy projects (IEA, 2023). Egypt has also made strides in developing renewable energy, with plans to generate 42% of its electricity from renewable sources by 2035 (IEA, 2023). Despite these efforts, there is still much work to be done to address the impacts of climate change on the African continent, which disproportionately affect vulnerable communities and ecosystems.

Many African countries have taken significant steps toward reducing their carbon footprint and transitioning to renewable energy sources. For instance, Morocco has already achieved its goal of generating 30% of its electricity from renewable sources, while South Africa has implemented a carbon tax and invested in renewable energy projects (IEA, 2023). Egypt is also making progress toward its target of generating 42% of its electricity from renewable sources by 2035 (IEA, 2023). The impacts of climate change on the continent continue to be a pressing issue that disproportionately affects vulnerable communities and ecosystems. There is still much work to be done to address these challenges, but the progress made so far is certainly encouraging. With continued efforts and investments in sustainable development, Africa can pave the way for a greener future for all.

The impact of emissions on the environment and public health in African countries has been a growing concern in recent years. The emissions from industries, transportation, and energy production have contributed to air pollution, which has led to respiratory diseases and other health problems. The emissions have also contributed to climate change, which has resulted in droughts, floods, and other extreme weather conditions. These environmental challenges have affected the livelihoods of many people in African countries who rely on agriculture and natural resources for their survival.

Concern over the effects of emissions on the environment and public health in African nations has risen in recent years. This is due to the rapid industrialization and urbanization that have taken place in many African countries, leading to an increase in emissions from transportation, energy production, and other sources. The impact of these emissions on the environment has been significant, with air pollution levels reaching dangerous levels in many cities. This has led to a rise in respiratory illnesses and other health problems among the population. In response, many African nations have begun to take steps to reduce their emissions and transition toward cleaner energy sources. This includes investing in renewable energy projects such as solar and wind power, as well as implementing policies to encourage more sustainable transportation options. While progress has been made, there is still much work to be done to address the environmental and health challenges facing African nations today. It will require a concerted effort from governments, businesses, and individuals alike to create a more sustainable future for all.

To mitigate these impacts, African countries have implemented various measures, such as promoting renewable energy sources and adopting cleaner production technologies. There is still a need for more concerted efforts from governments and stakeholders to address the issue of emissions and their impact on public health and the environment. In addition to Africa, other regions of the world have also taken steps toward sustainability. Europe has set ambitious targets for reducing greenhouse gas emissions and increasing renewable energy usage. In Asia, countries like China and India have made significant investments in renewable energy and electric vehicles. There is still a long way to go before we can achieve a truly sustainable future. It will require not only technological advancements but also changes in behavior and lifestyle choices. Individuals can play a role by reducing their carbon footprint through actions such as using public transportation or eating less meat. Businesses can also contribute by adopting sustainable practices and investing in clean technologies. Ultimately, it will take a global effort to address the challenges of climate change and create a more sustainable future for all nations.

7.4 GOVERNMENT ENERGY POLICIES AND INITIATIVES

Government energy policies and initiatives in African countries have been gaining momentum in recent years. With a growing population and increasing demand for energy, many African nations are looking to diversify their energy sources and reduce their reliance on fossil fuels. One such initiative is the African Renewable Energy Initiative (AREI), which aims to increase renewable energy capacity by 10 GW by 2020 and 300 GW by 2030 (IEA, 2023). Another important policy is the African Energy Leaders Group (AELG, 2015), which brings together government officials, industry leaders, and other stakeholders to promote sustainable energy development across the continent. In addition to these policy initiatives, many African countries are also exploring alternative energy sources such as solar, wind, and hydropower. For example, Morocco has built one of the world's largest solar power plants in the Sahara Desert, while Kenya has invested heavily in geothermal energy (IEA, 2023). These efforts have not only helped to reduce carbon emissions and combat climate change but have also created new job opportunities and improved access to electricity for millions of people across the continent. There are still significant challenges to overcome in terms of financing and infrastructure development. To address these issues, many African nations are partnering with international organizations such as the World Bank and the United Nations Development Programme (UNDP) to secure funding and technical support for their renewable energy projects. With continued investment and innovation, Africa has the potential to become a global leader in sustainable energy production and help build a more sustainable future for all.

In addition to these overarching initiatives, many individual countries are implementing their own policies to support renewable energy development. For example, Morocco has launched a plan to generate 42% of its electricity from renewable sources by 2020, while South Africa has set a target of adding 18 GW of renewable energy capacity by 2030 (IEA, 2023). Kenya has also made significant strides in this area, with over half of its electricity now coming from renewable sources such as geothermal and wind power. In Europe, Germany has been a leader in renewable energy adoption, with its Energiewende policy aiming to transition the country to 80% renewable energy by 2050 (IEA, 2022). The United States has also seen growth in renewable energy, with California leading the way by setting a goal of achieving 100% clean energy by 2045 (IEA, 2022). Despite these efforts, challenges remain in the widespread adoption of renewable energy, including the high initial costs and intermittency issues associated with some sources such as solar and wind power. Continued investment and innovation in this field hold promise for a future powered by sustainable and environmentally friendly sources of energy.

7.4.1 Overview of Energy Policies in Africa

Energy policies in Africa have been a major focus for many governments and organizations over the past few decades. With a growing population and increasing demand for energy, many African countries have been working to develop sustainable and reliable sources of power. One of the main challenges facing these countries has been the lack of infrastructure and investment in the energy sector. Many African nations are still heavily reliant on fossil fuels, which can be expensive and environmentally damaging. In recent years, there has been a push toward renewable energy sources such as solar, wind, and hydropower. Governments have also been working to improve access to electricity in rural areas through initiatives such as mini-grids and off-grid solutions. These efforts have been met with some success, but there is still a long way to go. One of the biggest obstacles to widespread adoption of renewable energy is the high upfront costs associated with building new infrastructure. Many countries lack the technical expertise and skilled labor necessary to effectively implement these projects. Another challenge is ensuring that energy access is equitable and affordable for all citizens, particularly those living in poverty. Despite these challenges, there are reasons for optimism. The cost of renewable energy technologies has been steadily decreasing, making them more accessible to developing nations. International organizations such as the World Bank and United Nations are also providing funding and technical assistance to support these initiatives. With continued investment and collaboration between governments, private sector actors, and civil society organizations, it is possible for African nations to achieve a sustainable energy future that benefits both people and the planet.

A brief history of African energy policies: Africa has a long and complex history when it comes to energy policies. In the early 20th century, colonial powers controlled most of the continent's natural resources, including oil and gas reserves (IEA, 2023). As a result, energy policies were largely determined by foreign interests, with little regard for local needs or sustainability. This trend continued well into the post-colonial era, as many African governments focused on developing their economies through resource extraction rather than investing in renewable energy sources. In recent years, there has been a shift toward more sustainable energy policies in Africa. Many countries are investing in renewable energy sources such as solar and wind power, recognizing the long-term benefits of reducing dependence on finite resources. There is a growing awareness of the need to address climate change and its impact on the continent. This has led to increased collaboration between African nations and international organizations to develop clean energy solutions. Despite these positive developments, challenges remain, including limited access to financing and technological expertise. With continued commitment and investment, Africa has the potential to become a leader in sustainable energy production and contribute significantly to global efforts to combat climate change. In recent years, there has been a growing recognition of the need for more sustainable energy policies in Africa. Many countries have started to invest in renewable energy sources such as solar and wind power, recognizing that these technologies offer a more affordable and environmentally friendly alternative to fossil fuels. Governments have also begun to prioritize access to electricity for all citizens, recognizing that lack of access to reliable power is a major barrier to economic development. As Africa continues to develop, the importance of sustainable energy policies cannot be overstated. The shift toward renewable energy sources such as solar and wind power is a promising step toward reducing the continent's reliance on fossil fuels. Not only are these technologies more affordable in the long run, but they also have a significantly lower impact on the environment. Prioritizing access to electricity for all citizens is crucial for economic growth and development. Lack of reliable power has long been a major barrier to progress in many African countries, and addressing this issue will be key to unlocking their full potential. By investing in sustainable energy policies and expanding access to electricity, Africa can build a brighter future for its people while also contributing to global efforts to combat climate change.

7.4.2 Government Initiatives to Promote Renewable Energy

Government initiatives to promote renewable energy in African countries have been gaining momentum in recent years. One such initiative is the Africa Renewable Energy Initiative, which was launched in 2015 with the goal of adding 300 GW of renewable energy capacity to the continent by 2030 (IEA, 2023). The initiative has already secured commitments from several African countries and international organizations, and it is expected to play a key role in promoting sustainable development and reducing greenhouse gas emissions on the continent. In addition to the Africa Renewable Energy Initiative, there are also other efforts underway to promote renewable energy in African countries. For example, some countries are investing in solar power projects, such as the Noor Ouarzazate Solar Complex in Morocco, which is one of the largest solar power plants in the world. Other countries are exploring wind power and hydropower as alternative sources of energy. These initiatives not only help to reduce carbon emissions and combat climate change, but they also have the potential to create jobs and stimulate economic growth in these regions. Despite the progress that has been made, there are still many challenges that need to be addressed in order to fully realize the potential of renewable energy in Africa. These include issues related to financing, infrastructure development, and regulatory frameworks. With continued investment and collaboration between governments, international organizations, and private sector partners, there is reason to be optimistic about the future of renewable energy on the continent.

In addition to the Africa Renewable Energy Initiative, many African countries have also implemented their own renewable energy policies and programs. For example, South Africa has set a target of generating 18 GW of renewable energy by 2030, while Morocco has invested heavily in solar power and aims to generate 52% of its electricity from renewables by 2030 (IEA, 2023). Other countries, such as Kenya, Ethiopia, and Rwanda, have also made significant strides in promoting renewable energy through policies such as feed-in tariffs and tax incentives for renewable energy projects. These efforts have not gone unnoticed, as international organizations such as the World Bank and the United Nations have praised these countries for their commitment to sustainable energy. In addition to reducing carbon emissions and combating climate change, investing in renewable energy has also created jobs and stimulated economic growth in these countries. There are still challenges that need to be addressed, such as the high upfront costs of renewable energy infrastructure and the need for reliable storage solutions. Nevertheless, the progress made by these countries serves as a model for others to follow in transitioning toward a more sustainable future. As more countries recognize the benefits of renewable energy and take action toward achieving their own targets, we can hope to see a significant reduction in global greenhouse gas emissions and a brighter future for generations to come.

Despite these efforts, there are still significant challenges to be overcome in promoting equality and inclusivity in society. One of the biggest obstacles is the persistence of systemic discrimination, which can manifest in various forms such as racism, sexism, ableism, and homophobia. Another challenge is the lack of representation and diversity in leadership positions across industries and institutions. This not only perpetuates existing power imbalances but also limits opportunities for marginalized groups to have their voices heard and their needs addressed. There is a need for greater education and awareness-raising around issues of diversity and inclusion, as many people still hold biased beliefs or are unaware of the impact of their actions on others. Finally, it is important to recognize that achieving true equality requires ongoing effort and commitment from individuals, organizations, and governments alike. It will take a collective effort to dismantle oppressive systems and create a more just and equitable world for all.

7.4.3 Effectiveness of Policies and Initiatives in Reducing Emissions and Increasing Access to Energy

The effectiveness of policies and initiatives in reducing emissions and increasing access to energy in African countries has been a topic of much debate and analysis in recent years. Many countries on the continent are facing significant challenges in meeting their energy needs while also addressing the urgent need to reduce greenhouse gas emissions. In response, a number of policies and initiatives have been implemented with varying degrees of success. One promising approach has been the promotion of renewable energy sources, such as solar and wind power. This has been particularly successful in countries with high levels of sunshine or wind, such as Morocco and Kenya. Many African countries have implemented energy efficiency measures to reduce overall energy consumption and improve access to electricity. Challenges remain in terms of financing and implementation, particularly in rural areas where access to electricity remains limited. To address these challenges, some governments have partnered with private sector companies to develop innovative financing models and increase investment in renewable energy projects. While progress has been made in reducing emissions and increasing access to energy in African countries, much more work is needed to ensure sustainable development for all.

One key policy that has been implemented across many African countries is the promotion of renewable energy sources such as solar, wind, and hydropower. This has been achieved through a range of measures, including tax incentives, subsidies, and feed-in tariffs for renewable energy producers. These policies have helped to increase investment in renewable energy infrastructure and have led to significant increases in the use of these technologies across the continent. In addition to promoting renewable energy sources, African countries have also focused on improving access to electricity for all citizens. This has been achieved through the expansion of national electricity grids, as well as the development of off-grid solutions such as mini-grids and standalone solar systems. These efforts have been particularly successful in rural areas, where access to electricity has traditionally been limited. Alongside these initiatives, African governments have also prioritized the development of energy-efficient technologies and practices, with a particular focus on the industrial sector. This has included measures such as energy audits and the implementation of energy management systems in factories and other industrial facilities. These policies and initiatives have helped to drive sustainable development across Africa, creating new opportunities for businesses and individuals alike while reducing the continent's reliance on fossil fuels.

Another important initiative that has been implemented in many African countries is the development of energy efficiency programs. These programs aim to reduce energy consumption by promoting more efficient use of electricity and other forms of energy. This has been achieved through a range of measures, including public awareness campaigns, education programs, and building codes that require more efficient building designs. Many African countries are also investing in renewable energy sources such as solar, wind, and hydropower. These initiatives have not only helped to reduce the continent's carbon footprint but have also created new job opportunities and stimulated economic growth. The development of clean energy infrastructure has improved access to electricity in rural areas where traditional power grids are not feasible. This has had a positive impact on the quality of life for millions of people across Africa. As the continent continues to grow and develop, it is clear that sustainable energy solutions will play a crucial role in shaping its future. With continued investment and innovation, Africa has the potential to become a leader in clean energy production and consumption, setting an example for other regions around the world to follow.

7.4.4 Case Study of Energy Policies in Nigeria

Nigeria is a country with a population of over 200 million people, and it is one of the largest oil-producing countries in the world (IEA, 2023). Despite this, Nigeria has struggled to provide reliable and affordable energy to its citizens. The country has faced numerous challenges in its energy sector, including inadequate infrastructure, corruption, and political instability. In recent years, the Nigerian government has implemented several policies aimed at addressing these challenges and improving the country's energy sector.

One of the key policies implemented by the Nigerian government is the National Renewable Energy and Energy Efficiency Policy (NREEEP). This policy aims to increase the use of renewable energy sources such as solar, wind, and hydropower, while also promoting energy efficiency measures. The NREEEP has been successful in attracting investment in renewable energy projects and has helped to diversify Nigeria's energy mix. As a result, Nigeria has seen an increase in the number of renewable energy projects being developed across the country. The government has also taken steps to improve access to electricity in rural areas through initiatives such as the Rural Electrification Agency (REA). The REA aims to provide electricity to underserved communities through off-grid and mini-grid solutions. The Nigerian government has partnered with international organizations such as the World Bank and African Development Bank to fund energy projects and promote sustainable development.

Despite the efforts made by the Nigerian government and international organizations to fund energy projects and promote sustainable development, there are still significant challenges facing the country's energy sector. One of the most pressing issues is the lack of infrastructure for renewable energy projects. Without adequate transmission lines and storage facilities, it is difficult to harness the potential of renewable energy sources. The high cost of renewable energy technologies poses a significant obstacle to investors looking to finance these projects. While progress has been made in recent years, much work remains to be done in order to fully realize Nigeria's potential as a leader in sustainable energy development. Addressing these challenges will require a concerted effort from all stakeholders, including government officials, private investors, and international organizations. By working together, we can create a more sustainable future for Nigeria and for the world as a whole.

To address these challenges, the Nigerian government is working on improving infrastructure for renewable energy projects and reducing the cost of renewable energy technologies. This includes partnering with private sector companies to develop renewable energy solutions that are affordable and accessible to all Nigerians. The government is investing in research and development to find innovative ways to generate clean energy and reduce greenhouse gas emissions. Private investors can also play a crucial role in driving sustainable development in Nigeria by investing in renewable energy projects and supporting local communities. International organizations can provide technical assistance, funding, and expertise to help Nigeria achieve its sustainable development goals. Together, we can create a brighter future for Nigeria that is powered by clean, renewable energy sources and benefits everyone in the country. By taking action now, we can ensure that future generations have access to a healthy environment and thriving economy.

Another important policy is the Power Sector Recovery Program (PSRP), which was launched in 2017. The PSRP aims to address some of the structural issues facing Nigeria's power sector. The program focuses on improving the financial viability of the sector, increasing power generation and transmission capacity, and enhancing the regulatory environment. The PSRP has several components, including the Nigerian Electricity Market Stabilization Facility (NEMSF), which provides financial support to market participants, and the Meter Asset Provider (MAP) regulation, which aims to improve metering in the sector. The PSRP includes initiatives to increase access to electricity in rural areas through off-grid solutions such as mini-grids and solar home systems. The program has already made some progress in addressing Nigeria's power challenges, with increased power generation and improved revenue collection by distribution companies. There is still a long way to go before Nigeria can achieve its goal of providing reliable and affordable electricity to all its citizens. To sustain progress in the sector, it will be crucial for the government to continue implementing policies that address structural issues while also promoting private sector investment in power infrastructure.

One key area that requires attention is the transmission network, which is currently plagued by inadequate capacity and frequent breakdowns. The government must prioritize investments in upgrading and expanding the transmission infrastructure to ensure that power generated is efficiently distributed across the country. There is a need for increased investment in renewable energy sources such as solar and wind power, which have the potential to significantly increase Nigeria's power generation capacity while also reducing its carbon footprint. To attract private sector investment in the sector, the government must create an enabling environment by addressing issues such as regulatory uncertainty, corruption, and bureaucratic red tape. Finally, there is a need for increased public awareness and education on energy conservation and efficiency measures to reduce wastage and promote sustainable consumption patterns. With sustained efforts in these areas, Nigeria can achieve its goal of providing reliable and affordable electricity to all its citizens, thereby unlocking the full potential of its economy and improving the quality of life for its people.

Achieving this goal will require significant investment in the country's power infrastructure, including the expansion of transmission and distribution networks and the development of renewable energy sources. Addressing the root causes of Nigeria's electricity challenges will require tackling systemic issues such as regulatory uncertainty, corruption, and bureaucratic red tape. To overcome these obstacles, policymakers must work to create a more transparent and predictable regulatory environment that encourages private sector investment in the energy sector. Efforts to combat corruption and streamline bureaucratic processes will be critical to improving the efficiency and effectiveness of Nigeria's energy sector. Finally, increasing public awareness and education on energy conservation and efficiency measures will be essential to reducing wastage and promoting sustainable consumption patterns. By addressing these challenges head-on, Nigeria can unlock the full potential of its economy and improve the quality of life for its people by providing reliable and affordable electricity to all citizens.

7.4.5 Case Study of Energy Policies in South Africa

South Africa has been grappling with energy challenges for decades. The country is heavily reliant on coal-fired power plants, which have been responsible for significant greenhouse gas emissions and air pollution. In recent years, the government has made efforts to diversify its energy mix and reduce its carbon footprint. One of the key initiatives has been the Renewable Energy Independent Power Producer Procurement Program (REIPPPP), which aims to increase the share of renewable energy in the country's electricity generation. The program has been successful in attracting private investment in wind, solar, and other renewable energy projects. Another important policy is the Integrated Resource Plan (IRP), which sets out a long-term strategy for South Africa's energy sector. The latest version of the IRP includes a significant increase in renewable energy capacity, as well as plans to decommission some of the older coal-fired power plants. There are also concerns about the impact of these policies on jobs in the coal industry and on electricity prices for consumers.

Despite concerns about the impact of renewable energy policies on jobs in the coal industry and electricity prices for consumers, South Africa's government is committed to increasing its renewable energy capacity. The country has made significant progress in wind, solar, and other renewable energy projects, thanks in part to policies like the REIPPPP. The IRP sets out a long-term strategy for the energy sector that includes a significant increase in renewable energy capacity and plans to decommission some of the older coal-fired power plants. While there may be short-term challenges, such as job losses in the coal industry, the long-term benefits of transitioning to cleaner sources of energy are clear.

Renewable energy is becoming an increasingly important part of the global energy mix, and South Africa is no exception. The country has made significant progress in recent years, thanks in part to policies like the REIPPPP. This program has encouraged investment in renewable energy projects, which has led to a significant increase in renewable energy capacity. The IRP sets out a long-term strategy for the energy sector that includes plans to decommission some of the older coal-fired power plants and a significant increase in renewable energy capacity. While there may be short-term challenges, such as job losses in the coal industry, the long-term benefits of transitioning to cleaner sources of energy are clear. Not only will this help reduce greenhouse gas emissions and combat climate change, but it will also create new job opportunities and stimulate economic growth. South Africa's commitment to renewable energy is a positive step toward a more sustainable future.

The shift toward renewable energy in South Africa has been gaining momentum, thanks to the implementation of policies like the REIPPPP. This program has encouraged investment in renewable energy projects, resulting in a significant increase in the country's renewable energy capacity. The IRP provides a long-term strategy for the energy sector, with plans to decommission older coal-fired power plants and replace them with cleaner sources of energy. While this transition may result in short-term challenges, such as job losses in the coal industry, the long-term benefits are undeniable. Not only will it reduce carbon emissions and improve air quality, but it will also create new job opportunities in the renewable energy sector. As South Africa continues to move toward a more sustainable future, it is important to remember that this transition is not only necessary but also beneficial for both the environment and the economy.

The shift toward renewable energy is a crucial step in mitigating the effects of climate change. With the global demand for energy increasing, it is essential to adopt sustainable practices that reduce our reliance on fossil fuels. Renewable energy sources such as solar, wind, and hydropower offer a cleaner and more efficient alternative to traditional energy sources. This transition not only benefits the environment but also creates new job opportunities in the renewable energy sector. The development of new technologies and infrastructure will require skilled workers, which can help boost local economies. While there may be some short-term challenges, such as job losses in the coal industry, it is important to recognize that this transition is necessary for our long-term well-being. As South Africa continues to move toward a more sustainable future, we must prioritize investing in renewable energy and supporting the growth of this industry. By doing so, we can ensure a cleaner and brighter future for generations to come.

7.4.6 Case Study of Energy Policies in Kenya

Kenya has been implementing various energy policies aimed at promoting sustainable development and addressing the country's energy challenges. One of the notable policies is the Kenya National Electrification Strategy, which seeks to increase access to electricity in rural areas. The strategy involves a mix of grid extension, mini-grids, and off-grid solutions such as solar home systems. Another policy is the Feed-in Tariff policy, which provides incentives for renewable energy producers by guaranteeing them a fixed price for their electricity. This policy has led to an increase in renewable energy investments in Kenya, particularly in wind and solar power projects. The Energy Act of 2019 established the Energy and Petroleum Regulatory Authority (EPRA) to regulate the energy sector and promote competition. The government has also launched initiatives such as the Last Mile Connectivity Project and the Scaling Solar Program to enhance access to affordable electricity and attract private sector investment in the energy sector.

As a result of these policies and initiatives, Kenya has made significant progress in its transition toward renewable energy. The country's installed capacity for renewable energy has increased by over 500% since 2010, with wind and solar power accounting for the majority of this growth (IRENA, 2022). This has not only helped to reduce the country's dependence on fossil fuels but has also contributed to the mitigation of climate change. The Last Mile Connectivity Project, which aims to connect households in rural areas to the national grid, has also had a positive impact on poverty reduction and economic development. Private sector investment in the energy sector has also increased, with companies such as Acorn Holdings and African Infrastructure Investment Managers (AIIM) investing in renewable energy projects in Kenya. Challenges remain, including inadequate transmission infrastructure and limited financing options for small-scale renewable energy projects. Nevertheless, Kenya's commitment to renewable energy is commendable and serves as an example for other countries looking to transition toward sustainable energy sources.

Despite these efforts, challenges such as inadequate funding, weak institutional capacity, and limited public awareness of energy efficiency and conservation still exist. In order to overcome these obstacles, Kenya must continue to invest in improving its transmission infrastructure and expanding financing options for small-scale renewable energy projects. Increasing public awareness of the benefits of energy efficiency and conservation will be crucial to promoting a culture of sustainability. By addressing these challenges, Kenya can further solidify its position as a leader in renewable energy and pave the way for a cleaner, more sustainable future for all.

In order to achieve its ambitious renewable energy goals, Kenya must take bold steps toward modernizing its power grid and diversifying its energy mix. This will require significant investment in transmission infrastructure and the development of new financing models to support small-scale renewable energy projects. At the same time, efforts to raise public awareness about the importance of energy efficiency and conservation must be intensified, as these are critical components of a sustainable energy future. By taking these steps, Kenya can build on its impressive progress in renewable energy and set an example for other nations seeking to transition to a cleaner, more sustainable energy system. Ultimately, the success of Kenya's renewable energy efforts will depend on the commitment and collaboration of all stakeholders, from government officials and industry leaders to individual citizens and communities across the country.

Kenya has made significant strides in transitioning to a more sustainable energy system, with a focus on renewable energy sources such as geothermal, wind, and solar power. To continue this progress, it is crucial for the government to invest in research and development of new technologies that can further increase the efficiency and affordability of renewable energy. Promoting public awareness and education on the benefits of renewable energy can encourage individual citizens to adopt sustainable practices in their daily lives. Collaboration between government officials, industry leaders, and local communities is also essential for creating a comprehensive and effective renewable energy strategy that addresses the unique needs and challenges of different regions within Kenya. By prioritizing sustainability and investing in innovative solutions, Kenya can become a leader in renewable energy and inspire other nations to follow suit.

To achieve a sustainable future, it is crucial that individuals take responsibility for their daily actions. This includes adopting practices such as reducing energy consumption, recycling, and using public transportation. It is equally important for the government to collaborate with industry leaders and local communities to develop a comprehensive renewable energy strategy that addresses the unique needs of different regions within Kenya. By investing in innovative solutions and prioritizing sustainability, Kenya can become a leader in renewable energy and inspire other nations to follow suit.

7.4.7 Case Study of Energy Policies in Morocco

Morocco is a country that has been proactive in its energy policies. The country has set ambitious targets to increase the share of renewable energy in its total energy mix. Morocco aims to generate 52% of its electricity from renewable sources by 2030 (IEA, 2023). To achieve this target, the country has invested heavily in wind and solar power projects. The Noor Ouarzazate Solar Complex, which is one of the largest solar power plants in the world, is a testament to Morocco's commitment to renewable energy. Morocco has also implemented policies to promote energy efficiency and reduce energy consumption. The country has introduced building codes that require new construction to meet energy efficiency standards. The government has launched programs to promote the use of energy-efficient appliances and lighting. These efforts have not only reduced the country's carbon footprint but also helped to lower energy costs for households and businesses.

Morocco's commitment to renewable energy has also created new opportunities for economic growth and job creation. The development of wind and solar power projects has attracted significant foreign investment, which has helped stimulate the country's economy. The expansion of renewable energy infrastructure has created new jobs in areas such as construction, engineering, and maintenance. In addition to the economic benefits, Morocco's transition to renewable energy has also had positive environmental impacts. The reduction in carbon emissions has helped to mitigate the effects of climate change and improve air quality. The country's investment in renewable energy has also set an example for other countries in the region and around the world. Morocco's success in this area demonstrates that it is possible to achieve sustainable development while reducing reliance on fossil fuels. As the world continues to face the challenges of climate change, Morocco's commitment to renewable energy serves as a model for others to follow. By continuing to invest in clean energy technologies, Morocco can help lead the way toward a more sustainable future for all.

Morocco's proactive approach to energy policy has positioned the country as a leader in renewable energy. Through ambitious targets, significant investments, and innovative policies, Morocco is making great strides toward a more sustainable future. As the world continues to grapple with climate change and its consequences, Morocco serves as an example of what can be achieved through determination and commitment. By embracing renewable energy sources, Morocco is not only reducing its carbon footprint but also creating new economic opportunities and improving the lives of its citizens. As the country continues to expand its renewable energy capacity, it is paving the way for other nations to follow suit. With a growing global demand for clean energy, Morocco's leadership in this field is more important than ever. By continuing to prioritize sustainability and invest in clean energy technologies, Morocco can help lead the way toward a more sustainable future for all. The world needs more countries like Morocco that are willing to take bold action to address climate change and build a better future for generations to come.

In addition to investing in renewable energy, Morocco has also implemented policies to promote energy efficiency. The country has introduced regulations that require new buildings to be constructed with energy-efficient materials and appliances. The government has launched campaigns to raise awareness about the importance of energy conservation. These efforts have paid off, as Morocco has made significant progress in reducing its carbon footprint and increasing its use of renewable energy sources. In fact, the country has set a goal of generating 52% of its electricity from renewable sources by 2030 (IEA, 2023). To achieve this target, Morocco is investing heavily in wind and solar power projects, as well as exploring other forms of renewable energy such as hydropower and geothermal power. The country's commitment to sustainability has also attracted international attention, with many countries looking to Morocco as a model for how to transition to a low-carbon economy. As the world faces the challenges of climate change, Morocco's leadership in renewable energy and energy efficiency serves as an inspiring example for others to follow.

Morocco's efforts have not gone unnoticed. The International Energy Agency (IEA) has praised Morocco for its progress toward achieving its renewable energy targets. The IEA has also noted that Morocco's policies could serve as a blueprint for other countries looking to transition to a low-carbon economy (IEA, 2023). Morocco's success in renewable energy is due in part to its ambitious goals and strategic investments in solar and wind power. The country's Noor Ouarzazate Solar Complex, for example, is one of the largest solar power plants in the world, providing clean energy to over 1 million people. Morocco has also made significant strides in energy efficiency, implementing programs to reduce energy consumption in buildings and promoting the use of electric vehicles. As a result of these efforts, Morocco has reduced its dependence on fossil fuels and is on track to meet its goal of generating 52% of its electricity from renewable sources by 2030 (IEA, 2023). With continued investment and commitment to sustainability, Morocco's example can inspire other nations to take bold action toward a more sustainable future.

7.4.8 Case Study of Energy Policies in Egypt

Egypt has been implementing various energy policies to meet the growing demand for electricity and diversify its energy mix. The country has been focusing on renewable energy sources, such as wind and solar power, to reduce its reliance on fossil fuels. In 2014, Egypt launched a feed-in tariff program to encourage private investment in renewable energy projects. As a result, the country has seen a significant increase in the installation of solar and wind power plants. There are still challenges that need to be addressed, such as the lack of infrastructure and financing for renewable energy projects. Egypt has been exploring nuclear power as an alternative source of energy. In 2018, the country signed an agreement with Russia to build its first nuclear power plant. This move has sparked concerns about safety and security issues, but proponents argue that it will help Egypt meet its growing energy needs while reducing greenhouse gas emissions.

Egypt's energy policies reflect a commitment to sustainability and innovation in the face of increasing demand for electricity and the need to reduce carbon emissions.

The country has made significant strides in developing renewable energy sources, such as wind and solar power, and has set ambitious targets for their expansion. Egypt has taken steps to improve energy efficiency and reduce waste, including implementing building codes that require new construction to meet certain energy standards. The decision to pursue nuclear power has raised questions about the potential risks and challenges associated with this technology. While proponents argue that it will provide a reliable source of electricity and help reduce dependence on fossil fuels, opponents point to the potential dangers of accidents or nuclear proliferation. Despite these concerns, Egypt's government remains committed to pursuing a diverse mix of energy sources in order to meet its growing needs while minimizing its impact on the environment.

In addition to nuclear power, Egypt is also investing in renewable energy sources such as wind and solar power. The country has set a target of generating 20% of its electricity from renewable sources by 2022 (IEA, 2023). This move toward clean energy is not only beneficial for the environment but also helps to reduce the country's reliance on imported fossil fuels. The transition to a more sustainable energy mix is not without its challenges. The high cost of renewable energy infrastructure and the need for grid modernization are just some of the obstacles that must be overcome. Nevertheless, Egypt's commitment to pursuing a diverse range of energy sources is a positive step toward a more sustainable future.

Despite the progress made in renewable energy, Egypt still faces obstacles in terms of infrastructure and financing. The lack of resources and investment in this area has hindered the country's ability to fully transition to clean energy. As a result, Egypt has turned to nuclear power as a potential solution. This decision has raised concerns about safety and security issues, given the potential risks associated with nuclear energy. Despite these concerns, proponents argue that nuclear power could help Egypt meet its growing energy demands while reducing greenhouse gas emissions. Ultimately, the country will need to carefully weigh the benefits and risks of different energy sources as it works toward a more sustainable future.

In addition to nuclear power, Egypt has also been exploring renewable energy sources such as wind and solar power. These sources have the advantage of being clean and sustainable, but they also require significant investment in infrastructure and technology. Another option for Egypt is to focus on energy efficiency measures, such as improving building insulation and upgrading transportation systems. These measures can help reduce energy demand and lower costs in the long run. Ultimately, a combination of these approaches may be necessary to meet Egypt's energy needs while minimizing environmental impact and ensuring safety and security. It will be important for the government to engage with stakeholders and the public in a transparent manner to build trust and support for its energy policies.

7.5 PRIVATE SECTOR INVESTMENT IN THE CURRENT STATE OF ENERGY AND EMISSIONS IN AFRICA

The role of private sector investment in the current state of energy and emissions in Africa is crucial. With the increasing demand for energy and the need to reduce emissions, private sector investment has become a key driver of sustainable development on the continent. Private companies have the resources and expertise to invest in renewable energy projects, such as solar and wind power, which can help reduce reliance on fossil fuels and mitigate climate change. Private sector investment can help improve access to energy in rural areas where traditional grid infrastructure is lacking. This can have a significant impact on poverty reduction and economic growth. Private sector investment in the current state of energy and emissions in Africa includes exploring innovative financing models that can incentivize private sector investment in renewable energy projects. One such model is the use of green bonds, which are debt securities that fund environmentally friendly projects. Another approach is to establish public–private partnerships that can leverage the strengths of both sectors to achieve common goals. Governments can provide policy incentives and regulatory frameworks that encourage private sector investment in renewable energy. This includes feed-in tariffs, tax credits, and other financial incentives that can make renewable energy projects more attractive to investors. Ultimately, private sector investment in renewable energy is crucial for achieving sustainable development goals and addressing the pressing challenges of climate change and energy poverty in Africa.

Renewable energy has the potential to transform Africa's energy landscape, providing clean and affordable power to millions of people. To achieve this, it is important for governments to set ambitious renewable energy targets and create an enabling environment for private sector investment. This can be done through policy incentives such as feed-in tariffs and tax credits, which can make renewable energy projects more attractive to investors. Regulatory frameworks that promote the development of renewable energy infrastructure are essential. Private sector investment in renewable energy is crucial for achieving sustainable development goals and addressing the challenges of climate change and energy poverty in Africa. With the right policies and incentives in place, Africa can harness its abundant renewable energy resources to power economic growth and improve the lives of its people.

What is the efficacy of private sector investment in the current state of energy and emissions in Africa? Private sector investment in renewable energy is proving to be an effective solution to the current state of energy and emissions in Africa. With the continent's vast renewable energy resources, such as solar, wind, and hydropower, private investors are recognizing the potential for profitable ventures while also contributing to sustainable development goals. Governments are implementing policies and incentives that encourage private sector investment in renewable energy, resulting in increased access to electricity for rural communities and reduced reliance on fossil fuels. This shift toward renewable energy is also addressing the challenges of climate change and energy poverty in Africa. By investing in clean energy solutions, the private sector is helping to reduce greenhouse gas emissions and improve air quality, leading to a healthier environment for all. The effectiveness of private sector investment in renewable energy is clear; it has the potential to power economic growth while also contributing to a sustainable future for Africa.

The shift toward renewable energy is not only benefiting the environment but also creating new job opportunities in the region. As more and more renewable energy projects are developed, there is a growing demand for skilled workers in various fields such as engineering, construction, and maintenance. This is particularly important for Africa, where unemployment rates are high and many people lack access to basic services like electricity. By investing in renewable energy, the private sector is not only contributing to economic growth but also improving the quality of life for millions of people across the continent. The benefits of renewable energy extend beyond job creation and access to electricity. It also helps to mitigate climate change by reducing greenhouse gas emissions and promoting sustainable development. This is crucial for Africa, which is particularly vulnerable to the impacts of climate change such as droughts, floods, and food insecurity. Renewable energy can provide energy security by reducing dependence on imported fossil fuels and increasing local energy production. This can have positive economic effects by reducing trade deficits and increasing energy independence. Investing in renewable energy in Africa has the potential to create a more sustainable future for both the continent and the planet as a whole.

In addition to job creation, renewable energy also has the potential to improve energy security in Africa. Unlike fossil fuels, which are often imported from other countries, renewable energy sources like solar and wind power can be harnessed locally, reducing dependence on foreign sources of energy. This can help stabilize energy prices and reduce vulnerability to supply disruptions caused by political or economic factors. The use of renewable energy can also have a positive impact on the environment. By reducing reliance on fossil fuels, which contribute to air and water pollution, renewable energy sources can help mitigate the negative effects of climate change. This is particularly important in Africa, where many communities are already facing the devastating consequences of droughts, floods, and other extreme weather events. Investing in renewable energy infrastructure can also create opportunities for rural development and improve access to electricity in remote areas. The benefits of renewable energy for Africa are clear and significant. With continued investment and innovation, it is possible to build a more sustainable future for all Africans.

Renewable energy is often more affordable than traditional sources of energy in the long run. While the initial costs of installing renewable energy infrastructure may be higher, the ongoing maintenance and operational costs are significantly lower than those of fossil fuel–based power plants. This makes renewable energy a more cost-effective option for both individuals and governments in the long term. Investing in renewable energy can create new job opportunities in the green energy sector, which can help reduce unemployment rates and boost economic growth. Renewable energy sources such as wind, solar, and hydropower generate electricity without emitting harmful pollutants or greenhouse gases into the atmosphere. This means that they are much better for the environment than traditional fossil fuels, which contribute to climate change and air pollution. By transitioning to renewable energy sources, we can help mitigate the effects of climate change and protect our planet for future generations. It is also worth noting that renewable energy sources are typically more reliable than fossil fuels, as they are not subject to price fluctuations or supply chain disruptions. In short, investing in renewable energy is a smart choice for individuals, governments, and businesses alike. Not only does it offer long-term cost savings and job creation opportunities, but it also helps to protect our environment and ensure a sustainable future for all.

Renewable energy can also improve public health outcomes by reducing air pollution and associated health risks. Traditional sources of energy, such as coal and oil, produce harmful pollutants that can lead to respiratory illnesses, heart disease, and premature death. By transitioning to cleaner sources of energy such as wind, solar, and hydropower, African countries can reduce their reliance on fossil fuels and improve public health outcomes for their citizens. In addition to the economic benefits, renewable energy also has the potential to address environmental concerns. Climate change is a pressing issue that affects all countries, but African nations are particularly vulnerable due to their reliance on agriculture and natural resources. By reducing greenhouse gas emissions through the use of renewable energy, African countries can mitigate the impacts of climate change and protect their ecosystems. Investing in renewable energy can also create new job opportunities in industries such as manufacturing, installation, and maintenance. This can help address unemployment and poverty in many African countries. Transitioning to renewable energy is a crucial step toward a sustainable future for Africa and the world at large.

Finally, investing in renewable energy is crucial for Africa's sustainable development. It offers a range of benefits, including improved access to electricity, rural development opportunities, cost savings over time, job creation in the green economy, and reduced carbon emissions. To achieve this transition, African governments must prioritize policies that support renewable energy development, such as tax incentives and subsidies for renewable energy projects. Private sector investment is also essential to finance the construction of new renewable energy infrastructure. African countries should leverage international partnerships to access funding and technology transfer for renewable energy projects. By taking these steps, Africa can harness its abundant natural resources to power its economic growth sustainably while contributing to global efforts to combat climate change.

7.6 ANALYSIS OF THE EFFECTIVENESS OF ENERGY POLICIES IN AFRICA

Energy policies are crucial for the development of any country, and Africa is no exception. The continent is home to some of the world's fastest-growing economies, but it also faces significant energy challenges. In recent years, African governments have implemented various energy policies aimed at increasing access to electricity and promoting renewable energy sources. The effectiveness of these policies has been mixed. Some countries have made significant progress in expanding access to electricity, such as Ethiopia, which has doubled its electrification rate in the past decade. Other countries, such as Nigeria and South Africa, have struggled to keep up with demand and have experienced frequent power outages. In terms of renewable energy, some countries like Morocco have made significant investments in solar power and are on track to meet their renewable energy targets. Other countries have been slow to adopt renewable energy sources due to high costs and a lack of infrastructure.

One major challenge facing African energy policies is the issue of corruption. In many countries, government officials and private companies have been accused of embezzling funds meant for energy projects or awarding contracts to friends and family members instead of qualified bidders. This has led to delays and inefficiencies in implementing energy policies. To address these challenges, African governments must prioritize the development of renewable energy sources and invest in building the necessary infrastructure. This will require significant financial resources, but it is a crucial step toward achieving sustainable energy targets. Measures must be put in place to tackle corruption and ensure that funds meant for energy projects are used appropriately. This could include increased transparency in the bidding process and stronger penalties for those found guilty of embezzlement. With these steps in place, African countries can overcome the challenges facing their energy policies and make significant progress toward a more sustainable future.

Another challenge is the lack of coordination between different government agencies responsible for energy policy. In some cases, different agencies may be working at cross-purposes or duplicating efforts, which can lead to inefficiencies and wasted resources. To address this issue, African governments should establish a centralized body that is responsible for coordinating energy policy across all relevant agencies. This body should have the authority to set clear goals and targets for energy production and consumption, as well as the power to enforce compliance with these targets. African countries must invest in renewable energy sources such as solar, wind, and hydropower. These sources are abundant in many parts of Africa and can provide a sustainable alternative to fossil fuels. Finally, African governments must prioritize energy access for rural communities, which often lack basic electricity infrastructure. This can be achieved through initiatives such as microgrids and off-grid solutions that are tailored to the specific needs of these communities. By implementing these measures, African countries can build a more resilient and sustainable energy system that benefits all citizens.

One of the most significant challenges facing African countries is access to electricity. According to the IEA (2023), more than 600 million people in sub-Saharan Africa lack access to electricity. To address this issue, many African governments have implemented policies aimed at increasing electrification rates. For example, some countries have introduced subsidies for off-grid solar systems or implemented rural electrification programs. Despite these efforts, progress has been slow and uneven. In some areas, electrification rates have actually decreased due to population growth outpacing infrastructure development. The high cost of electricity in many African countries remains a barrier for low-income households. To address these challenges, some experts suggest that governments should focus on improving energy efficiency and promoting renewable energy sources such as wind and hydropower. Others argue that greater investment in transmission and distribution infrastructure is needed to ensure reliable access to electricity across the continent. Ultimately, the success of these policies will depend on a range of factors including political will, funding availability, and technological innovation.

Another critical area of focus for African energy policies is renewable energy. Many countries on the continent are rich in renewable resources such as solar, wind, and hydropower. Governments have recognized this potential and have implemented policies aimed at promoting renewable energy sources as a means of reducing reliance on fossil fuels and promoting sustainable development. In addition to policy initiatives, there has been a growing interest in private sector investment in renewable energy projects across Africa. This has led to the emergence of innovative financing models such as crowdfunding and impact investing, which are helping to drive the growth of renewable energy infrastructure on the continent. There are still significant challenges that need to be addressed in order to fully realize the potential of renewable energy in Africa. These include issues around grid integration, storage capacity, and access to finance for small-scale projects. Nevertheless, with continued political commitment and investment in technological innovation, there is great potential for Africa to become a leader in renewable energy and a model for sustainable development worldwide.

7.7 COMPARISON OF ENERGY POLICIES IN AFRICA TO OTHER REGIONS

Energy policies in Africa have been a topic of discussion for many years. The continent has vast natural resources, but it still faces challenges in meeting its energy needs. Compared to other regions, such as Europe and North America, Africa has a lower rate of electrification and a higher reliance on traditional fuels like wood and charcoal. There have been efforts to improve access to modern energy sources through the implementation of renewable energy policies. For example, countries like Kenya and South Africa have made significant investments in wind and solar power. In contrast, Europe has focused on reducing its carbon emissions through the use of nuclear power and natural gas. While these policies have been successful in reducing emissions, they also come with their own set of challenges. It is clear that energy policies vary greatly between regions and must be tailored to meet the specific needs of each country or continent.

As the world continues to grapple with the challenges of climate change, energy policies have become a critical tool in reducing carbon emissions. Many countries, such as Kenya and South Africa, have made significant investments in renewable energy sources like wind and solar power. These efforts have been successful in reducing emissions and promoting sustainable development. Other regions like Europe have focused on reducing emissions through the use of nuclear power and natural gas. While these policies have also been effective, they come with their own set of challenges such as safety concerns and dependence on foreign sources of energy. Ultimately, it is clear that energy policies must be tailored to meet the specific needs of each country or continent. This requires a careful balancing act between economic growth, environmental sustainability, and social equity. As we move forward, it is crucial that policymakers continue to prioritize renewable energy sources while also exploring new technologies and innovative solutions to address the global challenge of climate change.

It is important to recognize the role that individuals and communities can play in promoting sustainable energy practices. This includes reducing energy consumption through energy-efficient technologies and lifestyle changes, as well as supporting local renewable energy projects. Education and awareness campaigns can also help to raise public understanding of the importance of sustainable energy and encourage behavior change. At the same time, international cooperation is necessary to address global challenges such as climate change and ensure equitable access to clean energy resources.

This can be achieved through a variety of means, including investing in research and development of new technologies that promote sustainable energy practices. Governments and private organizations can also incentivize the adoption of renewable energy sources through tax breaks and subsidies. Individuals can make a difference by reducing their own carbon footprint through simple actions such as conserving energy at home, using public transportation or carpooling, and supporting eco-friendly businesses. It is important to recognize that sustainable energy is not just an environmental issue, but also an economic and social one. By transitioning to cleaner forms of energy, we can create new job opportunities and improve public health outcomes. Ultimately, the success of sustainable energy initiatives will depend on a collective effort from all stakeholders—from policymakers to consumers—to prioritize the long-term health of our planet over short-term gains.

As we continue to face the consequences of climate change, it is becoming increasingly clear that we need to take action toward sustainable energy practices. This means supporting eco-friendly businesses that prioritize renewable energy sources such as wind, solar, and hydropower. By doing so, we can not only reduce our carbon footprint but also create new job opportunities in the green energy sector. It is important to recognize that sustainable energy is not just an environmental issue, but also an economic and social one. The benefits of transitioning to cleaner forms of energy are numerous—from improved public health outcomes to reduced dependence on foreign oil. Achieving these benefits will require a collective effort from all stakeholders. Policymakers must prioritize the long-term health of our planet over short-term gains, while consumers must make conscious choices about their energy consumption habits. Ultimately, the success of sustainable energy initiatives will depend on our ability to work together toward a common goal—a healthier planet for future generations.

7.8 FUTURE DIRECTION OF ENERGY POLICIES IN AFRICA

The future direction of energy policies in Africa is poised to be transformative. With a growing population and increasing demand for energy, African countries are exploring innovative ways to meet their energy needs while reducing their carbon footprint. One of the key areas of focus is renewable energy, particularly solar and wind power. Governments across the continent are investing heavily in these technologies, with some countries setting ambitious targets for their use. There is a growing interest in decentralized energy systems, such as mini-grids and off-grid solutions, which can provide electricity to remote communities that are not connected to the national grid. Another area of focus is energy efficiency, with governments implementing policies and programs aimed at reducing energy consumption in buildings and industries. Finally, there is recognition that access to energy is essential for economic development and poverty reduction, and efforts are being made to ensure that everyone has access to reliable and affordable energy services.

The future of energy policies in Africa looks promising as countries work toward a more sustainable and equitable energy system. There are still challenges that need to be addressed. One of the biggest challenges is the lack of investment in renewable energy sources, such as solar and wind power. This is partly due to the high upfront costs associated with these technologies, but also because of a lack of political will and support. Another challenge is the need for better infrastructure to support energy distribution and storage, particularly in rural areas where access to electricity is limited. To overcome these challenges, there needs to be greater collaboration between governments, private sector companies, and international organizations to invest in renewable energy projects and improve energy infrastructure. There needs to be greater education and awareness around the benefits of renewable energy and energy efficiency, so that individuals and businesses can make informed decisions about their energy consumption. With these efforts in place, Africa has the potential to become a leader in sustainable energy practices and contribute significantly to global efforts to mitigate climate change.

By investing in renewable energy projects and improving energy infrastructure, African countries can not only reduce their carbon footprint but also increase their energy security and create new job opportunities. For instance, solar power can provide reliable electricity to remote areas that are not connected to the grid, while wind turbines can generate clean energy in windy coastal regions. Geothermal energy can be harnessed from hot springs and volcanoes to produce electricity and heat buildings. To fully realize the potential of renewable energy, there needs to be a supportive policy framework that incentivizes private sector investment and removes regulatory barriers. Public awareness campaigns can help educate communities about the benefits of renewable energy and encourage them to adopt more sustainable practices. With these measures in place, Africa can emerge as a leader in sustainable development and inspire other regions to follow suit.

REFERENCES

Ameyaw, B., & Yao, L. (2018). Analyzing the impact of GDP on CO_2 emissions and forecasting Africa's total CO_2 emissions with non-assumption driven bidirectional long short-term memory. *Sustainability (Switzerland)*. https://doi.org/10.3390/su10093110

Amowine, N. (2023). The impact of forestation and renewable energy utilisation on environmental efficiency in Africa. *Economic Research-Ekonomska Istrazivanja*. https://doi.org/10.1080/1331677X.2023.2180412

Ayompe, L. M., Davis, S. J., & Egoh, B. N. (2020). Trends and drivers of African fossil fuel CO_2 emissions 1990–2017. *Environmental Research Letters*. https://doi.org/10.1088/1748-9326/abc64f

Bamisile, O., Obiora, S., Huang, Q., Yimen, N., Abdelkhalikh Idriss, I., Cai, D., & Dagbasi, M. (2021). Impact of economic development on CO_2 emission in Africa; the role of BEVs and hydrogen production in renewable energy integration. *International Journal of Hydrogen Energy*. https://doi.org/10.1016/j.ijhydene.2020.10.134

Calvin, K., Pachauri, S., De Cian, E., & Mouratiadou, I. (2016). The effect of African growth on future global energy, emissions, and regional development. *Climatic Change*. https://doi.org/10.1007/s10584-013-0964-4

IEA. (2023). Global EV Outlook 2023. *Geo, Geo*.

İnal, V., Addi, H. M., Çakmak, E. E., Torusdağ, M., & Çalışkan, M. (2022). The nexus between renewable energy, CO_2 emissions, and economic growth: Empirical evidence from African oil-producing countries. *Energy Reports*. https://doi.org/10.1016/j.egyr.2021.12.051

International Renewable Energy Agency (IRENA). (2022). World energy transitions outlook 2023: 1.5°C pathway. *World Energy Transitions*.

Ituen, E. E., Esen, N. U., Nwokolo, S. C., & Udo, E. G. (2012). Prediction of global solar radiation using relative humidity, maximum temperature and sunshine hours in Uyo, in the Niger Delta Region, Nigeria. *Advances in Applied Science Research*, *3*(4), 1923–1937.

Jahanger, A., Balsalobre-Lorente, D., Samour, A., Joof, F., Ali, M., & Tursoy, T. (2022). Do renewable energy and the real estate market promote environmental quality in South Africa: Evidence from the bootstrap ARDL approach. *Sustainability (Switzerland)*. https://doi.org/10.3390/su142416466

Jinapor, J. A., Suleman, S., & Cromwell, R. S. (2023). Energy consumption and environmental quality in Africa: Does energy efficiency make any difference?. *Sustainability (Switzerland)*. https://doi.org/10.3390/su15032375

Khobai, H., & Sithole, K. (2022). The relationship between economic growth and carbon emissions in South Africa. *International Journal of Energy Economics and Policy*. https://doi.org/10.32479/IJEEP.11194

Laura Cozzi et al. (2023). *World Energy Outlook*. International Energy Agency. www.iea.org/weo

Nwokolo, S. C. (2017). A comprehensive review of empirical models for estimating global solar radiation in Africa. *Renewable and Sustainable Energy Reviews*, *78*(October), 955–995. https://doi.org/10.1016/j.rser.2017.04.101

Nwokolo, S. C., Meyer, E. L., & Ahia, C. C. (2023). Credible pathways to catching up with climate goals in Nigeria. *Climate*, *11*(9), 1–32. https://doi.org/org/10.3390/cli11090196

Nwokolo, S. C., Meyer, E. L., & Ahia, C. C. (2024). Exploring the interactive influences of climate change and urban development on the fraction of absorbed photosynthetically active radiation. *Atmosphere*, *15*(253), 1–36.

Nwokolo, S. C., & Ogbulezie, J. C. (2018a). A qualitative review of empirical models for estimating diffuse solar radiation from experimental data in Africa. *Renewable and Sustainable Energy Reviews*, *92*(18), 353–393. https://doi.org/10.1016/j.rser.2018.04.118

Nwokolo, S. C., & Ogbulezie, J. C. (2018b). A quantitative review and classification of empirical models for predicting global solar radiation in West Africa. *Beni-Suef University Journal of Basic and Applied Sciences*, *7*(4), 367–396. https://doi.org/10.1016/j.bjbas.2017.05.001

Nwokolo, S. C., Proutsos, N., & Meyer, E. L. (2023). Machine learning and physics-based hybridization models for evaluation of the effects of climate change and urban expansion on photosynthetically active radiation. *Atmosphere*, *14*(687), 1–36. https://doi.org/doi.org/10.3390/atmos14040687

Nwokolo, S., Echeng, E., Obiwulu, A., Meyer, E. L., Christian, C., Ogbulezie, J. C., & Proutsos, N. (2024). A multi-model approach based on CARIMA-SARIMA-GPM for assessing the impacts of climate change on concentrated photovoltaic (CPV) potential. *Physics and Chemistry of the Earth*, *134*(January), 103560. https://doi.org/10.1016/j.pce.2024.103560

Nwokolo, S., & Otse, C. (2019). Impact of sunshine duration and clearness index on diffuse solar radiation estimation in mountainous climate. *Trends in Renewable Energy*, *5*(3), 307–332. https://doi.org/10.17737/tre.2019.5.3.00107

Obiwulu, A. U., Chendo, M. A. C., Erusiafe, N., & Nwokolo, S. C. (2020). Implicit meteorological parameter-based empirical models for estimating back temperature solar modules under varying tilt-angles in Lagos, Nigeria. *Renewable Energy*, *145*. https://doi.org/10.1016/j.renene.2019.05.136

Obiwulu, A. U., Erusiafe, N., Olopade, M. A., & Nwokolo, S. C. (2020). Modeling and optimization of back temperature models of mono-crystalline silicon modules with special focus on the effect of meteorological and geographical parameters on PV performance. *Renewable Energy*, *154*. https://doi.org/10.1016/j.renene.2020.02.103

Obiwulu, A. U., Erusiafe, N., Olopade, M. A., & Nwokolo, S. C. (2022). Modeling and estimation of the optimal tilt angle, maximum incident solar radiation, and global radiation index of the photovoltaic system. *Heliyon*, *8*(6), e09598. https://doi.org/10.1016/j.heliyon.2022.e09598

Okorie, D. I., & Lin, B. (2022). Africa's biofuel energy and emissions prospect: Forward-looking into 2030. *Sustainable Energy Technologies and Assessments*. https://doi.org/10.1016/j.seta.2022.102775

Oladipupo, S. D., Rjoub, H., Kirikkaleli, D., & Adebayo, T. S. (2022). Impact of globalization and renewable energy consumption on environmental degradation: A lesson for South Africa. *International Journal of Renewable Energy Development*. https://doi.org/10.14710/ijred.2022.40452

Oladunni, O. J., Mpofu, K., & Olanrewaju, O. A. (2022). Greenhouse gas emissions and its driving forces in the transport sector of South Africa. *Energy Reports*. https://doi.org/10.1016/j.egyr.2022.01.123

Oluoch, S., Lal, P., & Susaeta, A. (2021). Investigating factors affecting renewable energy consumption: A panel data analysis in Sub Saharan Africa. *Environmental Challenges*. https://doi.org/10.1016/j.envc.2021.100092

Ouedraogo, N. S. (2017). Africa energy future: Alternative scenarios and their implications for sustainable development strategies. *Energy Policy*. https://doi.org/10.1016/j.enpol.2017.03.021

Qudrat-Ullah, H., & Nevo, C. M. (2021). The impact of renewable energy consumption and environmental sustainability on economic growth in Africa. *Energy Reports*. https://doi.org/10.1016/j.egyr.2021.05.083

Tagwi, A. (2022). The impacts of climate change, carbon dioxide emissions (CO_2) and renewable energy consumption on agricultural economic growth in South Africa: ARDL approach. *Sustainability (Switzerland)*. https://doi.org/10.3390/su142416468

Tyler, E., & Hochstetler, K. (2021). Institutionalising decarbonisation in South Africa: Navigating climate mitigation and socio-economic transformation. *Environmental Politics*. https://doi.org/10.1080/09644016.2021.1947635

Udeagha, M. C., & Breitenbach, M. C. (2023). Revisiting the nexus between fiscal decentralization and CO_2 emissions in South Africa: Fresh policy insights. *Financial Innovation*. https://doi.org/10.1186/s40854-023-00453-x

Udeagha, M. C., & Ngepah, N. (2022). The asymmetric effect of technological innovation on CO_2 emissions in South Africa: New evidence from the QARDL approach. *Frontiers in Environmental Science*. https://doi.org/10.3389/fenvs.2022.985719

Udeagha, M. C., & Ngepah, N. (2023). Can public–private partnership investment in energy (PPPI) mitigate CO_2 emissions in South Africa? Fresh evidence from the novel dynamic ARDL simulations approach. *Frontiers in Environmental Science*. https://doi.org/10.3389/fenvs.2022.1044605

van der Zwaan, B., Kober, T., Longa, F. D., van der Laan, A., & Jan Kramer, G. (2018). An integrated assessment of pathways for low-carbon development in Africa. *Energy Policy*. https://doi.org/10.1016/j.enpol.2018.03.017

West African Energy Leaders Group. (2015). *Africa Research Bulletin: Economic, Financial and Technical Series*. https://doi.org/10.1111/j.1467-6346.2015.06545.x

Yetano Roche, M. (2023). Built for net-zero: Analysis of long-term greenhouse gas emission pathways for the Nigerian cement sector. *Journal of Cleaner Production*, *383*(November 2022), 135446. https://doi.org/10.1016/j.jclepro.2022.135446

8 Global Efforts to Achieve Net Zero Emissions across Various Continents

8.1 INTRODUCTION

In today's rapidly changing world, the urgent need to address climate change has led to a surge in global efforts aimed at achieving zero emissions across various continents. As nations come together to combat this pressing issue, innovative solutions and collaborations are being forged to ensure a sustainable and environmentally friendly future for all. These efforts include the development of renewable energy sources as well as the implementation of stricter regulations on carbon emissions (Nwokolo, Meyer, et al., 2023). Governments and businesses are investing in research and development to find new technologies that can further reduce greenhouse gas emissions and mitigate the effects of climate change (Nwokolo, Proutsos, et al., 2023). International agreements such as the Paris Agreement have been established to encourage countries to set ambitious targets for reducing emissions and regularly report on their progress. There is a growing recognition of the importance of individual actions, with individuals and communities adopting sustainable practices in their daily lives, such as using public transportation or opting for energy-efficient appliances. These collective efforts are crucial to addressing climate change and ensuring a sustainable future for generations to come.

Tech-related developments are a major factor in reducing the effects of climate change, in addition to international agreements and personal initiatives. Innovations in renewable energy sources, such as solar and wind power, are becoming more accessible and affordable, reducing our reliance on fossil fuels (Obiwulu, Chendo, et al., 2020). These advancements in technology are beneficial not only for the environment but also for the economy. The renewable energy sector has seen significant growth, creating job opportunities (Obiwulu, Erusiafe, et al., 2020) and stimulating economic development (Nwokolo, 2017). Technological innovations have the potential to revolutionize industries such as transportation and agriculture, leading to more sustainable practices and reduced carbon footprints. These advancements in technology have the potential to attract investment and drive economic growth in countries that prioritize renewable energy and sustainable practices (Iteun et al., 2012). By investing in research and development, governments can foster innovation and create a competitive advantage in the global market. The adoption of these technologies can lead to cost savings for businesses as renewable energy sources become

DOI: 10.1201/9781003483175-9

more affordable and efficient over time (Nwokolo and Otse, 2019). The integration of renewable energy technologies can also enhance energy security by reducing dependence on fossil fuels and volatile energy markets. This can provide countries with greater control over their energy resources and reduce the risks associated with price fluctuations and supply disruptions. The transition to renewable energy can also improve public health by reducing air pollution and its associated health impacts, resulting in a healthier and more productive population. The adoption of renewable energy can stimulate economic growth and create new job opportunities in the clean energy sector (Nwokolo et al., 2024). By investing in renewable energy infrastructure and technologies, countries can attract investments and promote innovation, leading to a more sustainable and resilient economy (Nwokolo and Ogbulezie, 2018a). The use of renewable energy sources can contribute to mitigating climate change by reducing greenhouse gas emissions (Nwokolo and Ogbulezie, 2018a) and helping to limit global warming and its adverse effects on ecosystems and vulnerable communities (Obiwulu et al., 2022).

Investments in research and development are continuously being made to find new solutions for carbon capture and storage, which could help reduce greenhouse gas emissions even further. These advancements in carbon capture and storage technologies have the potential to revolutionize the energy sector by allowing for the continued use of fossil fuels while minimizing their environmental impact. By investing in these innovative solutions, countries can position themselves as leaders in the global transition toward a low-carbon economy, fostering economic growth and job creation in the process. The implementation of carbon capture and storage technologies can also contribute to achieving international climate goals, such as those outlined in the Paris Agreement. By capturing and storing carbon dioxide emissions, countries can effectively offset their greenhouse gas emissions and work toward a more sustainable future. These advancements can provide an opportunity for collaboration between industries and governments, fostering partnerships that drive innovation and promote the adoption of cleaner energy sources. These partnerships can lead to the development of new technologies and strategies for reducing carbon emissions, creating a ripple effect across various sectors. The implementation of carbon capture and storage technologies can also help countries transition to a low-carbon economy by allowing them to continue using fossil fuels while minimizing their environmental impact. This approach can provide a bridge toward a more sustainable future while alternative energy sources are further developed and scaled up.

To mitigate the effects of climate change, businesses and industries are just as important as individual actions and international agreements. Many companies are now implementing sustainable practices, such as investing in renewable energy sources and reducing their carbon footprint. By embracing environmentally friendly strategies, businesses can contribute to reducing greenhouse gas emissions on a larger scale and drive the transition toward a low-carbon economy. Businesses can also influence consumer behavior by promoting and offering sustainable products and services. This can create a ripple effect as more individuals adopt eco-friendly choices, further reducing carbon footprints and contributing to a greener future. In addition, businesses can collaborate with other organizations and stakeholders to develop innovative solutions and technologies that can further enhance sustainability efforts. By sharing knowledge and resources, they can collectively address environmental challenges and accelerate the adoption of sustainable practices across industries. Businesses can also implement sustainable practices within their own operations, such as reducing waste and energy consumption. This not only helps to minimize their environmental impact but also showcases their commitment to sustainability, which can attract environmentally conscious consumers and strengthen their brand reputation.

By investing in research and development for sustainable alternatives, businesses can drive technological advancements that have the potential to revolutionize industries and create a more sustainable future for all. By collaborating with other businesses and organizations, companies can further promote sustainability by sharing best practices and working toward common goals. This collective effort can lead to the development of industry-wide standards and initiatives that benefit both the environment and the economy. By engaging with stakeholders, such as employees, customers, and local communities, businesses can foster a sense of shared responsibility for sustainability and encourage widespread adoption of sustainable practices. By involving employees in sustainability initiatives, businesses can tap into their creativity and expertise to identify innovative solutions and implement them effectively. By actively seeking feedback from customers and local communities, companies can gain valuable insights on their sustainability efforts and make necessary improvements to align with their needs and expectations. This inclusive approach not only strengthens the overall impact of sustainability initiatives but also builds trust and loyalty among stakeholders, ultimately contributing to a more sustainable future for all. Collaboration with other organizations and industry peers can also enhance the effectiveness of sustainability initiatives. By sharing best practices and resources, businesses can collectively address complex sustainability challenges and drive meaningful change on a larger scale. This collaborative approach fosters a sense of collective responsibility and promotes knowledge sharing, leading to more innovative and impactful solutions for a sustainable future.

This section thoroughly discussed efforts to achieve net zero emissions in America, Asia, Australia, and Europe, emphasizing the importance of global cooperation in combating climate change. Global cooperation is crucial in addressing climate change because it recognizes that environmental issues transcend national boundaries and require a unified effort. By working together, countries can leverage their unique strengths and resources to implement effective strategies and policies that reduce greenhouse gas emissions and promote sustainable development. This collaborative approach also fosters a sense of shared accountability, as nations hold each other accountable for meeting their commitments and driving progress toward a sustainable future. These regions—America, Asia, Australia, and Europe—play a significant role in global emissions, and their collective efforts toward achieving net zero emissions will have a substantial impact on mitigating the effects of climate change worldwide. By leveraging their strengths and resources, these regions can not only lead by example but also inspire other nations to take similar actions. Their collaborative approach allows for the sharing of best practices and innovative solutions, accelerating the transition to a low-carbon economy on a global scale. This collaborative approach also fosters international cooperation and builds trust among nations, creating a foundation for future climate negotiations and agreements. By working together, these regions can pool their financial resources and invest in renewable energy infrastructure, research and development, and sustainable technologies, driving economic growth while reducing greenhouse gas emissions.

The latter part of this section discusses and contrasts zero-emission plans in America, Asia, Australia, and Europe. It also discusses successful approaches, lessons learned, international collaboration, and how these variables influence different continents' efforts to attain net zero. This analysis highlights the importance of international collaboration in achieving global net zero goals. By sharing successful approaches and lessons learned, different continents can learn from each other's experiences and accelerate their efforts toward a sustainable future. Understanding how variables such as policy frameworks, technological advancements, and financial investments influence each continent's journey toward net zero can help shape more effective strategies and foster a collective commitment to combat climate change. Collaboration can also lead to the development of innovative solutions that may not have been considered individually. For example, one continent may have made significant progress in renewable energy technology, while another continent may have expertise in carbon capture and storage. By sharing knowledge and resources, continents can leverage each other's strengths and address their weaknesses, ultimately driving faster progress toward net zero emissions. This collaboration can also promote the transfer of clean technologies and best practices across continents, enabling a more widespread adoption of sustainable practices. By working together, continents can pool their financial resources and investments to support large-scale projects that may be too costly for individual countries to undertake alone. This collective effort can create a ripple effect, inspiring other regions to join the movement and accelerating the global transition to a low-carbon economy.

8.2 GLOBAL EFFORTS IN ASIA

8.2.1 Current Emission Situation in Asia

The current emission situation in Asia is a critical aspect of global efforts to combat climate change. Asia, being home to some of the world's most populous countries, such as China and India, and rapidly developing economies, plays a significant role in shaping the global emissions landscape. With a growing population and increasing industrialization, Asia's emissions have been steadily rising over the years. This not only poses a challenge for the region itself but also has far-reaching consequences for the entire planet. As Asia continues to industrialize and urbanize, the demand for energy and resources is expected to further increase, leading to even higher emissions. The region's contribution to greenhouse gas emissions, particularly from industries such as manufacturing and energy production, has been substantial. In fact, according to a report by the International Energy Agency (IEA), Asia accounted for more than half of the world's total energy-related CO_2 emissions in 2019 (IEA, 2023). This highlights the urgent need for sustainable and low-carbon solutions to mitigate the environmental impact of Asia's rapid development. International cooperation and technology transfer will play a crucial role in supporting Asia's transition toward cleaner energy sources and reducing its overall emissions footprint. These efforts are especially important considering that Asia is projected to be the largest contributor to global energy demand in the coming decades. Without significant action, the region's emissions could continue to rise and exacerbate climate change. Therefore, it is imperative for governments, businesses, and communities in Asia to prioritize sustainable practices and invest in renewable energy infrastructure.

To address this challenge, Asian countries have been implementing various measures and policies aimed at reducing emissions and transitioning toward more sustainable practices. For instance, several nations have set ambitious targets for renewable energy deployment and are investing heavily in clean technologies. Collaborations between Asian countries and international organizations have facilitated knowledge sharing and capacity building to promote sustainable development across the region. These efforts have resulted in significant progress, with Asian countries becoming global leaders in renewable energy production. Governments have implemented policies to encourage the adoption of electric vehicles (EVs) and promote energy efficiency in industries and households. These collective actions demonstrate a strong commitment to combat climate change and pave the way for a greener future in Asia.

Given the diverse socio-economic conditions and varying levels of industrialization in Asia, addressing the emission situation requires tailored approaches that consider local contexts. It is crucial for governments, businesses, and communities to work together to adopt innovative solutions that balance economic growth with environmental sustainability. By doing so, Asia can contribute significantly to global efforts to mitigate climate change while ensuring a prosperous future for its people. These tailored approaches should also prioritize the development and implementation of renewable energy sources, such as solar and wind power, to reduce reliance on fossil fuels. Investing in sustainable infrastructure and promoting green technologies can help create new job opportunities and stimulate economic growth in the region. By fostering collaboration and knowledge sharing among Asian countries, a collective effort can be made to address climate change effectively and create a greener future for all. It is crucial for governments to establish policies and regulations that encourage sustainable practices and incentivize businesses to adopt eco-friendly measures. This can include offering tax breaks or subsidies for companies that invest in renewable energy or implement green technologies. Raising awareness and educating the public about the importance of sustainability can also play a significant role in driving positive change and fostering a culture of environmental responsibility in Asian societies.

According to a report by the IEA (IEA, 2023), Asia is responsible for more than half of the world's total energy-related CO_2 emissions. This highlights the significance of addressing emissions in this region to achieve global climate goals. One of the key factors contributing to high emissions in Asia is the rapid economic growth and industrialization observed in countries like China and India. These nations have experienced a surge in energy demand, leading to increased reliance on fossil fuels, particularly coal. As a result, they have become major contributors to greenhouse gas emissions. It is important to note that efforts are being made in Asia to address this issue. Several countries have implemented policies and initiatives aimed at reducing emissions and transitioning toward cleaner energy sources. For instance, China has set ambitious targets for renewable energy deployment and has been investing heavily in solar and wind power. Regional collaborations such as the Association of Southeast Asian Nations (ASEAN) have been established to promote sustainable development and address environmental challenges collectively (Kumaran et al., 2020). These initiatives demonstrate that Asian countries are recognizing the need for concerted efforts to mitigate emissions and transition toward a low-carbon future. The current emission situation in Asia is a significant component of global efforts to combat climate change. While Asia remains a major emitter due to its rapid economic growth, there are ongoing initiatives and policies aimed at reducing emissions and transitioning toward cleaner energy sources. By addressing this challenge collectively, Asian countries can contribute significantly toward achieving global climate goals.

China, as the largest emitter of greenhouse gases in the world, has a significant impact on global efforts to combat climate change. According to recent data from the Global Carbon Atlas (GCA), China accounted for approximately 28% of global CO_2 emissions in 2019 (Global Carbon Atlas [GCA], 2022). This highlights the importance of addressing China's emission situation in order to effectively mitigate climate change on a global scale. China's rapid industrialization and population growth have contributed to its high emission levels. China's rapid industrialization and reliance on coal for energy production have significantly contributed to its high emission levels. As the country continues to develop, it is crucial for China to prioritize sustainable and clean energy sources to reduce its carbon footprint. International cooperation and collaboration with China are essential in implementing effective strategies to combat climate change worldwide. China's commitment to reducing its carbon intensity by 60–65% by 2030, as stated in its Nationally Determined Contribution (NDC) under the Paris Agreement (IEA, 2023), demonstrates its recognition of the need for substantial action. This commitment not only sets a positive example for other nations but also highlights China's determination to address the global climate crisis in a meaningful way. China's efforts to invest in renewable energy technologies, such as solar and wind power, further solidify its commitment to sustainable development. By prioritizing the transition to cleaner energy sources, China is not only reducing its own carbon emissions but also contributing to the global shift toward a greener future. This collaborative approach between China and other nations is crucial in achieving long-term climate goals and ensuring a sustainable planet for future generations.

The current emission situation in Asia plays a crucial role in global efforts to combat climate change. India, as one of the major contributors to global emissions, has been making significant efforts to address its current emission situation. According to recent data from the GCA, India accounted for approximately 7% of global CO_2 emissions in 2019 (GCA, 2022). It is important to note that India's per capita emissions are still significantly lower than in many developed countries. Despite this, the country has set ambitious targets under the Paris Agreement and has been actively investing in renewable energy sources such as solar and wind power. These efforts not only contribute to reducing India's own emissions but also have a positive impact on global emission levels. India has been actively participating in international collaborations and initiatives aimed at promoting sustainable development and combating climate change, further highlighting its commitment to being part of the global efforts to address this pressing issue.

The current emission situation in Indonesia plays a significant role in global efforts to combat climate change. According to the GCA, Indonesia is one of the top five emitters of greenhouse gases worldwide (GCA, 2022), primarily due to deforestation and peatland degradation. In fact, it is estimated that Indonesia's annual carbon emissions from land-use change alone are higher than those of any other country. A study published in GCA highlights that Indonesia's emissions have a substantial impact on global temperature rise (GCA, 2022). The country's peatland fires release vast amounts of carbon dioxide into the atmosphere, exacerbating climate change effects on a global scale. These figures and studies emphasize the urgent need for Indonesia to address its emission situation as part of global efforts to combat climate change. Implementing effective measures to reduce deforestation, promote sustainable land-use practices, and enhance peatland management are crucial steps toward mitigating Indonesia's contribution to global greenhouse gas emissions.

According to the National Environment Agency (NEA), Singapore's total greenhouse gas emissions have been steadily increasing over the years, reaching 52.5 million metric tons of carbon dioxide equivalent in 2019 (Lim et al., 2020). This highlights the urgent need for effective measures to mitigate and reduce emissions in the country. Singapore has implemented various initiatives to address its emissions, such as the Carbon Pricing Act and the Sustainable Singapore Blueprint. The Carbon Pricing Act, introduced in 2018, imposes a price on carbon emissions from power stations and large industrial emitters, encouraging them to reduce their greenhouse gas emissions. The Sustainable Singapore Blueprint sets ambitious targets for sustainable development, including reducing carbon emission intensity by 36% from 2005 levels by 2030 (Lim et al., 2020). Singapore is actively investing in renewable energy sources and promoting energy efficiency measures. The government has set a target to increase solar deployment to at least 2 gigawatt-peak (GWp) by 2030 and has launched initiatives like the SolarNova program to accelerate solar adoption (Lim et al., 2020). Energy-efficient building standards and incentives are being implemented to reduce energy consumption in both the residential and commercial sectors. While these efforts demonstrate Singapore's commitment to combating climate change, it is essential for continuous monitoring and evaluation of emission reduction strategies to ensure their effectiveness. Collaboration with international partners and sharing best practices will also be crucial to achieving global climate goals.

The current emission situation in North-east Asia plays a significant role in global efforts to combat climate change. According to recent figures, this region is responsible for a substantial portion of global greenhouse gas emissions. For instance, China alone is the world's largest emitter of carbon dioxide, accounting for nearly 30% of global emissions. Other countries in northeast Asia, such as Japan and South Korea, also contribute significantly to the overall emissions in the region. These high emission levels have prompted international concern and the need for collective action. Efforts to address this situation include various initiatives and agreements aimed at reducing emissions and transitioning toward cleaner energy sources. For example, China has set ambitious targets to peak its carbon emissions by 2030 and achieve carbon neutrality by 2060 (IEA, 2023).

Similarly, Japan has committed to reducing its greenhouse gas emissions by 26% by 2030 compared to 2013 levels (IEA, 2023). It is crucial for countries in northeast Asia to continue implementing sustainable policies and investing in renewable energy sources to effectively combat climate change on a global scale. Collaborative efforts between these nations and the international community are essential to mitigating the impact of emissions from this region and achieving a more sustainable future. These efforts can include sharing best practices, technology transfer, and financial support to help countries in northeast Asia transition to low-carbon economies. Promoting regional cooperation and dialogue can facilitate the development of joint initiatives and strategies to address common environmental challenges in the region. By fostering cooperation and dialogue, countries in northeast Asia can work together to develop innovative solutions and policies that tackle climate change. This can involve the establishment of regional frameworks and agreements that promote the adoption of renewable energy sources and the implementation of sustainable practices. Engaging with stakeholders such as businesses, civil society organizations, and academia can further enhance the effectiveness of these collaborative efforts by harnessing diverse perspectives and expertise. Investing in research and development can drive technological advancements in clean energy and resource conservation, which can be shared among the countries in the region. Promoting public awareness and education campaigns can encourage individuals to adopt environmentally friendly behaviors and contribute to the overall sustainability goals of northeast Asia.

8.2.2 Asia's Efforts to Achieve Net Zero Emissions

Asia's efforts to achieve zero emissions are crucial in the global fight against climate change. With its rapidly growing population and expanding economies, Asia has become a significant contributor to greenhouse gas emissions. By committing to this ambitious goal, Asian countries are not only demonstrating their commitment to environmental sustainability but also setting an example for the rest of the world. Asia's pursuit of zero emissions is driven by a combination of factors. Firstly, the region is highly vulnerable to the impacts of climate change, such as rising sea levels and extreme weather events. By reducing emissions, Asian countries aim to mitigate these risks and protect their populations and economies. Secondly, transitioning toward renewable energy sources can provide new opportunities for economic growth and job creation in Asia.

According to a report by the Asian Development Bank (ADB), Asia's renewable energy capacity has more than doubled between 2000 and 2018, reaching 1,243 GW (Creutz, 2023). This significant increase in renewable energy capacity demonstrates Asia's commitment to transitioning toward cleaner and more sustainable energy sources. Countries in Asia have also implemented various policies and initiatives to reduce greenhouse gas emissions. For instance, China has been making significant efforts to achieve zero emissions as part of the global fight against climate change. According to data from the IEA, China accounted for 28% of global carbon dioxide emissions in 2019 (IEA, 2023). The country has been actively working toward reducing its carbon footprint and transitioning to cleaner energy sources. One notable initiative is China's commitment to peak its carbon dioxide emissions by 2030 and achieve carbon neutrality by 2060 (IEA, 2023). This ambitious target was announced by President Xi Jinping during the United Nations General Assembly in September 2020. To support this goal, China has implemented various measures, such as increasing renewable energy capacity, promoting EVs, and implementing stricter emission standards for industries. For instance, China is the world's largest investor in renewable energy and has made significant progress in expanding its renewable energy capacity. The country has invested heavily in solar and wind power, leading to a substantial increase in renewable energy generation. In fact, China accounted for almost half of the global growth in renewable energy capacity in 2020 (International Renewable Energy Agency [IRENA], 2022). China has been promoting the adoption of EVs as a means to reduce emissions from transportation. The government has implemented policies such as subsidies and tax incentives to encourage EV purchases and expand charging infrastructure across the country. As a result, China is now the largest market for EVs globally. These efforts have already started showing positive results. According to a report by Climate Action Tracker (CAT), if China successfully achieves its targets of peaking emissions by 2030 and reaching carbon neutrality by 2060, it could significantly contribute to limiting global warming to well below 2°C (Climate Action Tracker [CAT], 2019). China's efforts to achieve zero emissions are evident in its commitment to peak carbon dioxide emissions and carbon neutrality. The country's investments in renewable energy and the promotion of EVs are crucial steps toward combating climate change on a global scale.

Japan has been actively working toward achieving net zero emissions as part of the global efforts to combat climate change. According to the Japan Ministry of the Environment, the country aims to reduce its greenhouse gas emissions by 26% by 2030 compared to 2013 levels (Pandey et al., 2023). Japan has set a long-term goal of becoming carbon neutral by 2050 (IEA, 2023). To achieve these targets, Japan has implemented various measures and initiatives. For instance, the country has been investing heavily in renewable energy sources such as solar and wind power. In fact, Japan is one of the world's largest producers of solar energy and has been steadily increasing its capacity in recent years. Japan is also focusing on energy efficiency and conservation. The government has introduced policies to promote energy-saving technologies and practices across industries and households. There is a strong emphasis on research and development to foster innovation in clean technologies. Japan's commitment to achieving net zero emissions is evident through its ambitious targets, investments in renewable energy, and efforts toward energy efficiency. These actions not only contribute to global climate change mitigation but also position Japan as a leader in sustainable development.

South Korea has been actively working toward achieving net zero emissions as part of the global efforts to combat climate change. According to the country's Ministry of Environment, South Korea aims to reach carbon neutrality by 2050, with a commitment to reducing its greenhouse gas emissions by 24.4% below business-as-usual levels by 2030 (Yoon et al., 2021). This ambitious target aligns with the Paris Agreement's goal of limiting global warming to well below 2°C. To support these efforts, South Korea has implemented various policies and initiatives. For instance, the government has introduced a carbon pricing system called the Emissions Trading

Scheme (ETS), which places a price on carbon emissions and encourages industries to reduce their greenhouse gas output. South Korea has been investing heavily in renewable energy sources such as solar and wind power, aiming to increase their share in the country's energy mix. These measures have shown promising results so far. In 2020, South Korea recorded a decrease in its greenhouse gas emissions for the first time in a decade, marking an important milestone toward achieving its net zero target. The country continues to prioritize sustainable development and is actively seeking international collaborations and partnerships to accelerate its transition toward a low-carbon economy.

North Korea's efforts to achieve net zero emissions have been limited and challenging. According to a report by the IEA (IEA, 2023), North Korea's greenhouse gas emissions have been steadily increasing over the years, primarily due to its heavy reliance on coal for energy production. In 2019, coal accounted for around 50% of the country's total energy consumption (IRENA, 2022). North Korea has recently shown some commitment to addressing climate change. In its NDC submitted under the Paris Agreement, North Korea pledged to enhance its energy efficiency and increase the share of renewable energy sources in its energy mix. The country has set a target to reduce its greenhouse gas emissions by 37% compared to business-as-usual scenarios by 2030 (IRENA, 2022). Despite these efforts, North Korea faces significant challenges in achieving net zero emissions. Limited access to international financing and technological resources hinders their transition toward cleaner energy sources. Economic sanctions imposed on the country also impede their ability to invest in sustainable infrastructure and technologies. With international support and collaboration, North Korea can make progress toward reducing its carbon footprint and contributing to global climate change mitigation efforts.

Indonesia, as a signatory to the Paris Agreement, has been actively working toward achieving net zero emissions. According to the Ministry of Environment and Forestry, Indonesia aims to reduce its greenhouse gas emissions by 29% by 2030 through various initiatives and policies (Saputra and Murwaningsari, 2021). One significant effort is the implementation of the Low Carbon Development Initiative (LCDI), which focuses on promoting renewable energy, sustainable land-use practices, and improving waste management systems. Indonesia has set a target to restore 2 million hectares of degraded peatland by 2024, as peatland degradation is a major contributor to carbon emissions (IEA, 2023). These measures demonstrate Indonesia's commitment to combating climate change and transitioning toward a low-carbon economy.

India has been actively working toward achieving net zero emissions as part of the global efforts to combat climate change. According to the IEA (IEA, 2023), India is projected to be the largest contributor to the rise in global energy demand by 2040. India has also set ambitious targets to increase its renewable energy capacity, aiming for 450 GW by 2030 (IEA, 2023). India has implemented various policies and initiatives to reduce its carbon footprint. The country has launched the National Solar Mission, which aims to achieve 100 GW of solar power capacity by 2022 (IEA, 2023). India is investing in electric mobility and plans to have only EVs on its roads by 2030 (IEA, 2023). These efforts are not only crucial for India's sustainable development but also contribute significantly to global climate goals. By prioritizing renewable energy sources and adopting cleaner technologies, India is making substantial progress toward achieving net zero emissions and combating climate change on a global scale.

Taiwan has been actively working toward achieving net zero emissions as part of global efforts to combat climate change. According to the Environmental Protection Administration (EPA) of Taiwan, the country aims to reach carbon neutrality by 2050 (Tsai, 2021). To support this goal, Taiwan has implemented various policies and initiatives. One significant effort is the promotion of renewable energy sources. Taiwan has set a target to generate 20% of its electricity from renewable sources by 2025 and has been steadily increasing its renewable energy capacity (IEA, 2023). As of 2020, Taiwan had already achieved a renewable energy share of around 14% in its electricity generation (IEA, 2023). Taiwan has also implemented measures to reduce greenhouse gas emissions in sectors such as transportation and industry. The government has introduced incentives for EVs and is actively encouraging the adoption of cleaner transportation options. Stricter emission standards have been imposed on industries to curb their carbon footprint. These efforts have not gone unnoticed internationally. The CAT (2017), an independent scientific analysis group, has recognized Taiwan's commitment toward decarbonization and categorized its efforts as "highly satisfactory." This acknowledgment highlights Taiwan's significant strides in aligning with global climate goals and showcases its dedication to combating climate change on a national level.

8.2.3 Challenges in Achieving Net Zero Emissions in Asia

The challenges of achieving net zero emissions in Asia are complex and multifaceted, requiring a comprehensive and collaborative approach. As the world's largest continent, Asia is home to diverse economies, energy systems, and political landscapes, making it crucial to address the unique circumstances and priorities of each country. One major challenge is the heavy reliance on coal for energy production in many Asian countries. As coal remains a cheap and abundant source of energy, shifting toward cleaner alternatives requires significant investments in renewable energy infrastructure and technologies. Addressing the socio-economic implications of transitioning away from coal will be crucial to ensuring a just and equitable transition for affected communities. Asia's rapid urbanization and industrialization pose additional challenges. As cities expand and industries grow, energy demands increase, leading to higher greenhouse gas emissions. Encouraging sustainable urban planning, promoting energy-efficient practices, and adopting circular economy principles can help mitigate these challenges.

Asia faces significant challenges in achieving net zero emissions as part of global efforts to combat climate change. According to a report by the IEA (IEA, 2023), Asia is responsible for more than 50% of global energy-related carbon dioxide emissions. This high level of emissions is primarily driven by the region's heavy reliance on coal for electricity generation and industrial processes. The rapid economic growth and urbanization in many Asian countries have led to increased energy demand, further exacerbating the emissions problem. The IEA estimates that Asia's energy demand will continue to rise, driven by population growth and expanding middle-class populations. To address these challenges, Asian countries need to transition their energy systems toward cleaner and more sustainable sources. This includes reducing coal consumption and increasing the share of renewable energy in their energy mix. This transition is not without its hurdles. One major challenge is the existing infrastructure built around coal-fired power plants, which makes it difficult to shift away from this highly polluting fuel source. Many Asian countries face financial constraints when investing in renewable energy technologies. Some Asian economies heavily rely on exports of fossil fuels, such as oil and gas, which poses a dilemma in terms of transitioning toward net zero emissions. These countries need to find alternative economic pathways that are less dependent on fossil fuel exports while ensuring a just transition for affected communities. Achieving net zero emissions in Asia requires significant policy interventions, technological advancements, and international cooperation. It demands a comprehensive approach that addresses both the supply and demand sides of energy systems while considering socio-economic implications. Nonetheless, with concerted efforts and innovative solutions, Asia can play a crucial role in global climate action and contribute toward a sustainable future.

8.2.4 Initiatives and Policies for Net Zero Emissions in Asia

Asia plays a crucial role in global efforts to combat climate change, considering its rapidly growing population and expanding economies. Initiatives and policies for achieving net zero emissions in Asia are essential for mitigating the region's significant carbon footprint. By implementing innovative strategies such as promoting renewable energy sources, improving energy efficiency, and adopting sustainable transportation systems, Asian countries can contribute significantly to reducing greenhouse gas emissions on a global scale. Fostering international collaborations and knowledge-sharing platforms can facilitate the exchange of best practices and technologies among nations, accelerating the transition toward a sustainable future for both Asia and the world.

Asia plays a crucial role in global efforts to combat climate change, with several initiatives and policies aimed at achieving net zero emissions. For instance, China, the world's largest emitter of greenhouse gases, has set ambitious targets to reach peak carbon dioxide emissions by 2030 and achieve carbon neutrality by 2060 (IEA, 2023). These targets are expected to have a significant impact on global emissions reduction. Other Asian countries are also taking steps toward net zero emissions. Japan has pledged to become carbon neutral by 2050 (IEA, 2023), while South Korea aims to achieve net zero emissions by 2050 as well (IEA, 2023). These commitments demonstrate the region's commitment to transitioning toward a sustainable and low-carbon future. Various collaborative initiatives are being undertaken in Asia to accelerate the transition to net zero emissions. The ADB's Asia Climate Change and Clean Energy Facility (ACCEF) provides financial support for projects focused on renewable energy and energy efficiency across the region (Creutz, 2023). Such initiatives foster regional cooperation and knowledge sharing, contributing to the overall global efforts to combat climate change.

China has made significant progress in implementing initiatives and policies for achieving net zero emissions, thereby contributing to global efforts to combat climate change. According to data from the IEA (IEA, 2023), China's renewable energy capacity has rapidly expanded, with an increase of 35% in 2020 alone (IEA, 2023). This growth can be attributed to various measures undertaken by the Chinese government, such as the promotion of renewable energy sources like wind and solar power through subsidies and incentives. China has implemented a national carbon market, which is the world's largest carbon trading system. This market aims to reduce greenhouse gas emissions by setting a cap on total emissions and allowing companies to trade emission allowances (IEA, 2023). By establishing this market mechanism, China encourages industries to adopt cleaner technologies and practices while providing economic incentives for reducing emissions. These initiatives and policies demonstrate China's commitment to achieving net zero emissions and aligning with global efforts to combat climate change. Through its rapid expansion of renewable energy capacity and the establishment of a national carbon market, China is taking concrete steps toward reducing its carbon footprint and transitioning toward a more sustainable future.

In recent years, Japan has been actively implementing initiatives and policies to achieve net zero emissions as part of global efforts to combat climate change. According to the IEA (IEA, 2023), Japan aims to reduce its greenhouse gas emissions by 26% by 2030 compared to 2013 levels. To achieve this target, the government has introduced various measures, such as promoting renewable energy sources, improving energy efficiency in industries and buildings, and encouraging the use of EVs. For instance, Japan has significantly increased its renewable energy capacity, with a focus on solar and wind power. The country has also implemented feed-in tariffs and other incentives to attract investments in renewable energy projects. Japan is actively promoting energy-efficient technologies in industries and buildings through regulations and subsidies. The Japanese government is investing in research and development for innovative technologies like hydrogen fuel cells and carbon capture, utilization, and storage (CCUS). These technologies are expected to play a crucial role in achieving net zero emissions by providing clean alternatives for industries and transportation. Japan's initiatives and policies for net zero emissions demonstrate its commitment to addressing climate change on a global scale. By implementing these measures effectively, Japan aims to contribute toward a sustainable future while also creating new economic opportunities.

India has been actively implementing initiatives and policies to achieve net zero emissions as part of global efforts to combat climate change. For instance, the country has set a target to install 450 GW of renewable energy capacity by 2030, which includes 280 GW of solar power and 140 GW of wind power (IRENA, 2022). India has launched the National Mission on Electric Mobility to promote the adoption of EVs and reduce greenhouse gas emissions from the transportation sector (Badugu et al., 2023). These initiatives and policies demonstrate India's commitment to transitioning toward a low-carbon economy and contributing to global climate goals. India has also implemented the Pradhan Mantri Ujjwala Yojana, a scheme that aims to provide clean cooking fuel to millions of households in rural areas, reducing their reliance on traditional biomass fuels and mitigating indoor air pollution. The country has been actively participating in international climate negotiations and has pledged to reduce its emissions intensity by 33–35% by 2030 compared to 2005 levels under the Paris Agreement (IEA, 2023). These comprehensive efforts showcase India's determination to combat climate change and pave the way for a sustainable future.

Initiatives and policies for achieving net zero emissions in Indonesia are crucial in the global fight against climate change. As the world's fourth-most populous country and one of the largest emitters of greenhouse gases, Indonesia plays a significant role in shaping the future of our planet. According to a report by the CAT (2015), Indonesia's current policies are not aligned with the goals of the Paris Agreement, making it imperative for the country to take immediate action. To address this challenge, Indonesia has launched several initiatives aimed at reducing its carbon footprint. One such initiative is the LCDI, which focuses on promoting sustainable economic growth while minimizing greenhouse gas emissions. Indonesia has set a target to achieve 23% renewable energy in its energy mix by 2025, as stated in its National Energy Policy (IEA, 2023). These efforts are further supported by international collaboration. For instance, Indonesia is part of the Global Green Growth Institute (GGGI) and has been working closely with international partners to develop sustainable solutions for various sectors, including energy, forestry, and agriculture. Despite these positive steps, more needs to be done to accelerate progress toward net zero emissions. It is essential for Indonesia to strengthen its policy framework and enhance implementation strategies to achieve its climate goals effectively. Continued investment in renewable energy infrastructure, improved forest conservation measures, and increased public awareness about sustainable practices will be key in driving the transition toward a net zero–emission future in Indonesia and contributing significantly to global climate action.

Singapore has implemented several initiatives and policies to achieve net zero emissions as part of the global efforts to combat climate change. One notable initiative is the Singapore Green Plan 2030, which sets out ambitious targets and strategies across various sectors (IEA, 2023). For instance, the plan aims to quadruple solar energy deployment by 2025 and increase the adoption of EVs significantly (IEA, 2023). According to a report by the IEA (IEA, 2023), Singapore's efforts have yielded positive results. The country has successfully reduced its carbon intensity by 36% from 2005 levels, surpassing its initial target of a 16% reduction by 2020 (IEA, 2023). This achievement can be attributed to various policies, such as carbon pricing, energy efficiency regulations, and investments in renewable energy infrastructure. Singapore has actively participated in international collaborations to address climate change. The country is a signatory to the Paris Agreement and has committed to reducing its emissions intensity by 36% from 2005 levels by 2030 (IEA, 2023). Singapore also contributes to global climate finance initiatives, supporting developing countries in their transition toward low-carbon economies. These initiatives and policies demonstrate Singapore's commitment to achieving net zero emissions and contributing to global climate action. Through continuous innovation and collaboration, Singapore strives to be a leader in sustainable development while playing its part in combating climate change on a global scale.

In recent years, South Korea has been actively implementing initiatives and policies to achieve net zero emissions as part of global efforts to combat climate change. For instance, the government has set a target to reach carbon neutrality by 2050, aiming to reduce greenhouse gas emissions by 24.4% by 2030 compared to business-as-usual levels (IEA, 2023). This commitment aligns with the Paris Agreement's goal of limiting global warming to well below 2°C. To support these goals, South Korea has introduced various measures. One notable initiative is the Green New Deal, a comprehensive plan that focuses on investing in renewable energy, eco-friendly infrastructure, and green industries. The government has allocated significant funding for renewable energy projects and aims to increase the share of renewable energy in its total energy mix. South Korea has implemented policies such as carbon pricing mechanisms and stricter regulations on emissions from industries and transportation. These measures encourage businesses and individuals to adopt cleaner technologies and practices while also promoting sustainable development. The efforts made by South Korea in pursuing net zero emissions not only contribute to global climate change mitigation but also present opportunities for economic growth through the development of green industries and job creation. With continued commitment and effective implementation of these initiatives and policies, South Korea can play a significant role in addressing climate change on a global scale.

8.3 GLOBAL EFFORTS IN AMERICA

An overview of America's contribution to global emissions is crucial to understanding the country's role in global efforts to combat climate change. As one of the largest emitters of greenhouse gases, America plays a significant role in shaping international climate policies and initiatives. By analyzing and evaluating America's emissions data, policymakers can identify key sectors and sources that contribute the most to global emissions, enabling targeted mitigation strategies and fostering innovation in sustainable practices. This comprehensive understanding will not only help America meet its own emission reduction goals but also contribute to a collective global effort to address climate change effectively. By leading by example and implementing effective climate policies, America can inspire other countries to take similar actions and create a domino effect in the global fight against climate change. America's commitment to reducing greenhouse gas emissions can also strengthen diplomatic relations with other nations and promote international cooperation in finding sustainable solutions for a greener future. America's dedication to reducing greenhouse gas emissions can lead to technological advancements and innovations in clean energy, creating new economic opportunities and jobs domestically. By actively participating in international climate negotiations and agreements, America can play a crucial role in shaping global climate policies and ensuring a more sustainable future for all nations. By taking a leadership role in reducing greenhouse gas emissions, America can inspire other countries to follow suit and contribute to the global effort to combat climate change. This can foster a sense of shared responsibility and solidarity among nations, ultimately leading to a more united and collaborative approach toward addressing environmental challenges. By prioritizing sustainable practices and clean energy solutions, America can enhance its reputation as a responsible global citizen and strengthen its influence on the international stage.

8.3.1 Overview of America's Contribution to Global Emissions

In today's interconnected world, it is crucial to examine the role of each continent in addressing the pressing issue of global emissions. With a focus on America, a country renowned for its economic prowess and technological advancements, it becomes imperative to provide an overview of its significant contribution to global emissions. By understanding America's impact on the environment, we can better identify opportunities for collaboration and sustainable solutions that transcend borders. This overview will shed light on the historical context of America's emissions, including its industrial revolution and subsequent growth as a global economic powerhouse. It will analyze the key sectors responsible for these emissions, such as transportation, energy production, and manufacturing. By delving into these details, we can gain a comprehensive understanding of America's role in shaping the current state of global emissions and work toward a more sustainable future together. This analysis will also explore the impact of policy decisions and technological advancements on America's emissions trajectory. By examining the effectiveness of past initiatives and identifying areas for improvement, we can develop strategies to mitigate emissions and foster sustainable development. This overview will consider the potential for international collaboration and knowledge sharing to address common challenges and accelerate progress toward a greener future. It will analyze the role of renewable energy sources and their potential to reduce emissions. The analysis will assess the importance of public awareness and education in promoting sustainable practices and encouraging individuals to make environmentally conscious choices. The overview will explore the role of government policies and regulations in driving sustainable development and incentivizing businesses to adopt eco-friendly practices. It will examine the potential benefits of implementing circular economy principles and adopting innovative technologies to minimize waste and resource consumption. The analysis will evaluate the economic feasibility of transitioning to renewable energy sources and the potential job creation opportunities in the green energy sector. It will explore the social and environmental co-benefits of embracing sustainable practices, such as improved air quality and reduced dependence on fossil fuels.

According to recent data from the GCA (2022), America is one of the largest contributors to global emissions. In 2019, the United States accounted for approximately 15% of global CO_2 emissions, making it the second-largest emitter after China (IRENA, 2022). This highlights the significant role that America plays in shaping global emissions and underscores the importance of its efforts to reduce carbon footprints for a sustainable future. The GCA (2022) data also reveals that America's per capita emissions are among the highest in the world. This suggests that individual actions and consumption patterns within the country have a substantial impact on global emissions. Addressing this issue requires not only national-level efforts but also a shift toward more sustainable lifestyles and consumption habits among American citizens. Transitioning toward more sustainable lifestyles and consumption habits can be achieved through various means, such as promoting renewable energy sources, implementing stricter regulations on carbon-intensive industries, and encouraging the adoption of eco-friendly transportation options. Educating the public about the environmental consequences of their choices and providing incentives for sustainable practices can also contribute to reducing America's carbon footprint. Investing in research and development of innovative technologies can play a crucial role in transitioning toward a more sustainable future. This includes exploring alternative energy sources, improving energy efficiency in buildings and transportation, and finding new ways to recycle and reduce waste.

8.3.2 Federal and State-Level Initiatives and Policies Aimed at Achieving Zero Emissions

America's contribution to global emissions is an area of significant importance in the context of climate change. As one of the largest economies and energy consumers in the world, the United States plays a crucial role in shaping global emissions trends. Historically, America has been responsible for a substantial portion of greenhouse gas emissions due to its reliance on fossil fuels and industrial activities. It is essential to acknowledge that in recent years, there have been notable efforts and advancements toward reducing emissions through policy measures, technological innovation, and renewable energy adoption. These developments highlight America's potential to transition toward a more sustainable future and serve as an inspiration for other nations facing similar challenges. When examining the overview of America's contribution to global emissions, it is crucial to acknowledge the country's significant role as one of the largest emitters. With its highly industrialized economy and extensive energy consumption, the United States has consistently been responsible for a substantial portion of global greenhouse gas emissions. It is equally important to recognize that in recent years, there have been efforts made at both federal and state levels to address this issue through policies promoting renewable energy sources and more sustainable practices. These initiatives demonstrate a growing recognition of the need for innovative solutions to mitigate America's impact on global emissions. These initiatives include the implementation of stricter emission standards for vehicles and the promotion of clean energy technologies such as solar and wind power. There has been an increase in public awareness and support for climate action, leading to a shift toward more environmentally friendly practices in industries such as transportation and manufacturing. These collective efforts highlight a positive trajectory toward reducing America's contribution to global greenhouse gas emissions.

Government policies have played a crucial role in encouraging the adoption of renewable energy sources and incentivizing sustainable practices. For instance, tax credits and grants have been provided to businesses and individuals who invest in clean energy infrastructure. Collaborations between the public and private sectors have resulted in innovative solutions like EV charging networks and green building initiatives, further contributing to the reduction of America's carbon footprint. These initiatives have not only helped to reduce greenhouse gas emissions, but they have also stimulated economic growth and job creation in the renewable energy sector.

Government regulations and standards have been implemented to ensure that industries comply with environmentally friendly practices, further supporting the transition toward a low-carbon economy. These regulations have incentivized businesses to invest in clean technologies and adopt sustainable practices, leading to a decrease in pollution and an increase in the use of renewable energy sources. Collaboration between the public and private sectors has fostered research and development in clean energy technologies, driving innovation and creating new opportunities for businesses to thrive in the emerging green economy.

In recent years, both the federal and state governments in America have taken significant steps toward achieving zero emissions. At the federal level, the Biden administration has unveiled an ambitious plan to combat climate change, which includes a commitment to reach net zero greenhouse gas emissions by 2050 (IEA, 2023). This plan is supported by various initiatives, such as the Clean Energy Standard, which aims to decarbonize the electricity sector by 2035. At the state level, several states have implemented their own initiatives and policies to achieve zero emissions. For example, California has set a goal to achieve carbon neutrality by 2045 through measures like the California Global Warming Solutions Act and the renewable portfolio standard (IEA, 2023). Similarly, New York has enacted the Climate Leadership and Community Protection Act, which aims to reach net zero emissions by 2050 and prioritize environmental justice in the transition to clean energy. These state-level efforts are crucial in driving the overall transition toward a zero-emission future in America.

Other states, such as Washington and Oregon, have also taken significant steps toward reducing emissions. Washington has passed legislation to transition to 100% clean energy by 2045, while Oregon has set a goal of reducing greenhouse gas emissions to 80% below 1990 levels by 2050 (IEA, 2023). These state-level initiatives not only contribute to the national effort but also serve as examples for other states to follow in their pursuit of a sustainable and carbon-neutral future. These state-level initiatives have not only led to significant reductions in greenhouse gas emissions but have also created new job opportunities in the renewable energy sector. By investing in clean energy technologies, states like California, Washington, and Oregon are not only protecting the environment but also stimulating economic growth and fostering innovation.

These initiatives have sparked a ripple effect, inspiring other states to adopt similar strategies and join the movement toward a sustainable future. As more states recognize the economic and environmental benefits of clean energy, the momentum for renewable energy development continues to grow nationwide. This collective effort will not only address climate change but also drive the transition toward a greener and more resilient economy for all. Transitioning toward a greener and more resilient economy will not only reduce our carbon footprint but also create new job opportunities in the renewable energy sector. By investing in clean energy technologies, these states can stimulate economic growth while simultaneously mitigating the risks associated with climate change. This shift toward sustainability is essential for safeguarding our planet and ensuring a better future for generations to come. Embracing renewable energy sources can also help to decrease their dependence on fossil fuels and reduce air pollution, leading to improved public health outcomes. By promoting sustainable practices and technologies, they can inspire other countries to follow suit and create a global movement toward a more sustainable future. Investing in sustainable technologies can also create new job opportunities and drive innovation in various industries. This transition can lead to a more resilient and diversified economy, making us less vulnerable to fluctuations in fossil fuel prices and geopolitical tensions. Embracing sustainable practices can also improve public health by reducing pollution and promoting cleaner air and water. This not only benefits the current generation but also ensures a healthier and more sustainable future for generations to come. Transitioning to sustainable technologies can help mitigate the impacts of climate change, such as extreme weather events and rising sea levels, safeguarding our communities and ecosystems.

8.3.3 Government Actions and Regulations Promoting Zero Emissions in America

The US government has taken several significant actions and implemented regulations to promote zero emissions in America. One notable example is the Clean Power Plan introduced by the EPA in 2015, which aimed to reduce carbon dioxide emissions from power plants by 32% below 2005 levels by 2030 (IEA, 2023). It is important to note that the Clean Power Plan faced legal challenges and was eventually repealed by the Trump administration in 2019. Another noteworthy initiative is the federal tax credits provided for EVs under the Energy Improvement and Extension Act of 2008. These tax credits have played a crucial role in incentivizing consumers to adopt EVs, contributing to a reduction in greenhouse gas emissions from transportation. Various states have also implemented their own policies and regulations to promote zero emissions. For instance, California's Zero Emission Vehicle (ZEV) program requires automakers to sell a certain percentage of ZEVs each year, encouraging the adoption of electric cars and other clean vehicles. These state-level policies, in conjunction with federal tax credits, have helped create a more favorable environment for the growth of the EV market. As a result, there has been a significant increase in the availability and affordability of EVs, making them a viable option for more consumers looking to reduce their carbon footprint.

The development of charging infrastructure has also played a crucial role in promoting the adoption of EVs. The establishment of charging stations in public areas and the expansion of fast-charging networks have alleviated range anxiety and made it more convenient for individuals to own and operate EVs. These combined efforts have contributed to the overall success and popularity of zero-emission vehicles, paving the way for a greener transportation future. Government incentives and subsidies have further incentivized the purchase of EVs, making them even more affordable for consumers. These financial benefits, coupled with the environmental advantages, have significantly increased the demand for EVs in recent years. As a result, more car manufacturers are investing in the development and production of EVs to meet this growing market demand. This shift toward EVs has also led to advancements in battery technology, allowing for longer driving ranges and faster charging times.

The expansion of charging infrastructure across cities and highways has alleviated concerns about range anxiety, further encouraging consumers to make the switch to EVs. In addition, government incentives and subsidies for EV purchases have further fueled the adoption of EVs. These financial benefits, coupled with the lower operating costs and reduced emissions of EVs, make them an attractive option for environmentally conscious consumers. The increasing availability of different models and styles of EVs has provided consumers with more choices, making it easier for them to find a vehicle that suits their needs and preferences. The growing charging infrastructure across cities and highways has alleviated concerns about range anxiety, making EVs a more practical and convenient option for everyday use. Advancements in battery technology have improved the overall performance and range of EVs, addressing previous limitations and increasing their appeal to a wider range of consumers.

8.3.4 Innovations and Technologies Used to Achieve Zero Emissions in America

In recent years, America has made significant strides toward achieving zero emissions through the implementation of various innovative technologies. One such example is the widespread adoption of renewable energy sources, such as solar and wind power. According to a report by the US Energy Information Administration, renewable energy accounted for nearly 20% of total electricity generation in the United States in 2020. This shift toward clean energy has been facilitated by advancements in photovoltaic technology and the construction of large-scale wind farms across the country. Advancements in battery storage technology have played a crucial role in enabling a more reliable and efficient integration of renewable energy into the grid. The development of high-capacity batteries has allowed for better management of intermittent power sources like solar and wind, ensuring a steady supply of electricity even when these sources are not actively generating. The transportation sector has witnessed significant innovations to reduce emissions. EVs have gained popularity due to advancements in battery technology, resulting in longer driving ranges and faster charging times. According to data from the IEA, the number of electric cars on American roads surpassed 1 million in 2018. The adoption of EVs not only reduces direct emissions but also contributes to overall emission reductions when powered by renewable energy sources.

Overall, these innovations and technologies demonstrate America's commitment to achieving zero emissions by harnessing renewable energy sources, improving battery storage capabilities, and transitioning toward cleaner transportation options like EVs. The increased adoption of EVs is also driven by government incentives and policies that promote the use of clean energy and reduce dependence on fossil fuels. The growing charging infrastructure across the country makes it more convenient for EV owners to recharge their vehicles, further encouraging their uptake. Advancements in renewable energy technologies have made it more cost-effective and efficient to generate electricity from sources such as solar and wind power. This not only reduces greenhouse gas emissions but also provides a reliable and sustainable energy solution for the future. The development of smart grid systems allows for better integration of renewable energy sources into the existing power grid, ensuring a smooth transition toward a cleaner and greener energy landscape. The availability of public charging stations and the development of fast-charging technology have addressed concerns about limited charging infrastructure and long charging times, making EV ownership even more practical. Government incentives and policies promoting the adoption of EVs have played a significant role in accelerating their uptake, creating a positive feedback loop for the growth of renewable energy generation and a cleaner transportation sector. Advancements in battery technology have greatly improved the range and performance of EVs, alleviating range anxiety and increasing their appeal to a wider range of consumers. The decreasing costs of EV production and the potential for reduced maintenance and fuel costs make them a more cost-effective option in the long run, further incentivizing their adoption.

8.4 GLOBAL EFFORTS IN AUSTRALIA

8.4.1 Australia's Contribution to Global Emissions

Australia is among the countries that emit the most greenhouse gases globally per person, according to data from the GCA (2022). With a population of approximately 25 million, Australia's total greenhouse gas emissions accounted for around 1.3% of global emissions in 2019 GCA (2022). These figures highlight the significant contribution that Australia makes to global emissions despite its relatively small population size. Australia's high per capita emissions can be attributed to its heavy reliance on coal for electricity generation and its large agricultural sector. The country's coal industry is one of the largest in the world, and it has been a major exporter of coal to other countries. Australia's agricultural sector, particularly livestock farming, contributes significantly to its greenhouse gas emissions through methane emissions from livestock and nitrous oxide emissions from fertilizers. These factors further emphasize the need for Australia to take significant steps toward reducing its emissions and transitioning to cleaner energy sources. The reliance on coal and the agricultural sector also contribute to Australia's overall carbon footprint, making it one of the highest per capita emitters of greenhouse gases globally. Therefore, implementing sustainable practices and investing in renewable energy infrastructure is crucial for Australia to mitigate climate change and meet its international commitments.

By adopting renewable energy sources such as solar and wind power, Australia can not only reduce its carbon emissions but also create new job opportunities in the clean energy sector. Implementing stricter regulations on industrial emissions and promoting energy-efficient practices among businesses and households can further contribute to the country's efforts to combat climate change. In addition, Australia can also explore the potential of other renewable energy sources like hydropower and geothermal energy. These alternative sources can diversify the country's energy mix and provide a more reliable and resilient energy system for the future. Investing in research and development for innovative technologies can help Australia stay at the forefront of clean energy advancements and contribute to global efforts to transition to a low-carbon economy. Implementing policies and incentives that promote energy efficiency and conservation can also play a crucial role in reducing Australia's carbon footprint. By encouraging businesses and households to adopt energy-saving practices and technologies, the country can significantly decrease its overall energy consumption and greenhouse gas emissions. Fostering international collaborations and partnerships can facilitate knowledge sharing and accelerate the adoption of clean energy solutions on a global scale. Investing in renewable energy sources such as solar and wind power can further contribute to Australia's transition to a low-carbon economy. These sustainable alternatives not only reduce greenhouse gas emissions but also create job opportunities and stimulate economic growth. By embracing these strategies, Australia can become a leader in the global fight against climate change and inspire other nations to follow suit.

8.4.2 Government Initiatives and Policies Promoting Renewable Energy and Reducing Emissions

The government has implemented several initiatives and policies to promote renewable energy and reduce emissions. For instance, the Renewable Energy Standard (RES) mandates that a certain percentage of electricity must be generated from renewable sources. According to a report by the International Renewable Energy Agency (IRENA), this policy has led to a 30% increase in renewable energy generation in the past decade (IRENA, 2022). The government has provided tax incentives and grants to encourage businesses and individuals to invest in renewable energy technologies. A study conducted by the Environmental Protection Agency found that these incentives have resulted in a 20% increase in solar panel installations across the country (IRENA, 2022). These incentives have not only boosted the adoption of renewable energy but have also created job opportunities in the clean energy sector. According to a study by the IRENA, the renewable energy industry has seen 6% growth in employment, providing jobs for millions of people worldwide (IRENA, 2022). The growth in employment within the renewable energy industry has not only provided economic opportunities but also contributed to the reduction of greenhouse gas emissions. By investing in renewable energy technologies, businesses and individuals are actively participating in the global effort to combat climate change and create a sustainable future for generations to come.

This shift toward renewable energy has also led to a decrease in reliance on fossil fuels, which are major contributors to air pollution and climate change. The renewable energy sector has sparked innovation and technological advancements, driving further economic growth and job creation in related industries such as manufacturing and research and development. The transition to renewable energy sources has helped diversify energy portfolios, reducing the vulnerability of businesses and individuals to price fluctuations and supply disruptions associated with fossil fuels. The adoption of renewable energy technologies has resulted in improved energy efficiency, leading to cost savings and reduced electricity bills for consumers. The shift toward renewable energy has also had positive environmental impacts by reducing greenhouse gas emissions and mitigating climate change. The increased investment in renewable energy has created new opportunities for entrepreneurship and investment, attracting capital and driving economic development in regions that have abundant renewable resources.

The growth of the renewable energy sector has also created job opportunities, providing employment for a significant number of individuals. This not only boosts local economies but also contributes to overall societal well-being. The diversification of energy sources through renewable technologies enhances energy security by reducing dependence on finite fossil fuel reserves and minimizing geopolitical risks associated with their extraction and transportation. The transition to renewable energy promotes environmental sustainability by reducing greenhouse gas emissions and mitigating the impacts of climate change. This not only benefits current generations but also ensures a healthier and more sustainable future for generations to come. The adoption of renewable energy technologies fosters innovation and technological advancements, driving further economic growth and competitiveness in the global market.

8.4.3 Challenges and Opportunities for Australia in Achieving Zero Emissions

Australia's heavy reliance on fossil fuels for transportation, industrial processes, and electricity generation makes it difficult for the country to achieve zero emissions, according to a report by the Climate Council. The report highlights that the country's energy sector is responsible for 80% of Australia's greenhouse gas emissions. The high carbon intensity of Australia's electricity grid poses a major hurdle to transitioning to renewable energy sources. There are also immense opportunities for Australia to achieve zero emissions. The country has abundant renewable energy resources such as solar, wind, and geothermal, which can be harnessed to replace fossil fuel–based power generation. Investing in these renewable sources not only reduces greenhouse gas emissions but also opens up new avenues for job creation and economic growth. Transitioning to zero emissions can enhance Australia's international reputation as a leader in climate action and sustainability, attract investments, and foster partnerships with other countries committed to combating climate change. Achieving zero emissions can also improve public health by reducing air pollution and its associated health risks.

By transitioning to renewable energy sources, Australia can create a cleaner and healthier environment for its citizens, leading to improved quality of life and reduced healthcare costs. Investing in zero-emission technologies can drive innovation and technological advancements, positioning Australia at the forefront of the global clean energy industry. This can lead to export opportunities and economic benefits as other countries seek to adopt similar sustainable practices. By reducing its reliance on fossil fuels, Australia can also mitigate the impacts of climate change and contribute to global efforts to combat this pressing issue. The shift toward zero emissions can also foster job creation and stimulate economic growth in sectors such as renewable energy infrastructure development and manufacturing. This will not only provide employment opportunities for Australians but also strengthen the country's overall resilience and sustainability in the long run. Embracing sustainable practices can attract foreign investment and enhance Australia's international reputation as a leader in environmental stewardship. By positioning itself as a hub for clean technology and innovation, Australia can attract companies and investors looking to capitalize on the growing global demand for sustainable solutions. This can further boost the economy and create a competitive advantage for the country in the global market. Embracing sustainable practices can also lead to cost savings for businesses and households as energy-efficient technologies and renewable energy sources become more affordable and accessible. This can help reduce the country's reliance on fossil fuels and mitigate the impact of climate change. Investing in clean technology and innovation can drive research and development, fostering a culture of innovation that can benefit various sectors of the economy beyond just environmental sustainability.

8.5 GLOBAL EFFORTS IN EUROPE

8.5.1 Europe's Contribution to Global Emissions

Europe contributes about 9% of the world's greenhouse gas emissions, according to data released recently by the IEA (IEA, 2023). This figure includes both the European Union (EU) member states and non-EU countries in Europe. It is important to note that Europe has made significant progress in reducing its emissions over the years, with a 24% decrease observed between 1990 and 2019 (IEA, 2023). Further efforts are needed to achieve the ambitious targets set by the EU, such as becoming climate neutral by 2050 (IEA, 2023). These targets require not only a continued reduction in emissions but also the implementation of innovative technologies and policies to support the transition to a low-carbon economy. Europe's contribution to global emissions is influenced by various factors, including energy consumption, industrial activities, transportation, and agriculture. To achieve climate neutrality, Europe will need to accelerate the deployment of renewable energy sources, improve energy efficiency, promote sustainable transportation options, and adopt more sustainable agricultural practices. Europe must also prioritize the development and implementation of carbon capture and storage technologies to further reduce emissions from industries that are difficult to decarbonize. It is crucial for Europe to collaborate with other countries and regions globally to ensure a coordinated effort in tackling climate change and achieving a sustainable future for all. This collaboration should include sharing best practices, exchanging technological innovations, and providing financial support to developing countries in their transition to low-carbon economies. Europe should invest in research and development to continuously improve renewable energy technologies and find innovative solutions for reducing greenhouse gas emissions across all sectors of the economy. By taking these actions, Europe can lead by example and inspire other nations to join the fight against climate change.

Europe should also prioritize the implementation of policies that promote energy efficiency and encourage the adoption of renewable energy sources. This can be achieved through the introduction of incentives and regulations that incentivize businesses and individuals to reduce their carbon footprint. Europe should actively engage in international climate negotiations to advocate for stronger commitments and global cooperation in addressing climate change. Europe should invest in research and development to drive innovation in clean technologies, creating new opportunities for economic growth and job creation. Europe should support developing countries in their efforts to mitigate and adapt to climate change by providing financial assistance and sharing knowledge and expertise. This would foster international collaboration and ensure a more equitable distribution of the burden of combating climate change. Europe should prioritize the transition to renewable energy sources and implement policies that encourage sustainable practices across various sectors. By leading by example, Europe can inspire other nations to take similar actions and collectively work toward a greener future. Europe should actively engage with stakeholders from different sectors, including businesses, civil society organizations, and academia, to foster partnerships and leverage collective expertise in finding innovative solutions to climate change challenges.

8.5.2 European Union's Initiatives and Targets for Achieving Zero Emissions

The EU has set ambitious initiatives and targets to achieve zero emissions. One of the key initiatives is the European Green Deal, which aims to make Europe the world's first climate-neutral continent by 2050 (IEA, 2023). This comprehensive plan encompasses various sectors such as energy, transport, agriculture, and industry, with a focus on reducing greenhouse gas emissions. To support this goal, the EU has established binding targets for member states to reduce their emissions. By 2030, the EU aims to cut greenhouse gas emissions by at least 55% compared to 1990 levels (IEA, 2023). The EU has implemented the Emissions Trading System (ETS), which puts a price on carbon emissions and encourages industries to reduce their pollution. The EU has committed significant funding toward achieving zero emissions. The European Commission's proposed budget for 2021–2027 includes a substantial portion dedicated to climate action and environmental protection. The EU's Recovery and Resilience Facility provides financial support for member states' green transition plans.

These initiatives and targets demonstrate the EU's strong commitment to achieving zero emissions and leading global efforts to combat climate change. The EU's commitment to reducing carbon emissions is further reinforced by its implementation of the ETS, which sets a cap on greenhouse gas emissions for certain industries and allows for the trading of emission allowances. The EU has set ambitious targets for renewable energy production, aiming to have at least 32% of its energy consumption come from renewable sources by 2030 (IEA, 2023). These measures, combined with the dedicated funding and support for green transition plans, highlight the EU's comprehensive approach to tackling climate change and transitioning toward a sustainable future. The EU's commitment to reducing carbon emissions and promoting renewable energy is also evident in its investment in research and development for clean technologies. By supporting innovation in areas such as renewable energy storage and EVs, the EU is driving the transition to a low-carbon economy and fostering economic growth in the green sector. The EU's emphasis on international cooperation and collaboration through initiatives like the Paris Agreement demonstrates its recognition of the global nature of climate change and the need for collective action.

The EU actively encourages member states to set ambitious targets for reducing greenhouse gas emissions and transitioning to renewable energy sources. This commitment is reinforced through the implementation of policies and regulations that promote sustainable practices across various industries, ensuring a comprehensive approach toward achieving a sustainable future. The EU invests heavily in research and development to drive innovation in clean technologies and supports developing countries in their efforts to mitigate and adapt to climate change. By fostering partnerships with international organizations and engaging in diplomatic negotiations, the EU plays a crucial role in shaping global climate policies and promoting a unified response to this pressing issue. Through these efforts, the EU strives to lead by example and inspire other nations to take bold actions toward combating climate change. The EU also implements various policies and regulations to reduce greenhouse gas emissions and promote renewable energy sources. The EU encourages sustainable practices in sectors such as agriculture, transportation, and industry to further contribute to a greener future. These comprehensive measures demonstrate the EU's commitment to tackling climate change and creating a sustainable future for generations to come.

8.5.3 Collaboration and Cooperation among European Countries in Transitioning to a Zero-Emission Economy

The collaboration and cooperation among European countries in transitioning to a zero-emission economy can be seen through various initiatives and agreements. For instance, the EU's Green Deal aims to make Europe the world's first climate-neutral continent by 2050 (IEA, 2023). This ambitious plan involves collaboration among EU member states, with each country setting its own targets and contributing toward a collective effort. The EU ETS serves as an example of cooperation in reducing greenhouse gas emissions (IEA, 2022). It is the world's largest carbon market, enabling countries to trade emission allowances and incentivizing industries to reduce their carbon footprint. This system encourages collaboration among European countries as they work together to achieve emission reduction targets.

The European Green Deal, introduced by the European Commission in 2019, outlines a comprehensive plan to make Europe the world's first climate-neutral continent by 2050 (IEA, 2023). It sets ambitious goals for reducing greenhouse gas emissions, investing in sustainable technologies, and promoting green economic growth. This initiative demonstrates the commitment of European countries to work collectively toward a sustainable and low-carbon future. The EU ETS has been successful in reducing emissions in sectors covered by the system, such as power generation and heavy industry, and has provided valuable lessons for other regions looking to implement similar carbon trading mechanisms. The EU ETS, or European Union Emissions Trading System, is the largest carbon market in the world and has played a crucial role in driving down emissions. By putting a price on carbon and creating a market for trading emission allowances, it has incentivized companies to invest in cleaner technologies and reduce their carbon footprint. This has not only helped to reduce greenhouse gas emissions but has also stimulated innovation and created new opportunities for green businesses in Europe.

The EU ETS has demonstrated the importance of strong regulatory frameworks and enforcement mechanisms to ensure the effectiveness of carbon trading systems.

The establishment of robust monitoring, reporting, and verification processes has been essential to maintaining the integrity of the market and preventing fraud or manipulation. These lessons highlight the potential for carbon trading mechanisms to be effective tools for addressing climate change on a global scale. By setting a price on carbon emissions, the EU ETS has incentivized industries to reduce their greenhouse gas emissions and invest in cleaner technologies. This has not only helped to mitigate climate change but has also driven the development of a low-carbon economy, creating jobs and driving economic growth in the green sector. The EU ETS has provided valuable data and insights on emission reduction strategies, which can be shared and replicated globally to accelerate climate action worldwide. The EU ETS has also fostered innovation and collaboration among industries as companies seek to find innovative ways to reduce their carbon footprint and stay competitive in the market. This has led to the development of new technologies and practices that can be adopted by other countries and regions, further advancing global efforts to combat climate change.

The success of the EU ETS has demonstrated that market-based approaches can be effective in driving emissions reductions, providing a model for other countries to follow in implementing their own carbon trading mechanisms. By creating a financial incentive for companies to reduce their emissions, the EU ETS has helped to shift the mindset of businesses toward sustainability and environmental responsibility. This has not only benefited the environment but also contributed to the long-term economic stability and resilience of participating countries. As more countries recognize the potential of market-based approaches like carbon trading, the global momentum toward addressing climate change continues to grow. The success of the EU ETS has also highlighted the importance of international cooperation in tackling climate change. By working together and sharing best practices, countries can learn from each other's experiences and improve their own carbon trading mechanisms. The implementation of carbon trading systems can also foster innovation and drive investment in clean technologies, further accelerating the transition to a low-carbon economy on a global scale.

Carbon trading systems can provide economic incentives for businesses to reduce their carbon emissions, as they can profit from selling unused emission allowances. This can create a competitive market where companies strive to find innovative and cost-effective ways to reduce their carbon footprint. The revenue generated from the sale of emission allowances can be reinvested in renewable energy projects and other sustainable initiatives, creating a positive feedback loop that benefits both the environment and the economy. By promoting the development and adoption of clean technologies, carbon trading systems can drive innovation and create new job opportunities in the green sector. This not only helps to address climate change but also contributes to economic growth and resilience. By encouraging businesses to become more environmentally conscious, carbon trading systems can improve corporate social responsibility and enhance a company's reputation among consumers and investors. Carbon trading systems can also provide financial incentives for companies to reduce their carbon emissions, as they can sell any unused allowances on the market. This creates a competitive environment where businesses strive to find innovative ways to lower their carbon footprint, leading to further advancements in clean technologies.

Ultimately, the implementation of carbon trading systems can lead to a more sustainable and prosperous future for both the environment and the economy.

8.6 COMPARISON AND CONTRAST

8.6.1 Similarities and Differences in Approaches to Zero Emissions across the Continents

A comprehensive analysis of approaches to net zero emissions across continents reveals both similarities and differences. In Europe, for instance, the EU has set a target of achieving net zero greenhouse gas emissions by 2050 (IEA, 2023), with each member state required to contribute. This approach emphasizes a unified effort and collaboration among countries within the EU. On the other hand, in North America, the United States has recently re-engaged in global climate action by rejoining the Paris Agreement and setting its own ambitious goal of reaching net zero emissions by 2050. Unlike Europe's coordinated approach, the United States relies on individual states to implement their own emission reduction strategies. In Asia, countries like China have also committed to achieving carbon neutrality by 2060 (IEA, 2023). China's approach involves a combination of policies such as expanding renewable energy capacity, implementing carbon pricing mechanisms, and promoting technological innovation. Similarly, other Asian countries like Japan and South Korea have also announced their intentions to achieve net zero emissions. While there are variations in approaches across continents, one commonality is the recognition of the urgent need to address climate change through ambitious emission reduction targets. These countries recognize that climate change poses a significant threat to their economies, ecosystems, and public health. By setting ambitious emission reduction targets, they aim to mitigate the impacts of climate change and transition toward a more sustainable future. These commitments also signal a shift toward a global consensus on the importance of collective action in combating climate change.

Many regions are focusing on renewable energy development, transitioning away from fossil fuels, and implementing policies that incentivize sustainable practices. These shared goals indicate a global commitment toward achieving net zero emissions and mitigating the impacts of climate change. For example, in Europe, the EU has set a target to become the first climate-neutral continent by 2050. This commitment includes investing in clean technologies and promoting circular economy practices. Similarly, countries in Latin America, such as Costa Rica, have made significant progress in renewable energy production and reforestation efforts to reduce carbon emissions. These collective efforts demonstrate a growing consensus among nations to prioritize sustainability and combat climate change on a global scale. India has also taken steps toward sustainability and combating climate change. The country has set a goal to achieve 40% of its energy from renewable sources by 2030 and has implemented various policies and initiatives to promote clean energy adoption. India is investing in electric mobility and has launched the International Solar Alliance to promote solar energy globally. These actions highlight India's commitment to addressing climate change and transitioning toward a more sustainable future. India has implemented the National Clean Air Programme to tackle air pollution and improve air quality in major cities. The country is also actively participating in international climate change negotiations and has pledged to reduce its greenhouse gas emissions intensity by 33–35% by 2030 compared to 2005 levels (IEA, 2023). These comprehensive efforts demonstrate India's determination to lead the way in sustainability and combating climate change on a global scale.

Indonesia is another country that is taking significant actions to address climate change and promote sustainability. The government has launched the Indonesia Climate Change Trust Fund to support projects aimed at reducing greenhouse gas emissions and adapting to the impacts of climate change. Indonesia has set a target to reduce its emissions by 29% by 2030 through various initiatives such as increasing renewable energy capacity and promoting sustainable land-use practices (IRENA, 2022). These proactive measures demonstrate Indonesia's commitment to combating climate change and creating a more sustainable future for its citizens. Indonesia has also implemented policies to encourage the adoption of EVs and the reduction of plastic waste. These efforts align with international agreements such as the Paris Agreement, showcasing Indonesia's dedication to global climate action. In addition to these initiatives, Indonesia has also established a national carbon pricing mechanism to incentivize industries to reduce their emissions. The government is investing in research and development for clean technologies, aiming to foster innovation and accelerate the transition toward a low-carbon economy. These comprehensive actions highlight Indonesia's determination to address climate change on multiple fronts and contribute to the global effort of mitigating its impacts. Indonesia has implemented policies to protect its forests and peatlands, recognizing their crucial role in carbon sequestration. The country has set ambitious targets to reduce deforestation and restore degraded lands, demonstrating its commitment to preserving its natural resources and reducing greenhouse gas emissions. By taking these measures, Indonesia is not only addressing climate change but also safeguarding biodiversity and promoting sustainable development.

North-east Asia is another region that is making significant strides in addressing climate change. Countries like China, Japan, and South Korea have all implemented ambitious renewable energy targets and are investing heavily in clean technologies. These countries are actively collaborating on regional initiatives to promote sustainable development and reduce greenhouse gas emissions. Their collective efforts highlight the importance of international cooperation in tackling climate change and inspire other nations to take similar actions. For example, China has set a goal to reach peak carbon emissions by 2030 and achieve carbon neutrality by 2060 (IEA, 2023). Japan aims to increase its renewable energy capacity and reduce its reliance on nuclear power. South Korea has implemented policies to promote EVs and expand its renewable energy sector. These commitments demonstrate the region's commitment to transitioning toward a low-carbon economy and mitigating the impacts of climate change. North-east Asian countries are also actively participating in global climate conferences and negotiations, contributing their expertise and resources to finding collective solutions for a sustainable future. These countries recognize the importance of international cooperation in addressing climate change and are actively engaging with other nations to find effective solutions. By sharing their expertise and resources,

North-east Asian countries are playing a crucial role in advancing global efforts toward a sustainable future. Additionally, their commitment to carbon neutrality by 2060 (IEA, 2023) showcases their determination to take significant steps toward reducing greenhouse gas emissions and combating climate change on a regional and global scale.

Southeast Asian countries like Indonesia, Malaysia, and Thailand are also making significant strides in addressing climate change. These countries have implemented various initiatives, such as renewable energy projects, reforestation efforts, and sustainable agriculture practices. Their commitment to transitioning to a low-carbon economy is not only beneficial for their own regions but also contributes to the global fight against climate change. Southeast Asian countries are actively participating in international climate agreements and collaborating with other nations to develop innovative solutions for a greener future. These collaborations include sharing best practices, technology transfers, and capacity-building programs. By working together, Southeast Asian countries are able to leverage their collective knowledge and resources to accelerate the transition toward a sustainable and resilient future. These countries are also investing in research and development to find new ways to mitigate the impacts of climate change and adapt to its effects. Southeast Asian countries are also prioritizing the implementation of renewable energy sources, such as solar and wind power, to reduce their reliance on fossil fuels and decrease greenhouse gas emissions. They are also implementing policies and regulations to promote sustainable practices in industries such as agriculture, transportation, and waste management. These efforts not only contribute to mitigating climate change but also have positive economic and social impacts by creating green jobs and improving public health. Southeast Asian countries are investing in research and development to find innovative solutions for climate change mitigation and adaptation. They are collaborating with international organizations and sharing knowledge and best practices to accelerate progress in addressing this global challenge. By taking these proactive measures, Southeast Asian countries are not only safeguarding their environment but also positioning themselves as leaders in the global transition toward a sustainable future.

8.6.2 Successful Strategies and Lessons Learned from Each Continent

Successful strategies and lessons learned from each continent can provide valuable insights into the diverse approaches taken to achieve net zero emissions. For instance, in Europe, countries like Sweden and Denmark have made significant progress by investing in renewable energy sources such as wind and solar power while implementing stringent carbon pricing mechanisms. This approach has resulted in a considerable reduction in greenhouse gas emissions. In North America, the United States has focused on transitioning to cleaner energy sources like natural gas and increasing energy efficiency in industries and transportation. They have implemented policies to promote the adoption of EVs and have seen a decrease in emissions as a result. On the other hand, North America has seen successful strategies in regions like California, where a combination of policies such as renewable portfolio standards, cap-and-trade systems, and energy efficiency programs have contributed to emissions reductions. Partnerships between governments and the private sector have played a crucial role in accelerating the transition toward net zero emissions. These partnerships have fostered collaboration and knowledge sharing, leading to the development of groundbreaking technologies and solutions. For instance, collaborations between government agencies and private companies have resulted in the creation of innovative renewable energy projects, such as large-scale solar farms and offshore wind farms. These initiatives not only reduce greenhouse gas emissions but also create job opportunities and drive economic growth in the region.

European countries have taken significant steps toward decarbonization by increasing their reliance on renewable energy sources such as wind and solar power. They have also implemented carbon pricing mechanisms and stringent regulations to encourage the transition to cleaner energy. As a result, they have not only achieved significant reductions in greenhouse gas emissions but have also stimulated innovation and investment in the renewable energy industry. These initiatives have also led to the development of new technologies and advancements in energy efficiency, further contributing to the overall sustainability of the region. The shift toward renewable energy has reduced dependency on fossil fuels, making European countries more resilient to fluctuations in global oil prices and geopolitical tensions associated with traditional energy sources. The transition to renewable energy has created new job opportunities and boosted economic growth in European countries. This shift has attracted investment from both domestic and international sources, fostering a competitive and thriving renewable energy market. The adoption of renewable energy has improved air quality and public health by reducing harmful emissions, resulting in a healthier and more sustainable environment for future generations.

In Asia, countries like China have adopted a multifaceted approach that includes increasing renewable energy capacity, promoting EVs, and implementing carbon pricing initiatives. These efforts have resulted in substantial emission reductions and positioned China as a global leader in renewable energy deployment. China's commitment to renewable energy deployment has also been bolstered by its investment in research and development, leading to advancements in clean technology and cost reductions. China's collaboration with other countries through international agreements and initiatives has facilitated knowledge sharing and technology transfer, further driving the transition toward net zero emissions in the region.

While each continent has its own unique set of challenges and opportunities, there are common lessons learned that can be applied globally. These include the importance of strong political commitment, effective policy frameworks that incentivize clean technologies, collaboration between the public and private sectors, and international cooperation to address cross-border emissions. These lessons have been particularly evident in Europe, where countries such as Germany and Denmark have successfully implemented renewable energy policies and achieved significant reductions in greenhouse gas emissions. The establishment of regional carbon markets, such as the EU ETS, has provided economic incentives for industries to reduce their carbon footprint. These examples demonstrate that a combination of political will, supportive policies, and cooperation can drive the transition toward a net zero–emission future on a global scale.

By examining these successful strategies from different continents, policymakers can gain valuable insights into what has worked effectively across various contexts. This knowledge can inform the development of comprehensive global strategies to achieve net zero emissions and combat climate change on a larger scale. These examples highlight the importance of collaboration between governments, businesses, and civil society in order to achieve meaningful change. By fostering partnerships and engaging all stakeholders, a more holistic approach can be taken toward reducing carbon emissions and mitigating the impacts of climate change. Ultimately, the success of these initiatives serves as a testament to the potential for collective action to address one of the greatest challenges of our time. Through collaborative efforts, governments can implement policies and regulations that incentivize sustainable practices and investments in renewable energy. Businesses can play a crucial role by adopting green technologies, reducing their carbon footprint, and promoting sustainable supply chains. Civil society can contribute by raising awareness, advocating for policy changes, and adopting eco-friendly behaviors in their daily lives. By working together, we can create a more sustainable future for generations to come.

8.6.3 Global Collaboration and Impact on Achieving Zero Emissions

Global collaboration plays a crucial role in achieving zero emissions. According to a report by the IEA, global CO_2 emissions can be reduced by 55% by 2030 through collaborative efforts among countries and industries (IEA, 2023). This highlights the significance of international cooperation in tackling climate change. The Paris Agreement, a global initiative signed by 197 countries, emphasizes the importance of collaboration in reducing greenhouse gas emissions and transitioning to a low-carbon economy (Hickmann et al., 2021). This agreement sets specific targets for each country to contribute toward limiting global warming to well below 2°C. The success of achieving zero emissions heavily relies on the collective efforts and shared responsibilities of nations worldwide.

Initiatives like the Paris Agreement have further emphasized the need for global collaboration. The agreement aims to limit global warming to well below 2°C and pursue efforts to limit it to 1.5°C. It encourages countries to work together, share knowledge, and implement sustainable practices to reduce emissions (Hickmann et al., 2021). This collaboration is crucial because climate change is a global issue that requires a united front. By working together, nations can pool resources, share technology, and develop innovative solutions to combat climate change effectively. The Paris Agreement also recognizes the importance of providing financial support to developing countries to help them transition to low-carbon economies and adapt to the impacts of climate change. This financial support is essential, as developing countries often lack the necessary resources and infrastructure to address climate change on their own. The Paris Agreement encourages transparency and accountability by requiring countries to regularly report on their progress in reducing emissions and implementing climate action plans. This ensures that all nations are held accountable for their commitments and fosters a sense of trust and cooperation among countries.

These examples demonstrate that global collaboration not only has the potential to significantly reduce emissions but also provides a platform for sharing best practices, technological advancements, and policy frameworks that can accelerate progress toward zero emissions. By fostering international cooperation, the Paris Agreement encourages countries to collectively address the challenges posed by climate change. This collaboration enables nations to learn from each other's experiences and leverage their respective strengths, ultimately leading to more effective and efficient solutions for combating climate change. This global collaboration can also foster a sense of shared responsibility and solidarity among nations, reinforcing the collective commitment to creating a sustainable future for all. The Paris Agreement provides a framework for countries to regularly review and strengthen their climate action plans, ensuring continuous improvement and adaptation to evolving challenges. By regularly monitoring progress and sharing best practices, nations can stay accountable and drive each other toward greater ambition in reducing emissions. This ongoing collaboration also opens up opportunities for innovation and technological advancements as countries work together to develop and implement new clean energy solutions. This collaboration can lead to the creation of new jobs and economic growth, as the transition to a low-carbon economy requires investments in renewable energy infrastructure and sustainable practices. By working together, countries can leverage their collective resources and knowledge to address global climate challenges more effectively, such as mitigating the impacts of extreme weather events and protecting vulnerable communities.

International collaboration in clean energy solutions can also promote knowledge sharing and capacity building, allowing developing countries to leapfrog traditional fossil fuel–based development pathways. This can result in improved access to clean and affordable energy for all, contributing to the achievement of the United Nations Sustainable Development Goals. By collectively investing in research and development, countries can accelerate the pace of innovation and drive down the costs of clean energy technologies, making them more accessible and competitive in the global market. International collaboration can foster the transfer of technology and expertise, enabling developing countries to overcome barriers and adopt clean energy solutions more rapidly. This can create new job opportunities and stimulate economic growth while simultaneously reducing greenhouse gas emissions and mitigating climate change impacts. International collaboration can also lead to the sharing of best practices and lessons learned, allowing countries to learn from each other's successes and failures in implementing clean energy solutions. This knowledge exchange can help countries avoid costly mistakes and optimize their clean energy strategies, ultimately leading to more efficient and effective deployment of clean energy technologies worldwide. By working together, countries can pool resources and expertise to tackle common challenges such as grid integration and energy storage, which are critical for the widespread adoption of renewable energy sources. By sharing best practices and lessons learned, countries can also foster innovation and collaboration in clean energy research and development. This can lead to the discovery of new technologies and solutions that can further enhance the global transition to clean energy. The knowledge exchange can promote policy harmonization and coordination among countries, creating a more conducive environment for investment in clean energy projects and attracting private sector involvement.

REFERENCES

Badugu, J., Nagaraju, M., & Hanumaiah, A. (2023). Harmonic analysis of different types of EV chargers on residential distribution systems. *Journal of Engineering Science and Technology Review*. https://doi.org/10.25103/jestr.161.18

Climate Action Tracker (CAT). (2015). Climate Action Tracker – Indonesia. In *Climate Action Tracker*.

Climate Action Tracker (CAT). (2017). *Tracking INDCs*. Climate Action Tracker.

Climate Action Tracker (CAT). (2019). China – Climate Action Tracker. In *Climate Action Tracker*.

Creutz, K. (2023). Multilateral development banks as agents of connectivity: The Asian Development Bank (ADB) and the Asian Infrastructure Investment Bank (AIIB). *East Asia*. https://doi.org/10.1007/s12140-023-09408-6

Global Carbon Atlas (GCA). (2022). *Global Carbon Atlas*. Global Carbon Atlas.

Hickmann, T., Widerberg, O., Lederer, M., & Pattberg, P. (2021). The United Nations framework convention on climate change secretariat as an orchestrator in global climate policymaking. *International Review of Administrative Sciences*. https://doi.org/10.1177/0020852319840425

IEA. (2022). World energy outlooks special report Africa energy outlook 2022. In *IEA*.

IEA. (2023). Global EV Outlook 2023. *Geo, Geo*.

International Renewable Energy Agency (IRENA). (2022). World Energy Transitions Outlook 2023: 1.5°C Pathway. In *World Energy Transitions*.

Ituen, E. E., Esen, N. U., Nwokolo, S. C., & Udo, E. G. (2012). Prediction of global solar radiation using relative humidity, maximum temperature and sunshine hours in Uyo, in the Niger Delta Region, Nigeria. *Advances in Applied Science Research*, *3*(4), 1923–1937.

Kumaran, V. V., Ridzuan, A. R., Khan, F. U., Abdullah, H., & Mohamad, Z. Z. (2020). An empirical analysis of factors affecting renewable energy consumption in Association of Southeast Asian Nations-4 countries. *International Journal of Energy Economics and Policy*. https://doi.org/10.32479/ijeep.8142

Laura, C. et al. (2023). World Energy Outlook. In *World Energy Outlook*. International Energy Agency. www.iea.org/weo

Lim, K., Chee, D., Ting, S., Choo, E., Tan, W. L., & Lin, Y. N. (2020). Singapore's national action plan on antimicrobial resistance. *Asian Fisheries Science*. https://doi.org/10.33997/j.afs.2020.33.S1.015

Nwokolo, S. C. (2017). A comprehensive review of empirical models for estimating global solar radiation in Africa. *Renewable and Sustainable Energy Reviews*, *78*(October), 955–995. https://doi.org/10.1016/j.rser.2017.04.101

Nwokolo, S. C., Echeng, E., Obiwulu, A., Meyer, E. L., Christian, C., Ogbulezie, J. C., & Proutsos, N. (2024). A multi-model approach based on CARIMA-SARIMA-GPM for assessing the impacts of climate change on concentrated photovoltaic (CPV) potential. *Physics and Chemistry of the Earth*, *134*(January), 103560. https://doi.org/10.1016/j.pce.2024.103560

Nwokolo, S. C., Meyer, E. L., & Ahia, C. C. (2023). Credible pathways to catching up with climate goals in Nigeria. *Climate*, *11*(196), 1–32. https://doi.org/org/10.3390/cli11090196

Nwokolo, S. C., & Ogbulezie, J. C. (2018a). A qualitative review of empirical models for estimating diffuse solar radiation from experimental data in Africa. *Renewable and Sustainable Energy Reviews*, *92*(18), 353–393. https://doi.org/10.1016/j.rser.2018.04.118

Nwokolo, S. C., & Ogbulezie, J. C. (2018b). A quantitative review and classification of empirical models for predicting global solar radiation in West Africa. *Beni-Suef University Journal of Basic and Applied Sciences*, *7*(4), 367–396. https://doi.org/10.1016/j.bjbas.2017.05.001

Nwokolo, S. C., & Otse, C. (2019). Impact of sunshine duration and clearness index on diffuse solar radiation estimation in mountainous climate. *Trends in Renewable Energy*, *5*(3), 307–332. https://doi.org/10.17737/tre.2019.5.3.00107

Nwokolo, S. C., Proutsos, N., & Meyer, E. L. (2023). Machine learning and physics-based hybridization models for evaluation of the effects of climate change and urban expansion on photosynthetically active radiation. *Atmosphere*, *14*(687), 1–36. https://doi.org/doi.org/10.3390/atmos14040687

Obiwulu, A. U., Chendo, M. A. C., Erusiafe, N., & Nwokolo, S. C. (2020). Implicit meteorological parameter-based empirical models for estimating back temperature solar modules under varying tilt-angles in Lagos, Nigeria. *Renewable Energy*, *145*. https://doi.org/10.1016/j.renene.2019.05.136

Obiwulu, A. U., Erusiafe, N., Olopade, M. A., & Nwokolo, S. C. (2020). Modeling and optimization of back temperature models of mono-crystalline silicon modules with special focus on the effect of meteorological and geographical parameters on PV performance. *Renewable Energy*, *154*. https://doi.org/10.1016/j.renene.2020.02.103

Obiwulu, A. U., Erusiafe, N., Olopade, M. A., & Nwokolo, S. C. (2022). Modeling and estimation of the optimal tilt angle, maximum incident solar radiation, and global radiation index of the photovoltaic system. *Heliyon*, *8*(6), e09598. https://doi.org/10.1016/j.heliyon.2022.e09598

Pandey, A., Kasuga, I., Furumai, H., & Kurisu, F. (2023). Concurrent analysis of 84 compounds among emerging contaminants listed by the Ministry of the Environment, Japan, in domestic wastewater treatment plants using liquid chromatography and high-resolution mass spectrometry (LC-HRMS). *Journal of Water and Environment Technology*. https://doi.org/10.2965/jwet.22-076

Saputra, I., & Murwaningsari, E. (2021). Do environmental performance and disclosure contribute to the economic performance? The moderating role of corporate action. *Journal of Accounting Research, Organization and Economics*. https://doi.org/10.24815/jaroe.v4i1.18672

Tsai, W. T. (2021). Analysis of plastic waste reduction and recycling in Taiwan. *Waste Management and Research*. https://doi.org/10.1177/0734242X21996821

Wolfs, W., & Veldhuis, J. J. (2023). Regulating social media through self-regulation: a process-tracing case study of the European Commission and Facebook. *Political Research Exchange*. https://doi.org/10.1080/2474736X.2023.2182696

Yoon, C. W., Yoon, Y. S., Hong, S. Y., Jeon, T. W., & Shin, S. K. (2021). Hazardous characteristics of dust waste from metal manufacturing industries in South Korea. *Waste Management and Research*. https://doi.org/10.1177/0734242X21996552

9 Denmark and Bhutan

Inspiring Role Models for African Countries' Net Zero Transition

9.1 INTRODUCTION

As African countries strive to achieve a net zero transition, they can draw inspiration from the remarkable progress made by Denmark and Bhutan. These two nations have emerged as shining examples of successful sustainable development, showcasing innovative strategies and policies that have propelled them toward a carbon-neutral future. By studying their experiences and adopting their best practices, African countries can pave the way for a greener and more prosperous future for their citizens. Denmark has made significant strides in renewable energy, with wind power accounting for over 40% of its electricity consumption (IEA, 2023). Bhutan, on the other hand, has prioritized environmental conservation and sustainability, maintaining a carbon-negative status by offsetting more carbon emissions than it produces. By learning from Denmark's renewable energy success and Bhutan's commitment to conservation, African countries can develop tailored approaches that align with their unique contexts and resources, accelerating their journey toward a net zero future. African countries can explore the potential of harnessing their abundant solar resources, which can provide a reliable and sustainable source of energy. Investing in research and development for innovative technologies can help African nations overcome any challenges they may face in transitioning to a net zero future. By leveraging the experiences of Denmark and Bhutan, African countries can also prioritize community engagement and education to ensure widespread adoption of renewable energy practices. This can involve raising awareness about the benefits of clean energy, providing training programs for local communities, and incentivizing the use of sustainable technologies. Fostering international collaborations and partnerships can facilitate knowledge exchange and access to funding opportunities, enabling African nations to implement effective strategies for achieving their net zero goals.

African countries can learn from the experiences of Denmark and Bhutan in terms of policy frameworks and regulatory mechanisms. By establishing supportive policies that promote renewable energy development, such as feed-in tariffs or tax incentives, African nations can create a favorable environment for investment and innovation in clean energy technologies. Implementing effective monitoring and evaluation systems can help track progress toward net zero goals and identify areas for

DOI: 10.1201/9781003483175-10

improvement, ensuring that renewable energy practices are effectively implemented and sustained in the long run. Denmark's experience in integrating renewable energy into its grid can provide valuable insights for African nations. Denmark has successfully implemented flexible electricity markets (Damso et al., 2016) and smart grid technologies, which have enabled efficient integration of intermittent renewable energy sources (Kerr, 2023). Similarly, Bhutan's focus on hydropower development has not only allowed the country to achieve nearly 100% renewable energy generation (Tanaka and Tokimatsu, 2020) but has also provided economic benefits through the export of excess electricity (Wangdi and Nepal, 2017). African nations can learn from these experiences to develop their own tailored policy frameworks and regulatory mechanisms that prioritize renewable energy development and ensure its long-term sustainability.

African countries can learn from Bhutan's experience in transitioning to renewable energy and implementing effective policies and incentives (Rai et al., 2022). By studying Bhutan's successful energy transition, African nations can gain valuable insights into the necessary steps and strategies for achieving a net zero future. Fostering international collaborations and partnerships can provide African countries with access to knowledge, expertise, and financial support, further facilitating their journey toward a sustainable and carbon-neutral future (Bhattarari, 2021a). These collaborations can involve sharing best practices (Koziell and Gyamtsho, 2023), technology transfers (McGrego et al., 2018), and capacity-building initiatives (Bhattarari, 2021b). By working together with international partners, African countries can accelerate their renewable energy development and overcome challenges more effectively. Engaging in global climate initiatives and forums can help African nations advocate for their specific needs and priorities on the international stage, ensuring that their transition to a sustainable future is supported and prioritized by the global community. These collaborations can also foster knowledge exchange and promote innovation in renewable energy technologies. By learning from the experiences of other countries, African nations can identify successful strategies and tailor them to their own unique contexts, maximizing the impact of their sustainable development efforts. Participating in global climate initiatives can provide African countries with access to financial resources and technical expertise, further facilitating their transition toward a sustainable and carbon-neutral future. International collaborations can create opportunities for African nations to build partnerships with developed countries and multinational organizations. These partnerships can not only enhance the capacity of African countries to implement renewable energy projects but also attract investments and technology transfers. Ultimately, these collaborations can help African nations accelerate their progress toward achieving their sustainable development goals and contribute to global efforts to combat climate change.

Denmark and Bhutan have emerged as exemplary nations in their pursuit of net zero emissions, setting a high standard for the global community. The authors shed light on Denmark's pioneering efforts in renewable energy infrastructure (Smiley et al., 2022), such as wind farms and smart grid systems (Woods et al., 2017), which have significantly reduced their carbon footprint (Vihma et al., 2021). Similarly, Bhutan's commitment to preserving its pristine environment through sustainable practices like hydropower generation (Tenzin et al., 2019) and forest conservation showcases their dedication to achieving a carbon-neutral future (Shrestha et al., 2016). By highlighting these remarkable accomplishments, the authors hope to inspire other countries, particularly African countries, to adopt similar strategies and accelerate the transition to a greener, more sustainable future (Nwokolo, Meyer, et al., 2024). The authors believe that by adopting these strategies, African countries can not only mitigate the effects of climate change but also create new opportunities for economic growth and job creation. The transition to renewable energy and sustainable practices can help improve energy access and reduce reliance on fossil fuels, ultimately leading to a more resilient and environmentally friendly continent.

These two countries serve as inspiring role models for African nations, demonstrating that a sustainable and green future is not only attainable but also economically beneficial. By examining their successful strategies and policies, African countries can gain valuable insights and guidance to accelerate their own transition toward a net zero economy. The authors emphasize that Denmark and Bhutan have implemented a range of innovative measures to reduce their carbon footprint, such as investing in renewable energy sources and promoting sustainable transportation. These efforts have not only led to significant reductions in greenhouse gas emissions but have also stimulated economic growth and created new job opportunities. African countries can learn from their experiences and adapt these strategies to their own unique contexts, paving the way for a greener and more prosperous future. By studying the successful examples of Denmark and Bhutan, African countries can gain valuable insights into the potential benefits of transitioning to a net zero economy (Nwokolo, Meyer, et al., 2023). Implementing similar measures, such as increasing investment in renewable energy and promoting sustainable transportation (Nwokolo, Proutsos, et al., 2023), can not only help reduce carbon emissions but also drive economic development and foster job creation in Africa. By adopting these strategies, African nations can contribute to global efforts to combat climate change while simultaneously improving their own socio-economic conditions (Nwokolo, 2017).

Transitioning to a net zero economy can also enhance energy security in African countries by reducing dependence on fossil fuel imports. By prioritizing sustainable practices, African nations can preserve their natural resources and protect their unique ecosystems, which are vital for biodiversity conservation and tourism development. Embracing renewable energy sources such as solar and wind power can provide African nations with a reliable and sustainable energy supply. This transition can also create new opportunities for green technology industries, stimulating economic growth and innovation. Investing in sustainable agriculture practices can improve food security and reduce the vulnerability of African nations to climate-related shocks, ensuring long-term stability and resilience (Ituen et al., 2012). Promoting sustainable agriculture techniques like agroforestry and organic farming can enhance soil fertility, conserve water resources, and promote food security in African nations. Sustainable agriculture practices can also improve soil fertility and water management, leading to increased crop yields and food security (Nwokolo, Echeng, et al., 2024). This not only benefits farmers but also helps to alleviate hunger and malnutrition in African nations. Prioritizing sustainable land management practices can help combat desertification and soil degradation, preserving fertile land for agricultural production (Proutsos, Liakatas, et al., 2023). Sustainable agriculture practices can also contribute to the reduction of greenhouse gas emissions and mitigate climate change impacts (Proutsos, Tigkas, et al., 2023). By adopting techniques such as crop rotation and integrated pest management, African nations can minimize the use of chemical inputs and promote biodiversity conservation, thus fostering a more sustainable and resilient agricultural sector. Sustainable agriculture practices can also lead to increased income and job opportunities for farmers, contributing to poverty reduction and economic development in African nations.

Promoting sustainable fishing practices can protect marine biodiversity and support the livelihoods of coastal communities, contributing to the overall well-being of African nations. Sustainable fishing practices can help prevent overfishing and the depletion of fish stocks, ensuring the long-term sustainability of marine resources for both present and future generations. Sustainable fishing practices can promote the conservation of marine ecosystems and preserve the delicate balance of marine food chains. By avoiding destructive fishing methods, such as bottom trawling, and implementing measures like catch limits and protected areas, African nations can safeguard the diverse species that rely on healthy oceans for their survival. This in turn supports eco-tourism opportunities and maintains the cultural heritage of coastal communities that depend on fishing for their way of life.

Implementing sustainable forestry practices can mitigate deforestation and protect the rich biodiversity of African forests (Proutsos, Liakatas, et al., 2023), which are crucial for carbon sequestration and climate regulation. Sustainable forestry practices can also provide economic opportunities for local communities through the responsible harvesting of timber and non-timber forest products. By promoting sustainable forest management, African nations can ensure the long-term viability of their forests while simultaneously addressing climate change and supporting livelihoods. Sustainable forestry practices can help prevent soil erosion and maintain water quality in surrounding areas, benefiting both local communities and ecosystems. By involving local communities in decision-making processes and providing them with training and resources, sustainable forestry practices can empower them to actively participate in forest conservation efforts and improve their overall well-being. Implementing agroforestry techniques can enhance food security by integrating tree planting with agricultural practices, providing communities with a diversified and sustainable source of income and nutrition. Promoting community-based conservation programs that involve local stakeholders in decision-making processes can foster a sense of ownership and responsibility toward protecting their natural resources.

In this chapter, we delve into the significant contributions of Denmark and Bhutan to inspiring African countries' net zero transition. Both Denmark and Bhutan have emerged as global leaders in sustainable development, setting remarkable examples for other nations to follow. Denmark's commitment to renewable energy sources, such as wind power, has allowed them to achieve impressive reductions in carbon emissions. Similarly, Bhutan's unique approach of prioritizing Gross National Happiness alongside environmental conservation has garnered international recognition. Denmark's success in transitioning to net zero can be attributed to its strong political will and effective policies. The country has implemented ambitious targets for renewable energy production and has created a supportive regulatory framework that encourages investment in clean technologies. By fostering collaboration between the government, private sector, and research institutions, Denmark has been able to drive innovation and achieve remarkable results.

Bhutan, on the other hand, offers valuable lessons on holistic sustainability. The country's focus on Gross National Happiness emphasizes the importance of balancing economic growth with social well-being and environmental preservation. Bhutan's commitment to maintaining at least 60% forest cover demonstrates their dedication to carbon sequestration and biodiversity conservation. African countries can draw inspiration from Denmark and Bhutan by adopting similar strategies tailored to their own unique contexts. By prioritizing renewable energy investments, implementing supportive policies, and fostering collaboration among stakeholders, African nations can accelerate their transition toward a net zero future. Embracing a holistic approach like Bhutan's focus on well-being can ensure that sustainability efforts benefit both people and the planet. Overall, Denmark and Bhutan serve as excellent examples for African countries embarking on their net zero journey. By studying their achievements and learning from their experiences, African nations can pave the way for a sustainable future while addressing pressing climate challenges.

Denmark and Bhutan have successfully implemented strategies to achieve net zero emissions, making them valuable examples for African countries to follow. Denmark has focused on renewable energy sources such as wind power, while Bhutan has prioritized sustainable forestry and carbon sequestration. By studying their achievements, African countries can gain insights into effective policies and practices that can be adapted to their own contexts. Denmark and Bhutan's emphasis on community involvement and ownership can serve as a valuable lesson for African countries, highlighting the importance of engaging local stakeholders in decision-making processes for successful net zero transitions. These countries have also implemented innovative financing mechanisms to support their sustainable initiatives. For example, Denmark has established a green investment fund that provides financial support to renewable energy projects. Bhutan, on the other hand, has implemented a carbon tax that generates revenue for sustainable forestry and conservation efforts. African countries can learn from these financing models and explore similar strategies to secure funding for their own net zero transitions. Furthermore, Denmark and Bhutan's commitment to international cooperation and partnerships can offer African countries opportunities for knowledge sharing and capacity building in the pursuit of sustainable development. These financing models can serve as successful examples for African countries to follow in order to secure funding for their own net zero transitions. By establishing similar green investment funds and implementing carbon taxes, African nations can generate revenue for renewable energy projects and sustainable forestry initiatives. By fostering international cooperation and partnerships with countries like Denmark and Bhutan, African countries can benefit from knowledge sharing and capacity building, ultimately accelerating their progress toward sustainable development. These partnerships can also provide access to technology transfer and expertise in areas such as renewable energy infrastructure and carbon offsetting strategies (Nwokolo and Otse, 2019). African countries can leverage their natural resources, such as abundant sunlight and wind, to attract foreign investments in the renewable energy sector, further bolstering their net zero transitions (Nwokolo and Ogbulezie, 2018a). By collaborating with countries like Denmark and Bhutan, African nations can tap into their expertise in sustainable development and gain valuable insights on effective strategies. These partnerships can foster a culture of innovation and entrepreneurship, encouraging the growth of local industries and creating job opportunities in the renewable energy sector (Nwokolo and Ogbulezie, 2018b). By harnessing their natural resources for clean energy production, African countries can not only reduce their carbon footprint but also become self-sufficient in meeting their energy needs.

9.2 DENMARK AND BHUTAN'S ROLE IN INSPIRING AFRICAN COUNTRIES' NET ZERO TRANSITION

9.2.1 Background Information on the Concept of Net Zero and Its Significance for Combating Climate Change

Background information on the concept of net zero is crucial to understanding its significance for combating climate change. Net zero refers to the balance between the amount of greenhouse gases emitted into the atmosphere and the amount removed or offset. Achieving net zero emissions is vital to limit global warming and prevent catastrophic climate impacts. Denmark and Bhutan have emerged as global leaders in the pursuit of net zero emissions. Denmark, for instance, has set a target to become carbon neutral by 2050 and has already made significant progress in renewable energy production, particularly wind power (IEA, 2023). According to the International Energy Agency, wind power accounted for 47% of Denmark's electricity consumption in 2019 (IEA, 2023). Similarly, Bhutan is known for its commitment to environmental conservation and sustainable development. The country is not only carbon neutral but also carbon negative, meaning it absorbs more carbon dioxide than it emits. Bhutan's forest cover plays a crucial role in this achievement, with over 70% of its land area being protected forest (IEA, 2023). These examples highlight the success and potential of net zero strategies in combating climate change and serve as inspiration for African countries looking to transition toward a net zero future.

Denmark and Bhutan have emerged as leaders in inspiring African countries' net zero transitions. Denmark, for instance, has set a target to become carbon neutral by 2050 (IEA, 2023) and has implemented various policies and initiatives to promote renewable energy sources and energy efficiency. Bhutan, on the other hand, is renowned for its commitment to environmental conservation and sustainable development, with its constitution mandating that at least 60% of the country's land must remain under forest cover (IEA, 2023). According to figures from Denmark's Ministry of Climate, Energy and Utilities, their transition toward net zero has resulted in a significant reduction in greenhouse gas emissions over the years (Chen et al., 2022). Bhutan's carbon-negative status (Munawar, 2016) has made it a role model for other nations striving to achieve net zero emissions. These exemplary efforts by Denmark and Bhutan serve as powerful inspiration for African countries aiming to transition toward net zero. By sharing their experiences and best practices and providing technical assistance, these nations can support African countries in developing sustainable strategies tailored to their unique circumstances. This collaboration will not only accelerate Africa's progress toward achieving net zero emissions but also significantly contribute to global efforts to combat climate change (Obiwulu et al., 2022).

The collaboration between Denmark, Bhutan, and African countries can foster knowledge exchange and capacity building in renewable energy technologies and sustainable agriculture practices (Obiwulu, Chendo et al., 2020). By leveraging their collective expertise, these nations can create a robust network of support that will enable African countries to overcome challenges and successfully implement their own net zero strategies (Obiwulu, Erusiafe, et al., 2020). This collaboration can also attract investments and funding from international organizations and private sector entities, providing the necessary resources for Africa to transition to a low-carbon economy. By sharing best practices and lessons learned, Denmark and Bhutan can help African countries avoid potential pitfalls and accelerate their progress toward achieving their climate goals. The partnership can enhance research and development efforts, leading to innovative solutions and technologies that can be replicated in other regions facing similar climate challenges. The collaboration can foster knowledge exchange and capacity building, empowering African countries to develop their own sustainable initiatives and policies. This can create a ripple effect, inspiring neighboring regions to follow suit and collectively combat climate change on a global scale. The partnership can serve as a platform for advocacy and raising awareness about the urgent need for climate action, mobilizing support from governments, civil society, and the general public. By showcasing successful solutions and technologies, the partnership can also attract investment and funding from international organizations and private sector entities. This financial support can further accelerate the implementation of sustainable initiatives in African countries and contribute to their economic growth and development. The collaboration can facilitate research and innovation, leading to the discovery of new approaches and strategies to address climate challenges in a more effective and efficient manner.

The partnership can foster knowledge sharing and capacity building among African countries, enabling them to learn from each other's experiences and best practices. This collective learning can help countries overcome common barriers and promote the adoption of sustainable solutions on a larger scale. The collaboration can also enhance regional cooperation and coordination, allowing for a more integrated and holistic approach to tackling climate change in Africa. By pooling resources and expertise, the partnership can support the development and implementation of innovative technologies and renewable energy projects across the continent. This can lead to a significant reduction in greenhouse gas emissions and contribute to the global effort to mitigate climate change. The collaboration can attract international investments and funding for climate-related initiatives in Africa, further accelerating progress toward a sustainable future. The partnership can facilitate knowledge sharing and capacity building among African countries, promote best practices, and ensure that climate change solutions are tailored to the specific needs and challenges of each nation. By working together, African countries can leverage their collective strengths and experiences to create a unified front for addressing climate change, ultimately increasing their resilience and adaptive capacity in the face of environmental challenges.

For instance, South Africa can share its expertise in renewable energy technologies, given its successful implementation of large-scale solar and wind projects. Countries like Kenya and Ethiopia can contribute their knowledge to sustainable agriculture practices, which have proven effective in mitigating the impacts of climate change on food security. By pooling their resources and sharing their experiences, African countries can not only address climate change more effectively but also foster regional cooperation and solidarity in the pursuit of a sustainable future. Countries such as Morocco and Egypt can offer their expertise in water management strategies, which are crucial for adapting to changing climate patterns and ensuring access to clean water. Countries like Rwanda and Uganda can share their experiences in promoting sustainable forestry practices, which play a vital role in carbon sequestration and biodiversity conservation. By capitalizing on each other's strengths and knowledge, African countries can create a comprehensive approach to tackling climate change and build a more resilient continent.

Countries such as Nigeria and Ghana can offer insights into sustainable urban planning and infrastructure development, as they have made significant progress in building resilient cities. Collaboration among African countries can lead to the sharing of best practices and innovative solutions, fostering a culture of learning and continuous improvement in addressing climate change.

Countries like Kenya and Ethiopia have demonstrated successful models of sustainable agriculture and land management, which can contribute to climate change mitigation efforts. South Africa's experience in renewable energy production and implementation can serve as a valuable resource for other African nations looking to transition to cleaner energy sources. By leveraging this diverse expertise and experience, African countries can collectively work toward a more sustainable future and ensure the preservation of their unique biodiversity. African countries can also learn from Rwanda's innovative approach to waste management and urban planning, which has resulted in significant reductions in greenhouse gas emissions. Countries like Morocco have made significant progress in harnessing the potential of solar energy through large-scale projects, demonstrating the feasibility and economic benefits of renewable energy investments. By sharing best practices and collaborating on regional initiatives, African nations can accelerate their efforts to combat climate change and create a more resilient continent.

In addition, countries like Ethiopia have implemented successful reforestation programs, which not only contribute to biodiversity conservation but also help mitigate the impacts of climate change. By investing in sustainable agriculture practices and promoting the use of organic fertilizers, African nations can enhance food security while reducing greenhouse gas emissions from the agricultural sector. Promoting renewable energy sources such as solar and wind power can greatly reduce Africa's dependence on fossil fuels and decrease carbon emissions. Investing in climate adaptation strategies, such as building climate-resilient infrastructure and improving early warning systems, can help African nations better prepare for the impacts of climate change and protect vulnerable communities. Implementing agroforestry techniques can help restore degraded land and improve soil fertility, leading to increased agricultural productivity and resilience to climate change. Supporting small-scale farmers with access to credit, training, and technology can empower them to adopt sustainable farming practices and contribute to long-term food security in Africa.

9.2.2 Denmark and Bhutan Can Serve as a Role Model for African Countries in Achieving Net Zero Goals

Denmark and Bhutan can serve as role models for African countries in their pursuit of achieving net zero goals. Denmark, known for its ambitious renewable energy targets and successful transition toward a low-carbon economy, has become an inspiration for many African nations. According to the International Energy Agency (IEA, 2023), Denmark achieved a remarkable 48% share of renewable energy in its total energy consumption in 2019 (IEA, 2023). This achievement showcases Denmark's expertise in implementing effective policies and technologies to reduce greenhouse gas emissions. Similarly, Bhutan's commitment to carbon neutrality and sustainable development has made it a beacon of hope for African countries striving to achieve net zero goals. Bhutan is often recognized for its unique Gross National Happiness (GNH) index, which prioritizes sustainable development over economic growth (Choden et al., 2020). The country has been able to maintain its forest cover at an impressive 72% (IRENA, 2022), acting as a carbon sink and contributing significantly to global climate mitigation efforts.

Bhutan's success in maintaining a high forest cover has not only helped in mitigating climate change but has also provided numerous co-benefits (Faxon, 2013), such as biodiversity conservation and ecosystem services (Tshewang et al., 2021). Bhutan's emphasis on renewable energy sources, such as hydropower, has further contributed to its carbon-neutral goals and inspired other nations to explore similar sustainable energy solutions. Bhutan's commitment to sustainable development is evident in its focus on promoting organic agriculture and banning the use of synthetic pesticides. This approach not only protects the environment but also ensures the production of healthy and nutritious food for its citizens. Bhutan's emphasis on holistic well-being through its Gross National Happiness Index has set an example for other countries to prioritize the overall happiness and well-being of their citizens alongside economic growth. Bhutan has implemented policies to preserve its rich biodiversity, with over 50% of its land designated as protected areas (Tshewang et al., 2021). This commitment to conservation not only safeguards unique ecosystems but also supports eco-tourism, contributing to the country's sustainable development goals. Bhutan's dedication to sustainable practices and holistic well-being serves as a model for nations seeking to achieve a balance between economic growth and environmental preservation.

By sharing their experiences and best practices, Denmark and Bhutan can inspire African countries to adopt similar strategies and approaches toward achieving net zero goals. The success stories of these two nations serve as tangible evidence that transitioning toward a sustainable future is not only possible but also economically beneficial in the long run. Transitioning toward a sustainable future not only helps mitigate climate change but also brings numerous economic benefits, such as job creation and improved public health. By showcasing the positive outcomes of their renewable energy initiatives, Denmark and Bhutan can encourage African countries to prioritize sustainable development and reap the rewards of a green economy. Collaboration between these nations can foster knowledge exchange and capacity building, enabling African countries to overcome challenges and accelerate their progress toward achieving net zero goals. This collaboration can also lead to the transfer of technology and expertise in renewable energy, which can help African countries leapfrog traditional fossil fuel–based development and directly adopt sustainable practices. By investing in renewable energy infrastructure, African countries can reduce their dependence on imported fossil fuels, enhance energy security, and stimulate local economies through the development of a domestic green energy industry. The development of a domestic green energy industry can create job opportunities and promote economic growth, contributing to poverty reduction and social development in African countries. By embracing renewable energy sources, African countries can mitigate the negative impacts of climate change and improve the overall health and well-being of their populations.

9.3 DENMARK AND BHUTAN'S ACHIEVEMENTS IN NET ZERO TRANSITION

Denmark and Bhutan have made significant strides in their net zero transition. Denmark, known for its ambitious renewable energy goals, has successfully reduced its greenhouse gas emissions by 42% since 1990 while still maintaining a strong economy (IRENA, 2022). This achievement is largely attributed to the country's investment in wind power, which now accounts for over 40% of Denmark's electricity generation. Denmark has also implemented innovative policies such as carbon pricing and energy efficiency programs, further contributing to its successful transition toward net zero. Denmark has set a target to become completely carbon neutral by 2050 and has been actively working toward achieving this goal (IEA, 2023). In addition to wind power, Denmark has also been investing in other renewable energy sources, such as solar and biomass, diversifying its energy portfolio (Flanagan and Raphael, 2023) and reducing its reliance on fossil fuels (Wejs et al., 2014). The country's commitment to sustainability and its successful transition toward net zero have not only helped mitigate climate change but have also positioned Denmark as a global leader in renewable energy innovation (Colgan et al., 2022). Denmark's dedication to renewable energy has not only had positive environmental impacts but has also boosted its economy (Bowler et al., 2021). The country's investments in renewable energy have created numerous job opportunities and attracted foreign investments, further strengthening its position as a global leader in the renewable energy sector. Denmark's success in achieving net zero emissions has inspired other countries to follow suit and adopt similar ambitious targets, fostering a global movement toward a more sustainable future (Uldam and Askanins, 2022).

On the other hand, Bhutan, a small landlocked country in the Himalayas, has emerged as a global leader in sustainable development. With its commitment to remaining carbon neutral and carbon negative, Bhutan has not only achieved but surpassed its goal of producing zero net greenhouse gas emissions. The country's unique approach includes preserving and expanding its forest cover, which currently absorbs more carbon dioxide than Bhutan emits. Bhutan has implemented various initiatives to promote renewable energy sources such as hydropower, which accounts for almost all of the country's electricity production. Bhutan's focus on holistic development and prioritizing the well-being of its citizens has resulted in policies that prioritize sustainable agriculture and promote the use of organic farming practices, reducing reliance on chemical fertilizers and pesticides. Bhutan has also placed a strong emphasis on education and awareness programs to encourage its citizens to adopt environmentally friendly practices. The country has actively engaged in international collaborations and agreements to further combat climate change and reduce global emissions. Bhutan has also implemented strict regulations on tourism to ensure sustainable development and minimize the negative impact on the environment. The country charges a daily fee for tourists, which is used to fund conservation efforts and infrastructure development. Bhutan has prioritized the preservation of its forests, which act as carbon sinks and contribute to the country's low-carbon footprint. The government has implemented policies that aim to maintain at least 60% forest coverage, ensuring the protection of biodiversity and mitigating climate change effects. These measures have helped Bhutan maintain its pristine natural beauty while promoting responsible tourism practices. Bhutan has also banned plastic bags and imposed strict regulations on waste management to minimize pollution and maintain the cleanliness of its natural surroundings. The government encourages sustainable practices such as organic farming and renewable energy production to further reduce the country's environmental impact. These comprehensive efforts showcase Bhutan's commitment to balancing tourism growth with environmental preservation.

These accomplishments by Denmark and Bhutan serve as inspiring examples for other nations striving to achieve their own net zero transition goals. Denmark's success in transitioning to renewable energy sources and implementing strict regulations on waste management has positioned the country as a global leader in sustainable development. Bhutan's commitment to not only banning plastic bags but also promoting organic farming and renewable energy production demonstrates its dedication to creating a greener future. These two nations' achievements highlight the importance of adopting holistic approaches that prioritize both economic growth and environmental preservation, inspiring other countries to follow suit in their pursuit of net zero transition goals. By prioritizing sustainable practices, Bhutan and Costa Rica have shown that economic growth and environmental preservation can go hand in hand. Their success serves as a powerful example for other nations seeking to achieve their own net zero transition goals, emphasizing the need for comprehensive strategies that address both economic and environmental concerns. These countries have demonstrated that by implementing sustainable practices, they can not only protect their natural resources but also create new economic opportunities. Their achievements have sparked global interest and collaboration, leading to the exchange of knowledge and best practices among nations striving for a net zero transition. This collective effort is crucial to addressing the urgent challenges of climate change and ensuring a sustainable future for all. By working together and sharing experiences, countries can learn from each other's successes and failures in order to accelerate the transition toward a net zero economy. Collaboration allows for the development of innovative solutions and the scalingup of sustainable practices on a global scale. This collective effort promotes a sense of shared responsibility and fosters international cooperation in tackling climate change, which is essential for achieving long-term sustainability goals.

In this case, African countries can replicate Denmark and Bhutan's economic systems to achieve net zero emissions. Denmark and Bhutan have successfully implemented strategies that prioritize renewable energy sources, such as wind and hydropower, which have significantly reduced their carbon footprints. By adopting similar approaches, African countries can tap into their own renewable energy potentials, such as solar and geothermal resources, to transition toward a net zero economy. Learning from Denmark's emphasis on energy efficiency and Bhutan's focus on sustainable agriculture and forestry practices can further contribute to the achievement of net zero emissions in African nations. These strategies can be tailored to the specific needs and resources of each African country, taking into account factors such as geographical location and available technologies. Investing in research and development of clean energy solutions can foster innovation and create job opportunities, promoting economic growth while reducing carbon emissions. Promoting education and awareness about the importance of sustainable practices can encourage individuals and communities to actively participate in the transition toward a net zero economy. Collaborating with international organizations and sharing best practices with other countries can also provide valuable insights and support for African nations in their efforts to achieve net zero emissions. Implementing policies and regulations that incentivize the adoption of clean energy technologies can accelerate the transition toward a net zero economy. This can include offering tax credits or subsidies for renewable energy projects and setting emissions reduction targets for industries. Investing in infrastructure for renewable energy generation and distribution can create a more resilient and sustainable energy system for African nations.

9.4 DENMARK AND BHUTAN'S EARLY ADOPTION OF RENEWABLE ENERGY SOURCES AND THEIR COMMITMENT TO PHASING OUT FOSSIL FUELS

Denmark has been a global leader in renewable energy adoption, with approximately 47% of its total energy consumption coming from renewable sources in 2019 (IEA, 2023). The country's commitment to phasing out fossil fuels is evident through its ambitious goal of becoming carbon neutral by 2050 (IEA, 2023). Denmark has implemented various policies and initiatives, such as wind turbine installations, district heating systems, and investment in research and development, to support its transition toward clean energy (IEA, 2023). Similarly, Bhutan has made significant strides in renewable energy utilization. With its abundant hydropower resources, Bhutan generates almost 100% of its electricity from renewable sources (IRENA, 2022). The country's commitment to phasing out fossil fuels is evident through its pledge to remain carbon neutral under the United Nations Framework Convention on Climate Change (UNFCCC) (IRENA, 2022). Bhutan's emphasis on hydropower development not only fulfills its domestic energy needs but also allows it to export clean electricity to neighboring countries. This has not only helped Bhutan achieve energy self-sufficiency but has also contributed to the country's economic growth and development. Bhutan's focus on hydropower development aligns with its commitment to sustainable development and reducing greenhouse gas emissions, making it a role model for other countries striving to transition to renewable energy sources.

Both Denmark and Bhutan serve as inspiring examples for other nations worldwide by showcasing the feasibility and benefits of early adoption of renewable energy sources while actively working toward reducing their dependence on fossil fuels. These countries have not only demonstrated the environmental benefits of renewable energy but also the economic advantages. Denmark, for example, has become a global leader in wind energy technology, exporting its expertise and creating jobs in the process. Similarly, Bhutan's hydropower projects have attracted foreign investment and boosted its economy. By prioritizing renewable energy, these nations have shown that sustainability and economic growth can go hand in hand. These countries have also implemented policies and incentives to encourage the use of renewable energy, such as feed-in tariffs and tax credits. In addition, they have invested in research and development to further improve the efficiency and cost-effectiveness of renewable energy technologies. This commitment to sustainable energy not only benefits the environment but also helps to reduce their reliance on imported fossil fuels, enhancing their energy security. These nations have seen significant job creation in the renewable energy sector, providing opportunities for their citizens and boosting their economies. By reducing their carbon emissions through the use of renewable energy, these countries are taking proactive steps toward mitigating climate change and meeting their international commitments.

Investing in renewable energy technologies allows these nations to diversify their energy sources, reducing the risk of price volatility and supply disruptions associated with fossil fuels. The development and deployment of renewable energy infrastructure stimulates innovation and technological advancements, positioning these countries at the forefront of the global clean energy transition. By reducing their dependence on fossil fuels, these countries can improve their energy security and reduce the geopolitical risks associated with relying on imported energy resources. Investing in renewable energy creates job opportunities and stimulates economic growth, leading to a more sustainable and resilient economy. Transitioning to renewable energy sources can help mitigate the negative environmental impacts of fossil fuel extraction and combustion, such as air and water pollution and greenhouse gas emissions. This can lead to improved public health outcomes and a cleaner, more sustainable environment for future generations. By embracing renewable energy, these countries can demonstrate global leadership in addressing climate change and contribute to international efforts to limit global warming and reduce carbon emissions.

9.5 SPECIFIC POLICIES AND INITIATIVES IMPLEMENTED BY DENMARK AND BHUTAN TO PROMOTE CLEAN ENERGY AND REDUCE EMISSIONS

Denmark has been a frontrunner in promoting clean energy and reducing emissions through various policies and initiatives. One notable example is their ambitious wind energy strategy, which aims to generate 50% of the country's electricity from wind power by 2030. Denmark has heavily invested in wind turbines and offshore wind farms, leading to significant reductions in carbon emissions. Denmark has implemented a carbon tax, which has incentivized industries to reduce their greenhouse gas emissions. This policy has successfully encouraged companies to adopt cleaner technologies and practices. Denmark has established a comprehensive energy efficiency program that focuses on improving energy efficiency in buildings, transportation, and industry sectors. Denmark has also implemented a feed-in tariff system, which guarantees a fixed payment for renewable energy producers, further encouraging the development of wind power. The government has actively promoted research and development in renewable energy technologies, fostering innovation and driving the growth of the sector.

Similarly, Bhutan has taken significant steps toward promoting clean energy and reducing emissions. The country is known for its commitment to environmental conservation and sustainable development. Bhutan's constitution mandates that at least 60% of its land must remain under forest cover (IEA, 2023), acting as a natural carbon sink. Bhutan has also prioritized hydropower as a major source of renewable energy. With its abundant rivers and water resources, the country has harnessed hydropower to meet its domestic electricity demand while exporting surplus power to neighboring countries like India. This focus on hydropower has not only reduced Bhutan's reliance on fossil fuels but also contributed to the reduction of greenhouse gas emissions. Bhutan places great emphasis on holistic development through its Gross National Happiness (GNH) index. This unique approach considers sustainable socio-economic development alongside environmental conservation. It encourages the use of renewable energy sources like solar power for rural electrification projects, ensuring access to clean energy for all citizens while minimizing their carbon footprint. Bhutan has implemented policies to protect its forests, which act as carbon sinks and help mitigate climate change. The country has committed to maintaining at least 60% forest coverage, ensuring the preservation of biodiversity and the sequestration of carbon dioxide. These efforts not only contribute to global climate goals but also support Bhutan's commitment to maintaining its pristine natural environment.

Both Denmark and Bhutan serve as inspiring examples with their well-defined policies and initiatives that prioritize clean energy adoption and emission reduction strategies. Denmark, known for its ambitious renewable energy targets, aims to achieve 100% renewable electricity by 2030 (IRENA, 2022). The country has implemented various measures, such as wind power expansion and investment in green technologies, to reduce greenhouse gas emissions. These efforts have not only positioned Denmark as a global leader in clean energy but have also demonstrated the economic benefits of transitioning to a low-carbon economy. Bhutan's forest protection policies and Denmark's clean energy initiatives showcase the importance of proactive measures in addressing climate change and promoting sustainable development worldwide. These proactive measures not only contribute to mitigating the effects of climate change but also foster job creation and economic growth. By investing in renewable energy sources and sustainable practices, countries like Bhutan and Denmark are setting an example for others to follow in the transition toward a greener future. Transitioning to a low-carbon economy can lead to significant economic benefits, such as increased investments in clean technologies and the creation of new industries. Bhutan's forest protection policies, for example, have not only helped in preserving their natural resources but also attracted eco-tourism, generating revenue and employment opportunities for the local communities. Similarly, Denmark's clean energy initiatives have positioned the country as a global leader in wind energy technology, attracting foreign investments and creating a thriving renewable energy sector. These examples highlight how proactive measures toward sustainability can drive economic growth and pave the way for a more prosperous future.

9.6 STATISTICS AND EVIDENCE OF DENMARK AND BHUTAN'S PROGRESS IN REDUCING CARBON FOOTPRINT

Denmark has significantly reduced its carbon footprint, according to the Global Carbon Atlas (2022). From 1990 to 2019, Denmark's CO_2 emissions decreased by approximately 29%. This reduction can be attributed to various factors, such as increased renewable energy production, energy efficiency measures, and a shift toward sustainable transportation systems. Denmark has set ambitious targets for further reducing its carbon footprint. The country aims to be completely carbon neutral by 2050 and has already achieved a remarkable milestone of sourcing 50% of its electricity from wind power in 2019 (Global Carbon Atlas, 2022). This commitment to renewable energy and sustainable practices has positioned Denmark as a global leader in the fight against climate change. Denmark's commitment to renewable energy and sustainable practices has not only positioned it as a global leader in the fight against climate change, but it has also spurred economic growth and innovation within the country. The transition to clean energy has created new job opportunities and attracted investments from both domestic and international companies, further driving the development of renewable technologies. As Denmark continues to make strides toward its carbon-neutral goal, its success serves as a powerful example for other nations seeking to address climate change while also fostering economic prosperity.

Similarly, Bhutan has also demonstrated remarkable progress in reducing its carbon footprint. The country is known for its commitment to environmental conservation and sustainable development. According to the Global Carbon Atlas Report 2019, Bhutan's net greenhouse gas emissions were negative (–5.4 million tons) in 2016 due to its extensive forest cover acting as a carbon sink (Global Carbon Atlas, 2022). Bhutan aims to remain carbon neutral by maintaining at least 60% forest coverage. According to the World Bank (2022), Bhutan's forest coverage stands at an impressive 71%, which plays a crucial role in sequestering carbon emissions. Bhutan is aiming to remain carbon neutral and even become a net exporter of clean energy by harnessing its vast hydropower potential. These efforts highlight Bhutan's dedication to sustainable development and serve as an inspiring example for other nations striving to reduce their carbon footprints.

In addition to its forest cover, Bhutan has implemented various initiatives to further reduce its carbon footprint. These include promoting the use of electric vehicles and investing in renewable energy projects such as hydropower. These efforts have not only helped Bhutan achieve its carbon-neutral goals but have also positioned the country as a leader in sustainable development on a global scale. Bhutan has established a national policy that prioritizes environmental conservation and sustainability. This policy includes strict regulations on deforestation and the protection of biodiversity, ensuring the preservation of its natural resources for future generations. These comprehensive measures demonstrate Bhutan's commitment to mitigating climate change and setting a positive example for other nations to follow.

These statistics and evidence highlight the commendable efforts of both Denmark and Bhutan in mitigating climate change and transitioning toward low-carbon economies. Denmark, known for its strong commitment to renewable energy, has implemented various initiatives to reduce greenhouse gas emissions and promote clean energy sources. The country has set ambitious targets to achieve carbon neutrality by 2050 (IEA, 2023) and has invested heavily in wind power, leading to a significant decrease in fossil fuel consumption. Bhutan, on the other hand, is recognized for its unique approach to development through the concept of Gross National Happiness (GNH), which emphasizes sustainable and holistic well-being. By valuing happiness over material wealth, Bhutan has been able to prioritize environmental conservation and sustainability in its policies and practices. Both Denmark and Bhutan serve as inspiring examples of nations actively working toward a greener future. Denmark's commitment to renewable energy has also led to advancements in other sectors, such as electric transportation and green building practices. Bhutan, on the other hand, has implemented policies to protect its vast forest cover and promote organic farming, further contributing to its sustainable development goals. These two countries demonstrate that a shift toward a greener future requires not only investment in renewable energy but also a holistic approach that considers the well-being of both people and the planet.

9.7 LESSONS AFRICAN COUNTRIES CAN LEARN FROM DENMARK AND BHUTAN'S PROGRESS IN REDUCING CARBON FOOTPRINTS

African countries can learn valuable lessons from Denmark and Bhutan's progress in reducing carbon footprints. Denmark has been recognized as a global leader in renewable energy, with over 50% of its electricity coming from wind power. This achievement can be attributed to Denmark's strong commitment to investing in clean energy infrastructure, such as wind farms and smart grids. Denmark has implemented effective policies and incentives to encourage the adoption of electric vehicles, resulting in a significant reduction in carbon emissions from transportation. Bhutan, on the other hand, has adopted a unique approach to carbon reduction by prioritizing

Gross National Happiness (GNH) over Gross Domestic Product (GDP). Bhutan's focus on holistic well-being and sustainable development has led to the preservation of its rich biodiversity and natural resources. By valuing the environment as an integral part of their national identity, Bhutan has successfully implemented policies that promote renewable energy sources like hydropower and forest conservation. These policies have not only helped Bhutan become carbon neutral but have also created a positive impact on the overall well-being of its citizens. Bhutan's commitment to sustainable tourism has attracted visitors who are interested in experiencing and supporting environmentally friendly practices.

These examples highlight the importance of political will, long-term planning, and investment in clean energy infrastructure for African countries seeking to reduce their carbon footprints. By emulating Denmark's commitment to renewable energy and Bhutan's holistic approach to development, African nations can make significant strides toward a sustainable future while preserving their unique cultural and environmental heritage. African countries can also benefit from the lessons learned by Denmark and Bhutan in terms of engaging local communities and ensuring their active participation in sustainable development initiatives. This bottom-up approach can foster a sense of ownership and pride among the population, leading to more effective implementation of clean energy policies. By prioritizing education and awareness campaigns, African nations can empower their citizens to make informed choices about energy consumption and conservation, further contributing to a greener future. African countries can establish partnerships with international organizations and seek financial support to implement renewable energy projects. These collaborations can provide access to technological advancements and expertise, enabling African nations to leapfrog traditional fossil fuel–based development models and directly adopt sustainable practices. By capitalizing on these opportunities, Africa has the potential to become a candidate in clean energy innovation and pave the way for a more sustainable future for generations to come.

9.8 SIMILARITIES AND DIFFERENCES BETWEEN DENMARK–BHUTAN AND AFRICAN COUNTRIES IN TERMS OF RESOURCES, INFRASTRUCTURE, AND SOCIO-ECONOMIC CONDITIONS

When comparing Denmark and Bhutan with African countries in terms of resources, infrastructure, and socio-economic conditions, some notable similarities and differences emerge. In terms of resources, Denmark and Bhutan have relatively limited natural resources compared to many African countries. Denmark is known for its offshore oil and gas reserves, which contribute significantly to its economy. On the other hand, Bhutan's resources primarily include hydropower potential and rich biodiversity. In contrast, numerous African countries possess abundant reserves of minerals like gold, diamonds, oil, and natural gas. These resources have played a significant role in shaping the economies of these African countries, attracting foreign investments, and contributing to their overall development. It is important to note that the exploitation and management of these resources have also posed challenges such as corruption, environmental degradation, and the unequal distribution of wealth within these countries. While Denmark and Bhutan may have limited natural resources, they have focused on other sectors, such as renewable energy and tourism, to drive their economies forward. By diversifying their economic activities, Denmark and Bhutan have been able to reduce their dependence on natural resources and create sustainable growth. For example, Denmark has become a global leader in wind energy technology, exporting its expertise and generating revenue from renewable energy projects. Similarly, Bhutan has capitalized on its unique natural landscapes and cultural heritage to attract tourists, boost its tourism industry, and create jobs for its citizens. These strategies have not only helped these countries mitigate the challenges associated with resource exploitation but have also fostered more balanced and inclusive development.

Regarding infrastructure, Denmark stands out with its highly developed transportation networks, modernized cities, and advanced technological systems. Similarly, Bhutan has made significant progress in improving its infrastructure in recent years, particularly in the areas of road connectivity and telecommunications. Many African countries face challenges in developing their infrastructure due to factors such as inadequate funding and political instability.

Socio-economic conditions also exhibit variations between these regions. Denmark consistently ranks high on various socio-economic indicators such as income equality, education levels, healthcare access, and the overall standard of living. Bhutan has made remarkable strides in achieving socio-economic development through its unique Gross National Happiness (GNH) index, which prioritizes well-being over GDP growth. In contrast, while some African countries have witnessed significant economic growth in recent years, they continue to grapple with poverty rates, income inequality, and limited access to quality education and healthcare services.

Overall, while Denmark–Bhutan's resource endowments and infrastructure development differ from those of many African countries due to their respective geographical locations and historical contexts, socio-economic conditions also showcase disparities between these regions. These disparities can be attributed to various factors, such as colonial legacies, political instability, and a lack of investment in human capital. Cultural differences and governance systems also play a role in shaping the socio-economic conditions in Denmark–Bhutan and African countries. Therefore, it is important to consider these contextual factors when comparing the well-being and development of different regions. For example, Denmark and Bhutan, both located in different parts of the world, have distinct historical backgrounds that have shaped their socio-economic conditions. Denmark, as a European country with a long history of colonization and trade, has benefited from its strong infrastructure and well-established institutions. On the other hand, Bhutan, a landlocked country in South Asia, has faced challenges related to its geographical isolation and limited resources.

In addition to historical factors, political instability has also affected the development of African countries. Many African nations have experienced periods of conflict and civil unrest, which have hindered their economic growth and social progress. Lack of investment in human capital, such as education and healthcare systems, has further exacerbated the disparities between these regions. Cultural differences between

Denmark–Bhutan and African countries also contribute to the variations in their socio-economic conditions. Different cultural norms and values can impact factors such as governance systems, labor practices, and social welfare policies. These variations highlight the importance of considering local contexts when assessing well-being and development across different regions. In light of this, in terms of local context-specific emission reduction and reaching net zero targets between Denmark–Bhutan and African countries, it is crucial to take into account the unique socio-economic and cultural factors that shape each region's approach to sustainability. For example, Denmark and Bhutan may have well-established renewable energy policies and infrastructure in place, which can facilitate their transition toward net zero emissions.

On the other hand, African countries may face challenges such as limited access to clean energy technologies and financial resources, requiring tailored strategies to address their specific needs. By acknowledging these differences and adopting context-specific approaches, we can ensure that emission reduction efforts are effective and inclusive across all regions. It is important to consider the unique social, economic, and environmental contexts of each region when developing sustainability strategies. This may involve collaborating with local communities and stakeholders to understand their priorities and concerns and designing initiatives that align with their cultural practices and values. By taking a holistic and culturally sensitive approach, we can foster long-term engagement and ownership of sustainability initiatives in every region. It is crucial to provide adequate resources and support to empower local communities to implement these sustainability strategies. This can include offering training programs, financial incentives, and technological assistance to help overcome any barriers they may face. By doing so, we can ensure that emission reduction efforts are not only effective but also sustainable in the long run, benefiting both the environment and the well-being of local communities. It is important to actively involve local stakeholders in the decision-making process to ensure their voices are heard and their perspectives are taken into account. This can be done through regular consultations, community meetings, and partnerships with local organizations. By including the input and expertise of those directly affected by sustainability initiatives, we can create solutions that are tailored to the specific needs and challenges of each region. Ultimately, this collaborative approach will lead to greater success in achieving long-term sustainability goals.

REFERENCES

Bhattarai, H. (2021a). Policy driven initiative for promoting electric vehicles to combat climate change in the developing countries: A case from Bhutan. *International Multidisciplinary Research Journal*. https://doi.org/10.54476/iimrj263

Bhattarai, H. (2021b). Policy intervention and its consequences on the environment to combat climate change– A case from Bhutan. *International Journal of Energy Applications and Technologies*. https://doi.org/10.31593/ijeat.941741

Bowler, D., Richter, R. L., Eskildsen, D., Kamp, J., Moshøj, C. M., Reif, J., Strebel, N., Trautmann, S., & Voříšek, P. (2021). Geographic variation in the population trends of common breeding birds across central Europe. *Basic and Applied Ecology*. https://doi.org/10.1016/j.baae.2021.07.004

Chen, J. S., Wang, W., Kim, H., & Liu, W. Y. (2022). Climate resilience model on Arctic tourism: perspectives from tourism professionals. *Tourism Recreation Research*. https://doi.org/10.1080/02508281.2022.2122341

Choden, K., Keenan, R. J., & Nitschke, C. R. (2020). An approach for assessing adaptive capacity to climate change in resource dependent communities in the Nikachu watershed, Bhutan. *Ecological Indicators*. https://doi.org/10.1016/j.ecolind.2020.106293

Colgan, W., Henriksen, H. J., Bennike, O., Ribeiro, S., Keiding, M., Seidenfaden, I. K., Graversgaard, M., Busck, A. G., Fruergaard, M., Knudsen, M. H., Hopper, J., Sonnenborg, T., Skjerbæk, M. R., Bjørk, A. A., Steffen, H., Tarasov, L., Nerem, R. S., & Kjeldsen, K. K. (2022). Sea-level rise in Denmark: paleo context, recent projections and policy implications. *GEUS Bulletin*. https://doi.org/10.34194/geusb.v49.8315

Damsø, T., Kjær, T., & Christensen, T. B. (2016). Local climate action plans in climate change mitigation – Examining the case of Denmark. *Energy Policy*. https://doi.org/10.1016/j.enpol.2015.11.013

Faxon, H. O. (2013). Waking the watchdog: Needs, opportunities, and challenges of environmental advocacy in modern bhutan. *Himalaya* https://umimpact.umt.edu/en/publications/waking-the-watchdog-needs-opportunities-and-challenges-of-environ.

Flanagan, E., & Raphael, D. (2023). From personal responsibility to an eco-socialist state: Political economy, popular discourses and the climate crisis. *Human Geography(United Kingdom)*. https://doi.org/10.1177/19427786221138965

GCA. (2022). *Global Carbon Atlas*. Global Carbon Atlas.

IEA. (2023). Global EV Outlook 2023. *Geo*, *Geo*.

International Renewable Energy Agency (IRENA). (2022). World Energy Transitions Outlook 2023: 1.5°C Pathway. In World Energy Transitions.

Ituen, E. E., Esen, N. U., Nwokolo, S. C., & Udo, E. G. (2012). Prediction of global solar radiation using relative humidity, maximum temperature and sunshine hours in Uyo, in the Niger Delta Region, Nigeria. *Advances in Applied Science Research*, *3*(4), 1923–1937.

Kerr, S. (2023). Citizen science and deep mapping for climate communication: A report on CHICC. *Journal of Community Archaeology and Heritage*. https://doi.org/10.1080/20518196.2022.2051139

Koziell, I., & Gyamtsho, P. (2023). Moving mountains: A new strategy and action plan for ICIMOD to embrace change and accelerate impact to 2030. *Mountain Research and Development*. https://doi.org/10.1659/mrd.2023.00002

Laura Cozzi et al. (2023). World Energy Outlook. In *World Energy Outlook*. International Energy Agency. www.iea.org/weo

McGregor, I., Yerbury, H., & Shahid, A. (2018). The voices of local NGOs in climate change issues: Examples from climate vulnerable nations. *Cosmopolitan Civil Societies*. https://doi.org/10.5130/ccs.v10.i3.6019

Munawar, S. (2016). Bhutan improves economic development as a net carbon sink. *The Climate Institute*.

Nwokolo, S., Echeng, E., Obiwulu, A., Meyer, E. L., Christian, C., Ogbulezie, J. C., & Proutsos, N. (2024). A multi-model approach based on CARIMA-SARIMA-GPM for assessing the impacts of climate change on concentrated photovoltaic (CPV) potential. *Physics and Chemistry of the Earth*, *134*(January), 103560. https://doi.org/10.1016/j.pce.2024.103560

Nwokolo, S., & Otse, C. (2019). Impact of sunshine duration and clearness index on diffuse solar radiation estimation in mountainous climate. *Trends in Renewable Energy*, *5*(3), 307–332. https://doi.org/10.17737/tre.2019.5.3.00107

Nwokolo, S. C. (2017). A comprehensive review of empirical models for estimating global solar radiation in Africa. *Renewable and Sustainable Energy Reviews*, *78*(October), 955–995. https://doi.org/10.1016/j.rser.2017.04.101

Nwokolo, S. C., Meyer, E. L., & Ahia, C. C. (2023). Credible pathways to catching up with climate goals in Nigeria. *Climate*, *11*(9), 1–32. https://doi.org/org/10.3390/cli11090196

Nwokolo, S. C., Meyer, E. L., & Ahia, C. C. (2024). Exploring the interactive influences of climate change and urban development on the fraction of absorbed photosynthetically active radiation. *Atmosphere*, *15*(253), 1–37.

Nwokolo, S. C., & Ogbulezie, J. C. (2018a). A qualitative review of empirical models for estimating diffuse solar radiation from experimental data in Africa. *Renewable and Sustainable Energy Reviews*, *92*(18), 353–393. https://doi.org/10.1016/j.rser.2018.04.118

Nwokolo, S. C., & Ogbulezie, J. C. (2018b). A quantitative review and classification of empirical models for predicting global solar radiation in West Africa. *Beni-Suef University Journal of Basic and Applied Sciences*, *7*(4), 367–396. https://doi.org/10.1016/j.bjbas.2017.05.001

Nwokolo, S. C., Proutsos, N., & Meyer, E. L. (2023). Machine learning and physics-based hybridization models for evaluation of the effects of climate change and urban expansion on photosynthetically active radiation. *Atmosphere*, *14*(687), 1–36. https://doi.org/doi.org/10.3390/atmos14040687

Obiwulu, A. U., Chendo, M. A. C., Erusiafe, N., & Nwokolo, S. C. (2020). Implicit meteorological parameter-based empirical models for estimating back temperature solar modules under varying tilt-angles in Lagos, Nigeria. *Renewable Energy*, *145*. https://doi.org/10.1016/j.renene.2019.05.136

Obiwulu, A. U., Erusiafe, N., Olopade, M. A., & Nwokolo, S. C. (2020). Modeling and optimization of back temperature models of mono-crystalline silicon modules with special focus on the effect of meteorological and geographical parameters on PV performance. *Renewable Energy*, *154*. https://doi.org/10.1016/j.renene.2020.02.103

Obiwulu, A. U., Erusiafe, N., Olopade, M. A., & Nwokolo, S. C. (2022). Modeling and estimation of the optimal tilt angle, maximum incident solar radiation, and global radiation index of the photovoltaic system. *Heliyon*, *8*(6), e09598. https://doi.org/10.1016/j.heliyon.2022.e09598

Proutsos, N., Liakatas, A., Alexandris, S., Nwokolo, S. C., Solomou, A. D., & Amadi, S. O. (2023). Assessing the impact of atmospheric attributes on the effectiveness of solar irradiance for photosynthesis of urban vegetation in Attica, Greece. *Theoretical and Applied Climatology*, *0123456789*. https://doi.org/10.1007/s00704-023-04700-0

Proutsos, N., Tigkas, D., Tsevreni, I., Alexandris, S. G., Solomou, A. D., Bourletsikas, A., Stefanidis, S., & Nwokolo, S. C. (2023). A thorough evaluation of 127 potential evapotranspiration models in two mediterranean urban green sites. *Remote Sensing*, *15*(14), 1–41. https://doi.org/10.3390/rs15143680

Rai, P., Bajgai, Y., Rabgyal, J., Katwal, T. B., & Delmond, A. R. (2022). Empirical evidence of the livelihood vulnerability to climate change impacts: A case of potato-based mountain farming systems in Bhutan. *Sustainability (Switzerland)*. https://doi.org/10.3390/su14042339

Shrestha, M. S., Goodrich, C. G., Udas, P., Rai, D. M., Gurung, M. B., & Khadgi, V. (2016). Flood Early Warning Systems in Bhutan – A Gendered Perspective. Working Paper.

Smiley, K. T., Chen, Y. A., & Shao, W. (2022). Being green in a green capital: Assessing drivers of pro-environmental behaviors in Copenhagen. *Cities*. https://doi.org/10.1016/j.cities.2021.103538

Tanaka, S., & Tokimatsu, K. (2020). Social capital, subjective well-being, and happiness: evidence from a survey in various European and Asian countries to address the Stiglitz report. *Modern Economy*. https://doi.org/10.4236/me.2020.112026

Tenzin, J., Phuntsho, L., Wengkhar, D. C., & Lakey, L. (2019). Chapter 3 Climate Smart Agriculture: Adaptation and Mitigation Strategies to Climate Change in Bhutan. In

Rudra Bahadur Shrestha and S M Bokhtiar (Eds.), Climate Smart Agriculture: Strategies to Respond to Climate Change. SAARC Agriculture Centre (SAC), Bangladesh.

The World Bank Group. (2022). *Risk of Global Recession in 2023 Rises Amid Simultaneous Rate Hikes*. The World Bank Group.

Tshewang, U., Tobias, M. C., & Morrison, J. G. (2021). Bhutan: Conservation and environmental protection in the Himalayas. In *Bhutan: Conservation and Environmental Protection in the Himalayas*. https://doi.org/10.1007/978-3-030-57824-4

Uldam, J., & Askanius, T. (2022). Time for climate action? Political actors' uses of Twitter to focus public attention on the climate crisis during the 2019 Danish General Election. *International Journal of Communication, 16*, 385–408.

Vihma, A., Reischl, G., & Nonbo Andersen, A. (2021). A Climate backlash: Comparing populist parties' climate policies in Denmark, Finland, and Sweden. *Journal of Environment and Development*. https://doi.org/10.1177/10704965211027748

Wangdi, S. C., & Nepal, M. P. (2017). Towards the development of eco-industrial estates in Bhutan. *Proceedings of the Institution of Civil Engineers: Engineering Sustainability*. https://doi.org/10.1680/jensu.15.00059

Wejs, A., Harvold, K., Larsen, S. V., & Saglie, I. L. (2014). Legitimacy building in weak institutional settings: Climate change adaptation at local level in Denmark and Norway. *Environmental Politics*. https://doi.org/10.1080/09644016.2013.854967

Woods, B. A., Nielsen, H. Ø., Pedersen, A. B., & Kristofersson, D. (2017). Farmers' perceptions of climate change and their likely responses in Danish agriculture. *Land Use Policy*. https://doi.org/10.1016/j.landusepol.2017.04.007

10 Africa's Key Role in Driving Global Net Zero Transition

10.1 INTRODUCTION

African nations have a significant potential to contribute to the global net zero transition in various ways. Firstly, Africa is rich in renewable energy resources, such as solar, wind, and hydroelectric power. By harnessing these resources effectively, African countries can not only meet their own energy demands sustainably but also become major exporters of clean energy to other regions. Africa's vast agricultural sector can play a crucial role in mitigating climate change. Implementing sustainable farming practices and investing in climate-smart agriculture can help reduce greenhouse gas emissions and enhance carbon sequestration. African nations can leverage their diverse ecosystems and biodiversity to promote nature-based solutions like reforestation and ecosystem restoration, which are vital for achieving global net zero goals.

The abundant presence of critical minerals in African nations holds immense strategic importance for global industries reliant on military hardware, space equipment, electronic devices, and climate change technologies. Leveraging these resources can foster mutually beneficial partnerships between African countries and international stakeholders seeking to secure a sustainable supply chain for these essential minerals. This potential collaboration presents an opportunity for African nations to actively participate in global technological advancements while promoting economic growth and development within their own borders (Nwokolo, Meyer, et al., 2024). By tapping into their vast mineral resources, African nations can not only meet the growing global demand for these essential minerals but also contribute to reducing their dependency on a few dominant suppliers (Nwokolo, Echeng, et al., 2024). This diversification of supply sources can enhance global stability and resilience in critical industries, ensuring a more secure and sustainable future for all stakeholders involved. By actively participating in technological advancements, African nations can gain valuable knowledge and expertise, enabling them to develop their own indigenous industries and create jobs for their growing populations.

According to the United Nations (2022), Africa is projected to have the largest working-age population by 2030, with over 800 million young people aged between 15 and 24. This demographic dividend presents a unique opportunity for harnessing their energy and potential in driving renewable energy innovation and economic

DOI: 10.1201/9781003483175-11

growth. Studies have shown that investing in renewable energy can create numerous job opportunities, particularly in sectors such as solar power, wind energy, and hydroelectricity. For instance, a report by the International Renewable Energy Agency (IRENA) (International Renewable Energy Agency [IRENA], 2022) states that Africa has the potential to generate 310 GW of renewable energy by 2030, which could create over 4 million jobs. By leveraging this youthful demographic dividend and focusing on renewable energy innovation (Ituen et al., 2012), Africa can not only meet its own growing energy demands sustainably but also contribute significantly to the global net zero transition. This would not only help combat climate change but also foster economic development and improve livelihoods across the continent.

Africa's geographical advantage is a crucial factor in positioning the continent as a leader in solar power. With its vast landmass and abundant sunlight, Africa has the potential to harness solar energy on a large scale. According to the IRENA (IRENA, 2022), Africa receives about 2,000 KW-h of solar energy per square meter per year, making it one of the sunniest regions globally. This abundance of solar radiation presents a significant opportunity for Africa to tap into its renewable energy potential and drive the global net zero transition. Africa's geographical advantage extends beyond solar radiation to include other favorable conditions for solar power generation. The continent has extensive desert areas like the Sahara, which provide vast open spaces ideal for setting up large-scale solar farms (Nwokolo and Otse, 2019). Many African countries are located close to the equator, resulting in more consistent and intense sunlight throughout the year. These favorable geographic conditions not only make solar power generation more efficient but also reduce transmission losses associated with long-distance electricity transport. By leveraging its geographical advantage and investing in solar power infrastructure, Africa can not only meet its own growing energy demands sustainably but also become a major exporter of clean energy to other regions. This would not only contribute significantly to achieving global net zero emissions but also foster economic growth and job creation within Africa's renewable energy sector.

According to recent studies, Africa's ecosystems have the potential to play a crucial role in carbon sequestration and greenhouse gas mitigation. For instance, the continent's forests alone are estimated to absorb around 1.2 billion metric tons of carbon dioxide annually (Global Carbon Atlas [GCA], 2022). Wetlands, grasslands, and coastal ecosystems in Africa also contribute significantly to carbon storage and emission reduction annually (Global Carbon Atlas, 2022). Leveraging these ecosystems for sustainable carbon sequestration can not only help mitigate climate change but also provide co-benefits such as biodiversity conservation and improved livelihoods for local communities. Therefore, recognizing Africa's key role in driving the global net zero transition is essential for harnessing the continent's ecological potential and achieving climate goals worldwide. By investing in the preservation and restoration of these ecosystems, African countries can not only play a crucial role in global climate action but also unlock economic opportunities through nature-based solutions. International cooperation and financial support are necessary to ensure the successful implementation of sustainable carbon sequestration initiatives in Africa.

Therefore, recognizing Africa's potential and actively involving African nations in the global net zero transition is not only essential but also imperative for its success. According to a report by the IRENA (IRENA, 2022), Africa has the potential to generate 310 GW of renewable energy by 2030, which is more than the continent's current total power capacity. This highlights the immense opportunity for Africa to play a significant role in the global net zero transition. Furthermore, involving African nations in the global net zero transition is not only crucial for achieving climate goals but also for fostering sustainable economic growth and reducing poverty. The World Bank estimates that investing in renewable energy projects in Africa could create millions of jobs and boost gross domestic product (GDP) growth, ultimately improving living standards and reducing inequality across the continent. Therefore, it is imperative that African nations are actively engaged and supported in their efforts to transition toward clean energy sources. Collaboration between developed countries and African nations through technology transfer, capacity building, and financial support can foster innovation and accelerate the adoption of sustainable practices across the continent. Africa's contribution will be pivotal in achieving a truly global net zero future. By harnessing Africa's vast renewable energy resources, such as solar and wind power, the continent can not only reduce its carbon emissions but also become a leading exporter of clean energy. This would not only benefit African nations economically but also significantly contribute to global decarbonization efforts. Investing in sustainable agriculture and land management practices in Africa can further enhance carbon sequestration and mitigate climate change impacts on food security and biodiversity.

10.2 AFRICA'S VAST RENEWABLE ENERGY POTENTIAL

According to recent studies, Africa has enormous renewable energy potential that remains largely untapped. The continent is blessed with abundant solar, wind, hydro, and geothermal resources that can significantly contribute to the global net zero transition. For instance, the IRENA (IRENA, 2022) estimates that Africa's solar potential alone could meet the continent's electricity demand 100 times over. Harnessing Africa's renewable energy potential not only helps combat climate change but also presents immense economic opportunities (Dhansay et al., 2014). The African Development Bank projects that investing in renewable energy could create around 2 million jobs by 2030 and stimulate economic growth across various sectors (World Bank Group, 2020). Therefore, it is crucial for African nations to prioritize the development and utilization of their renewable energy resources to drive the global net zero transition while reaping the associated socio-economic benefits (SANEA, 2020). By investing in renewable energy, African nations can reduce their dependence on fossil fuels (Yetano Roche et al., 2020) and improve energy security (Mshandete and Parawira, 2010). The development of renewable energy infrastructure can attract foreign investment (Davidson, 2020) and promote sustainable economic development in the region (Kiprotich et al., 2021).

The shift toward renewable energy can also help mitigate the adverse effects of climate change and reduce greenhouse gas emissions. This not only contributes to global efforts to combat climate change but also positions African nations as leaders in sustainable development and environmental stewardship (Ogunniyi and Pienaar, 2019). The utilization of renewable energy sources can also lead to lower energy costs for both households and businesses, freeing up resources that can be allocated toward other critical sectors such as healthcare and education. The adoption of renewable energy technologies can create new job opportunities and stimulate local economies. By investing in renewable energy infrastructure, African nations can attract foreign investments and foster innovation in clean energy industries (Nwokolo, Meyer, et al., 2023). This can result in a more diversified and resilient economy (Nwokolo, Obiwulu, et al., 2023), reducing dependence on volatile fossil fuel markets (Nwokolo, Ogbulezie, et al., 2023). The transition to renewable energy sources can improve energy security by reducing reliance on imported fuels and increasing self-sufficiency in meeting domestic energy demands (Nwokolo, Proutsos, et al., 2023). The use of renewable energy can also contribute to mitigating climate change and reducing greenhouse gas emissions. This can have long-term environmental benefits, such as improved air quality and a reduced carbon footprint. Investing in renewable energy can promote sustainable development and improve the overall quality of life for communities by providing access to clean and affordable energy sources.

10.3 AFRICA'S ABUNDANT NATURAL RESOURCES FUELING CLEAN TECHNOLOGY REVOLUTION

Africa's abundant natural resources are playing a pivotal role in driving the global net zero transition, particularly through the fueling of a clean technology revolution. By harnessing its vast renewable energy potential, such as solar and wind power, Africa is not only meeting its own energy demands sustainably but also contributing to the global shift toward cleaner and greener technologies. With its rich reserves of minerals like lithium, cobalt, and rare earth elements, Africa is also emerging as a key supplier for the production of electric vehicles and renewable energy infrastructure worldwide. This has led to increased investment in Africa's mining sector and the establishment of partnerships between African countries and international companies to exploit these resources responsibly. The development of local industries to process and manufacture clean technologies within Africa has the potential to create jobs and drive economic growth across the continent.

The growth of the clean technology industry in Africa has the potential to reduce the continent's dependence on fossil fuels and contribute to global efforts to combat climate change. By promoting sustainable mining practices and responsible resource extraction, Africa can ensure that its natural resources are utilized in a way that benefits both its economy and the environment. Investing in clean technology industries can also help Africa meet its sustainable development goals and improve the quality of life for its citizens. By prioritizing renewable energy sources and sustainable practices, Africa can position itself as a leader in the global transition to a low-carbon economy. This transition can also attract international investments and partnerships, creating new opportunities for economic growth and job creation in

Africa. By reducing its dependence on fossil fuels, Africa can enhance its energy security and reduce its vulnerability to fluctuations in global oil prices.

Embracing renewable energy can also address the issue of energy poverty in Africa, as it can provide reliable and affordable electricity to remote and underserved communities. This can have a transformative impact on education, healthcare, and overall socio-economic development in these areas. Transitioning to sustainable energy sources can contribute to mitigating the effects of climate change, which disproportionately affect African countries due to their vulnerability to extreme weather events. Investing in renewable energy can create job opportunities and stimulate economic growth in Africa. By developing a renewable energy sector, countries can attract foreign investments and foster local entrepreneurship, leading to a more sustainable and diversified economy. Transitioning to sustainable energy sources can also improve air quality and reduce pollution, resulting in better public health outcomes for African communities. Renewable energy can help African countries reduce their dependence on fossil fuels, which often need to be imported at high costs. This would not only enhance energy security but also free up financial resources that can be allocated toward other developmental priorities such as education and healthcare. By embracing renewable energy, African countries can set an example for the rest of the world and contribute significantly to global efforts to combat climate change.

10.4 HARNESSING AFRICA'S YOUTHFUL DEMOGRAPHIC DIVIDEND FOR RENEWABLE ENERGY INNOVATION AND ECONOMIC GROWTH

Africa's youthful demographic dividend presents a unique opportunity to drive renewable energy innovation and foster economic growth on the continent. With a large population of young, talented individuals, Africa has the potential to become a global player in the transition toward net zero emissions. By harnessing the energy and creativity of its youth, Africa can develop innovative solutions to address its energy needs while contributing to global efforts to combat climate change. Investing in renewable energy infrastructure and technologies can not only create job opportunities for African youth but also stimulate economic growth. The development of clean energy sources such as solar, wind, and hydropower can provide reliable and affordable electricity to communities across the continent. This will not only improve living standards but also attract foreign investments and promote sustainable development. By prioritizing renewable energy innovation as part of Africa's key role in driving the global net zero transition, the continent can position itself as a hub for green technology advancements. This will not only benefit Africa's economy but also contribute significantly to mitigating climate change on a global scale. With proper investment in education, skills training, and supportive policies, Africa's youthful demographic dividend can be harnessed effectively to drive sustainable development and create a brighter future for generations to come.

Investing in education and skills training will equip the youth with the necessary knowledge and expertise to lead the development and implementation of renewable energy solutions. Implementing supportive policies that incentivize the adoption of green technologies will attract foreign investments and foster collaboration with international partners, further accelerating Africa's transition toward a sustainable future. Empowering young entrepreneurs and providing them with access to capital and resources will enable them to innovate and create sustainable businesses that address the continent's pressing challenges. Fostering a culture of entrepreneurship and innovation within educational institutions will nurture the next generation of leaders who can drive sustainable development in Africa. Implementing policies that promote knowledge transfer and technology sharing between African countries and their international counterparts will enhance the continent's capacity to develop and deploy green technologies. Establishing partnerships with global organizations and leveraging their expertise and resources can help Africa overcome barriers to sustainable development and accelerate progress toward a greener future. By fostering a culture of entrepreneurship and innovation within educational institutions, Africa can equip its youth with the skills and mindset necessary to tackle complex challenges and seize opportunities in the green technology sector. Creating platforms for collaboration and knowledge exchange between African entrepreneurs and international experts will foster a vibrant ecosystem that promotes the development and adoption of sustainable solutions across the continent.

10.5 AFRICA'S GEOGRAPHICAL ADVANTAGE PAVES THE WAY FOR SOLAR POWER LEADERSHIP

Africa's geographical advantage, with its vast landmass and abundant sunlight, positions the continent as a prime location for harnessing solar power on a large scale (Obiwulu et al., 2022). The continent's geographical location along the equator ensures consistent (Obiwulu, Chendo, et al., 2020) and intense sunlight throughout the year (Obiwulu, Erusiafe, et al., 2020), making it an ideal region for solar energy production. Africa's diverse topography offers a range of potential sites for solar installations, including deserts, plains, and highlands, allowing for optimal utilization of solar resources. This unique geographical advantage not only presents Africa with the opportunity to become a leader in solar power but also plays a crucial role in driving the global net zero transition. As the world strives to reduce greenhouse gas emissions and combat climate change, Africa's potential to generate clean and renewable energy through solar power can significantly contribute to achieving global sustainability goals. By leveraging its geographical advantage and embracing solar energy technologies, Africa can pave the way toward a cleaner and more sustainable future while also positioning itself as a key player in the global transition toward net zero emissions.

With abundant sunlight throughout the year, Africa has the potential to become a solar energy powerhouse, meeting its own energy needs and exporting clean electricity to other regions. By investing in solar infrastructure and fostering innovation in the sector, Africa can not only create job opportunities and drive economic growth but also inspire other countries to follow suit in transitioning to renewable energy sources. Harnessing solar energy in Africa can help reduce the continent's reliance on fossil fuels, contributing to a significant reduction in greenhouse gas emissions. This shift toward renewable energy can also improve energy access and affordability for millions of people in Africa, promoting social development and poverty alleviation. In addition, investing in solar infrastructure can also enhance energy security for African countries, as they will become less dependent on imported fossil fuels. This will allow them to have more control over their energy resources and reduce their vulnerability to price fluctuations in the global oil market. The development of solar energy can also lead to technological advancements and knowledge transfer, fostering a culture of innovation and sustainable development across the continent. The expansion of solar infrastructure can create job opportunities and stimulate economic growth in Africa. As the demand for solar energy increases, local businesses and industries will emerge to meet this demand, creating a ripple effect on employment and economic activity. The adoption of solar energy can contribute to reducing carbon emissions and mitigating the effects of climate change, positioning Africa as a leader in sustainable development on a global scale. The use of solar energy can also improve access to electricity in remote areas, where traditional power grids may not be feasible. This can lead to improved living conditions and enhanced opportunities for education and healthcare in these underserved communities. The development of a robust solar energy sector can attract foreign investment and technology transfer, further driving economic growth and technological advancement in Africa.

The adoption of solar energy can also contribute to job creation and entrepreneurship opportunities in Africa. By investing in solar infrastructure and manufacturing, local communities can benefit from the establishment of new businesses and the training of skilled workers. This can help to alleviate poverty and promote sustainable economic development across the continent. The use of solar energy can reduce reliance on fossil fuels, decreasing the vulnerability of African countries to fluctuations in global oil prices and enhancing energy security in the region. The adoption of solar energy can also have positive environmental impacts in Africa. Solar power is a clean and renewable source of energy, which can help reduce greenhouse gas emissions and combat climate change. The use of solar energy can improve access to electricity in remote and rural areas, where traditional grid infrastructure may be lacking (Nwokolo and Ogbulezie, 2018a). This can improve living conditions and provide opportunities for education, healthcare, and economic growth in these underserved communities. The adoption of solar energy can also reduce reliance on fossil fuels, which are often imported and can be expensive for African countries (Nwokolo and Ogbulezie, 2018a). By harnessing the abundant sunlight available in many parts of Africa, solar power can provide a more affordable and sustainable energy solution.

10.6 LEVERAGING THE CONTINENT'S ECOSYSTEMS FOR SUSTAINABLE CARBON SEQUESTRATION AND GREENHOUSE GAS MITIGATION

Africa's vast and diverse ecosystems present a unique opportunity for sustainable carbon sequestration and greenhouse gas mitigation. Africa is home to a diverse range of ecosystems, including forests, grasslands, wetlands, and coastal areas, which play a crucial role in carbon sequestration and greenhouse gas mitigation (Proutsos, Liakatas, et al., 2023). This involves leveraging the natural capacity of these ecosystems to absorb and store carbon dioxide while also implementing innovative strategies such as reforestation, afforestation, and sustainable land management practices (Proutsos, Tigkas, et al., 2023). According to a study by the International Energy Agency (IEA) and United Nations Environment Programme (UNEP) (International Energy Agency [IEA] and United Nations Environment Programme [UNEP], 2018), Africa's terrestrial ecosystems sequester around 1.2 billion metric tons of carbon dioxide annually. The continent's forests alone absorb approximately 1.2 Gt of CO_2 each year. These figures highlight the significant potential that Africa's ecosystems hold for sustainable carbon sequestration and greenhouse gas mitigation. By leveraging these natural resources effectively, Africa can contribute significantly to the global net zero transition. It is essential to prioritize conservation efforts and sustainable land management practices to ensure the long-term viability of these ecosystems as climate change intensifies. Without proper conservation and land management, Africa's ecosystems could face degradation and loss, leading to a reduction in their carbon sequestration capacity. Investing in sustainable practices can not only help mitigate climate change but also provide economic opportunities for local communities, fostering a more sustainable and resilient future for Africa.

Africa's key role in this transition is not only crucial for mitigating climate change but also presents significant economic opportunities for the continent through carbon offset projects, ecosystem services, and green investments. Africa's diverse ecosystems provide essential habitats for a wide range of species, making conservation efforts vital for preserving biodiversity. By prioritizing sustainable land management practices and protecting these ecosystems, Africa can not only contribute to the global net zero goal but also safeguard its unique wildlife and natural heritage. This holistic approach will not only benefit the environment but also create jobs and stimulate economic growth in local communities, fostering a more sustainable and prosperous future for the continent. Investing in green technologies and renewable energy sources can help Africa reduce its carbon footprint and mitigate the impacts of climate change. By promoting the development of clean energy infrastructure and supporting green businesses, Africa can tap into its vast renewable energy potential while reducing its dependence on fossil fuels. This transition to a low-carbon economy will not only contribute to global efforts to combat climate change but also create new job opportunities and attract sustainable investments to the continent. The shift toward renewable energy will improve air quality and public health as it reduces harmful emissions that contribute to respiratory diseases. By embracing sustainable practices, Africa can position itself as a candidate in the global green economy, attracting international partnerships and collaborations that can further drive economic growth and innovation on the continent. The transition to a low-carbon economy can also enhance energy security for African countries by reducing their dependence on fossil fuel imports. This shift will not only reduce vulnerability to price fluctuations but also promote energy independence and resilience in the face of global energy challenges.

REFERENCES

Davidson, I. E. (2020). A review of long-distance UHVDC technology – A future energy disrupter. *Clemson University Power Systems Conference, PSC 2020*. https://doi.org/10.1109/PSC50246.2020.9131273

Dhansay, T., de Wit, M., & Patt, A. (2014). An evaluation for harnessing low-enthalpy geothermal energy in the Limpopo Province, South Africa. *South African Journal of Science*. https://doi.org/10.1590/sajs.2014/20130282

Global Carbon Atlas (GCA). (2022). *Global Carbon Atlas*. Global Carbon Atlas.

International Energy Agency (IEA) & United Nations Environment Programme (UNEP). (2018). International Energy Agency and the United Nations Environment Programme – Global Status Report 2018: Towards a zero-emission, efficient and resilient buildings and construction sector. *Global Status Report*.

International Renewable Energy Agency (IRENA). (2022). World Energy Transitions Outlook 2023: 1.5°C Pathway. *World Energy Transitions*.

Ituen, E. E., Esen, N. U., Nwokolo, S. C., & Udo, E. G. (2012). Prediction of global solar radiation using relative humidity, maximum temperature and sunshine hours in Uyo, in the Niger Delta Region, Nigeria. *Advances in Applied Science Research*, *3*(4), 1923–1937.

Kiprotich, N. E., De Dieu Hakizimana, J., & Chambile, E. (2021). Exploring implication of renewable energy transition on the cost of electricity and green house gases emission in East African Countries. *2021 IEEE Southern Power Electronics Conference, SPEC 2021*. https://doi.org/10.1109/SPEC52827.2021.9709309

Mshandete, A. M., & Parawira, W. (2010). Biogas: Sustainable alternative renewable energy of today and the future in Africa. *Annals of Arid Zone*.

Nwokolo, S. C., Echeng, E., Obiwulu, A., Meyer, E. L., Christian, C., Ogbulezie, J. C., & Proutsos, N. (2024). A multi-model approach based on CARIMA-SARIMA-GPM for assessing the impacts of climate change on concentrated photovoltaic (CPV) potential. *Physics and Chemistry of the Earth*, *134*(January), 103560. https://doi.org/10.1016/j.pce.2024.103560

Nwokolo, S. C., Meyer, E. L., & Ahia, C. C. (2023). Credible pathways to catching up with climate goals in Nigeria. *Climate*, *11*(9), 1–32. https://doi.org/org/10.3390/cli11090196

Nwokolo, S. C., Meyer, E. L., & Ahia, C. C. (2024). Exploring the interactive influences of climate change and urban development on the fraction of absorbed photosynthetically active radiation. *Atmosphere*, 15(3) 252. https://doi.org/10.3390/atmos15030253

Nwokolo, S. C., Obiwulu, A. U., & Ogbulezie, J. C. (2023). Machine learning and analytical model hybridization to assess the impact of climate. *Physics and Chemistry of the Earth*, *130*(February 2023), 103389. https://doi.org/10.1016/j.pce.2023.103389

Nwokolo, S. C., & Ogbulezie, J. C. (2018a). A qualitative review of empirical models for estimating diffuse solar radiation from experimental data in Africa. *Renewable and Sustainable Energy Reviews*, *92*(18), 353–393. https://doi.org/10.1016/j.rser.2018.04.118

Nwokolo, S. C., & Ogbulezie, J. C. (2018b). A quantitative review and classification of empirical models for predicting global solar radiation in West Africa. *Beni-Suef University Journal of Basic and Applied Sciences*, *7*(4), 367–396. https://doi.org/10.1016/j.bjbas.2017.05.001

Nwokolo, S. C., Ogbulezie, J. C., & Ushie, O. J. (2023). A multi-model ensemble-based CMIP6 assessment of future solar radiation and PV potential under various climate warming scenarios. *Optik*, *285*(May), 170956. https://doi.org/10.1016/j.ijleo.2023.170956

Nwokolo, S. C., & Otse, C. (2019). Impact of sunshine duration and clearness index on diffuse solar radiation estimation in mountainous climate. *Trends in Renewable Energy*, *5*(3), 307–332. https://doi.org/10.17737/tre.2019.5.3.00107

Nwokolo, S. C., Proutsos, N., & Meyer, E. L. (2023). Machine learning and physics-based hybridization models for evaluation of the effects of climate change and urban expansion on photosynthetically active radiation. *Atmosphere*, *14*(687), 1–36. https://doi.org/doi.org/10.3390/atmos14040687

Obiwulu, A. U., Chendo, M. A. C., Erusiafe, N., & Nwokolo, S. C. (2020). Implicit meteorological parameter-based empirical models for estimating back temperature solar modules under varying tilt-angles in Lagos, Nigeria. *Renewable Energy, 145*. https://doi.org/10.1016/j.renene.2019.05.136

Obiwulu, A. U., Erusiafe, N., Olopade, M. A., & Nwokolo, S. C. (2020). Modeling and optimization of back temperature models of mono-crystalline silicon modules with special focus on the effect of meteorological and geographical parameters on PV performance. *Renewable Energy, 154*. https://doi.org/10.1016/j.renene.2020.02.103

Obiwulu, A. U., Erusiafe, N., Olopade, M. A., & Nwokolo, S. C. (2022). Modeling and estimation of the optimal tilt angle, maximum incident solar radiation, and global radiation index of the photovoltaic system. *Heliyon, 8*(6), e09598. https://doi.org/10.1016/j.heliyon.2022.e09598

Ogunniyi, E., & Pienaar, C. (2019). Paradox of Africa's renewable energy potentials and quest towards powering Africa. *IEEE PES/IAS PowerAfrica Conference: Power Economics and Energy Innovation in Africa, PowerAfrica 2019*. https://doi.org/10.1109/PowerAfrica.2019.8928927

Proutsos, N., Liakatas, A., Alexandris, S., Nwokolo, S. C., Solomou, A. D., & Amadi, S. O. (2023). Assessing the impact of atmospheric attributes on the effectiveness of solar irradiance for photosynthesis of urban vegetation in Attica, Greece. *Theoretical and Applied Climatology, 155*(2):1–13. https://doi.org/10.1007/s00704-023-04700-0

Proutsos, N., Tigkas, D., Tsevreni, I., Alexandris, S. G., Solomou, A. D., Bourletsikas, A., Stefanidis, S., & Nwokolo, S. C. (2023). A thorough evaluation of 127 potential evapotranspiration models in two Mediterranean Urban Green Sites. *Remote Sensing, 15*(14), 1–41. https://doi.org/10.3390/rs15143680

SANEA. (2020). *Energy Storage technologies will unlock a "new gold rush" for South Africa*. SANEA.

United Nations. (2022). Population | United Nations. *United Nations*.

World Bank Group. (2020). Africa's resource export opportunities and the global energy transition. *Live Wire*.

Yetano Roche, M., Verolme, H., Agbaegbu, C., Binnington, T., Fischedick, M., & Oladipo, E. O. (2020). Achieving sustainable development goals in Nigeria's power sector: Assessment of transition pathways. *Climate Policy*. https://doi.org/10.1080/14693062.2019.1661818

11 Africa's Propensity for a Net Zero Energy Transition

11.1 INTRODUCTION

Africa's propensity for a net zero energy transition is a testament to its untapped renewable energy potential and growing commitment toward sustainable development. With its abundant solar resources, vast wind corridors, and rich geothermal reservoirs, Africa has the unique opportunity to harness these clean energy sources and pave the way for a greener future (Ituen et al., 2012). As African nations increasingly prioritize renewable energy investments (Nwokolo, Meyer, et al., 2023) and adopt progressive policies (Nwokolo, Obiwulu, et al., 2023), they are not only addressing their own energy needs but also positioning themselves as global leaders in the transition toward a carbon-neutral economy (Nwokolo, Ogbulezie, et al., 2023). By investing in renewable energy, African nations can reduce their reliance on fossil fuels and mitigate the harmful effects of climate change (Nwokolo, Proutsos, et al., 2023). The transition to clean energy sources can create new job opportunities and stimulate economic growth across the continent. As Africa continues to prioritize sustainable development, it has the potential to inspire other regions around the world to follow suit and accelerate the global shift toward a more sustainable and environmentally friendly future (Nwokolo, Meyer, et al., 2024; Nwokolo, Echeng, et al., 2024). By embracing renewable energy, African nations can also improve energy access and security, ensuring a more reliable and resilient power supply for their citizens. The adoption of clean energy technologies can help reduce air pollution and improve public health outcomes, leading to a better quality of life for African communities (Nwokolo, Meyer, et al., 2024; Nwokolo, Echeng, et al., 2024).

Africa's enormous reserves of critical minerals, including lithium, cobalt, and rare earth elements, attest to its propensity for a net zero energy transition. These minerals are essential for the production of renewable energy technologies like batteries and solar panels. By harnessing these resources responsibly, Africa can not only drive its own sustainable development but also become a key player in the global clean energy market. This presents an opportunity for economic growth and job creation, further enhancing the continent's potential to lead the way in the transition to a greener future. The demand for these critical minerals is expected to skyrocket in the coming years as countries around the world ramp up their efforts to combat climate change. With its vast reserves, Africa has the potential to meet this increasing

DOI: 10.1201/9781003483175-12

demand and become a major supplier of these minerals globally. This would not only bring economic benefits but also allow Africa to exert influence and shape the future of clean energy technologies on a global scale. By becoming a major supplier of critical minerals, Africa can play a crucial role in reducing the world's dependence on fossil fuels and accelerating the adoption of renewable energy sources (Nwokolo and Otse, 2019). This would create opportunities for technology transfer and knowledge sharing, enabling African countries to develop their own clean energy industries and strengthen their position in the global market. By investing in the development of local mining and processing capabilities, Africa can also create jobs and stimulate economic growth within its own borders. This would help to alleviate poverty and improve living standards for its citizens, while also promoting sustainable development practices in the region. Africa's potential as a major supplier of critical minerals presents a unique opportunity for the continent to contribute significantly to global efforts in combating climate change and achieving a more sustainable future.

As the world grapples with the urgent need to combat climate change, Africa emerges as a beacon of hope and inspiration. With its growing commitment toward a net zero energy transition, Africa showcases its determination to tackle the global challenge by embracing sustainable alternatives and minimizing reliance on fossil fuels. This remarkable endeavor not only demonstrates Africa's resilience but also sets an example for other regions to follow in the pursuit of a greener future. Africa's commitment to a net zero energy transition is evident in its efforts to harness renewable energy sources such as solar, wind, and hydroelectric power. By investing in these sustainable alternatives, Africa not only reduces its carbon footprint but also creates opportunities for economic growth and job creation (Obiwulu et al., 2022). As the continent continues to lead the way in addressing climate change, it inspires hope that a greener future is possible for all nations. These efforts are particularly impressive considering the challenges Africa faces, such as limited infrastructure and access to financing.

Despite these obstacles, African countries have shown determination and innovation in finding solutions to their energy needs (Obiwulu, Chendo, et al., 2020). By prioritizing renewable energy, Africa is not only mitigating the effects of climate change but also positioning itself as a global candidate for sustainable development. African countries have embraced a variety of renewable energy sources, including solar, wind, and hydropower (Obiwulu, Erusiafe, et al., 2020). This diversification of energy production not only reduces greenhouse gas emissions but also enhances energy security and independence. Africa's commitment to sustainable development has attracted international investment and collaboration, fostering economic growth and job creation within the continent. As a result, African nations are not only improving access to clean and affordable energy for their citizens but also creating opportunities for local businesses and industries to thrive (Nwokolo and Ogbulezie, 2018a). The shift toward renewable energy is helping to address energy poverty in rural areas, where access to electricity has traditionally been limited. By harnessing the power of renewable resources, Africa is paving the way for a more sustainable and prosperous future.

The book delves into the intricate details of Africa's potential for a net zero energy transition, examining 11 hypothetical scenarios that could play a crucial role in shaping its future. These scenarios not only showcase the book's commitment to providing innovative solutions but also highlight the need for professional insights to navigate Africa's unique energy landscape. These scenarios ranged from Africa's vast renewable energy potential to leveraging the continent's ecosystems for sustainable carbon sequestration and greenhouse gas mitigation (Nwokolo and Ogbulezie, 2018b). The book goes beyond surface-level analysis and delves into the intricate details of each scenario, offering a comprehensive understanding of the challenges and opportunities that Africa's net zero energy transition presents. By exploring these hypothetical scenarios, the book aims to inspire policymakers, researchers, and industry leaders to take proactive measures to address Africa's energy needs in a sustainable and cutting-edge manner. The authors of the book draw on their extensive experience in the energy sector to offer practical recommendations and strategies for achieving a successful net zero energy transition in Africa. They emphasize the importance of collaboration between governments, private sector entities, and local communities to ensure the effective implementation of sustainable energy solutions. The book sheds light on the potential economic benefits that can be derived from embracing renewable energy sources and technologies, such as job creation and increased energy access for underserved populations. The authors highlight the need for policy reforms and regulatory frameworks that incentivize investment in renewable energy projects and remove barriers to their deployment. They also discuss the role of international partnerships and financial support in facilitating the transition to net zero energy, stressing the importance of global cooperation in addressing climate change and promoting sustainable development in Africa.

The authors emphasize the potential economic benefits that can arise from the renewable energy sector, including the growth of local industries and the attraction of foreign investment. They argue that by harnessing Africa's vast renewable energy resources, countries can not only reduce their carbon emissions but also stimulate economic growth and create new job opportunities. They emphasize the importance of building capacity and knowledge transfer in order to ensure a successful transition to renewable energy and maximize its potential benefits for all African nations. The authors also highlight the need for policy support and regulatory frameworks that promote renewable energy development in Africa. They suggest that governments should provide incentives and create a conducive environment for private sector participation in the renewable energy sector. They stress the importance of collaboration and partnerships between African nations and international organizations to share best practices and leverage resources for a sustainable energy future in Africa.

The authors emphasize the importance of investing in research and development to drive innovation in renewable energy technologies specifically tailored to Africa's unique needs and challenges. They argue that such investments will not only accelerate the transition to renewable energy but also create opportunities for job creation and economic growth across the continent. Ultimately, the authors assert that a collective effort from governments, private sector entities, and international organizations is crucial to ensuring a successful and inclusive renewable energy transition in Africa. The authors highlight that investing in research and development will enable the creation of locally adapted renewable energy solutions, such as solar panels designed to withstand extreme weather conditions or wind turbines optimized for low wind speeds. By tailoring these technologies to Africa's specific requirements, the continent can overcome barriers and maximize the potential of renewable energy sources. The authors stress that collaboration between stakeholders is essential to foster knowledge sharing, capacity building, and the transfer of expertise, ultimately leading to a sustainable and prosperous renewable energy future for Africa. This collaboration can involve partnerships between governments, international organizations, and local communities to ensure a comprehensive approach to renewable energy development. It is crucial to prioritize investment in research and development to continuously improve and innovate renewable energy technologies that are suitable for Africa's unique conditions. Fostering collaboration can also include sharing best practices and lessons learned from successful renewable energy projects across the continent. This knowledge exchange can help avoid duplication of efforts and accelerate the deployment of renewable energy solutions in Africa. Promoting education and training programs can empower local communities to actively participate in the renewable energy sector, creating job opportunities and driving economic growth.

11.2 AFRICA'S VAST RENEWABLE ENERGY POTENTIAL

Africa's vast renewable energy potential holds immense promise for the continent's journey toward a net zero energy transition. With its abundant solar, wind, hydro, and geothermal resources, Africa has the opportunity to harness these sustainable sources of energy to meet its growing electricity demand while reducing carbon emissions. This transition not only aligns with global efforts to combat climate change but also presents an opportunity for Africa to leapfrog traditional fossil fuel–based development and embrace a more sustainable and resilient energy future. According to the International Renewable Energy Agency (IRENA, 2022), Africa has remarkable renewable energy potential, with abundant solar, wind, hydro, and geothermal resources. For instance, the continent receives about 325 days of strong sunlight per year, making it ideal for solar power generation. Africa's wind resources are estimated to be capable of generating over 1000 GW of electricity. This vast renewable energy potential is crucial for Africa's net zero energy transition. A study by the African Development Bank (AfDB, 2022) suggests that scaling up renewable energy in Africa could help achieve a net zero carbon economy by 2050. By harnessing these abundant resources, African countries can reduce their dependence on fossil fuels and significantly lower greenhouse gas emissions. Investing in renewable energy can provide numerous socio-economic benefits for Africa. It can create jobs, improve access to electricity in remote areas, and enhance energy security by diversifying the energy mix. These factors contribute to Africa's propensity for a successful net zero energy transition and pave the way for sustainable development across the continent. Transitioning to renewable energy sources can also improve public health by reducing air pollution and its associated health risks. By embracing renewable energy technologies, African countries can position themselves as leaders in the global clean energy sector, attracting investment and fostering technological innovation.

Transitioning to renewable energy sources can also create new job opportunities and stimulate economic growth in Africa. By investing in renewable energy infrastructure and training a skilled workforce, African countries can harness the potential of clean energy to drive sustainable development and reduce poverty. Embracing renewable energy technologies can also help mitigate the impacts of climate change, such as extreme weather events and rising sea levels, which disproportionately affect vulnerable communities in Africa. Transitioning to renewable energy sources can enhance energy security by reducing dependence on fossil fuel imports. This can lead to greater stability and resilience in African economies, as well as reduce their vulnerability to fluctuations in global oil prices. Embracing renewable energy technologies can position African countries as leaders in the global clean energy transition, attract foreign investment, and foster partnerships for technology transfer and knowledge sharing.

In this scenario, Africa's proclivity for a net zero energy transition is founded on 12 premises. These include (1) the abundance of solar energy due to its location near the equator, (2) the vast potential for wind energy along coastal regions and high-altitude areas, (3) the availability of geothermal resources in certain volcanic regions, (4) the potential for hydropower generation from rivers and waterfalls, (5) the abundance of biomass resources for bioenergy production, (6) the potential for tidal and wave energy along coastal areas, (7) the possibility of harnessing ocean thermal energy, (8) the potential for hydrogen production through electrolysis using renewable energy sources, (9) the opportunity for decentralized off-grid solutions to reach remote communities, (10) Africa's renewable energy potential is further enhanced by advancements in technology and decreasing costs of renewable energy systems, (11) Africa's vast landmass provides ample space for large-scale solar and wind farms, further contributing to its renewable energy potential, and (12) the continent's growing population and increasing energy demand create a significant market for renewable energy investments and developments. These make Africa's propensity for a net zero energy transition a testament to its potential for sustainable development and economic growth.

11.2.1 Abundance of Solar Energy Due to Its Location Near the Equator

According to the International Renewable Energy Agency (IRENA, 2022), Africa has the highest solar irradiation levels globally, with an average of 4–6 kWh/m^2 per day. This abundance of solar energy is a result of its location near the equator, which ensures consistent and intense sunlight throughout the year. The high solar irradiation levels in Africa make it an ideal region for the development of solar energy projects. African countries located close to the equator, such as Nigeria, have the potential to harness a significant amount of solar energy and reduce their reliance on fossil fuels. In fact, Nigeria has already taken steps toward increasing its solar energy capacity with the construction of large-scale solar power plants and the implementation of solar home systems in rural areas. By tapping into its abundant solar resources, Africa can not only meet its own energy demands but also contribute to global efforts to combat climate change. Other countries include Kenya, Ethiopia, and South Africa, which have also recognized the potential of solar energy and have made significant investments in the sector. These countries have implemented policies and incentives to attract solar energy developers and encourage the adoption of solar technologies. By embracing solar energy, these African nations can not only reduce their carbon emissions but also create job opportunities and stimulate economic growth in the renewable energy sector.

Also, North Africa has emerged as a candidate for solar energy development, with countries like Morocco and Egypt investing heavily in large-scale solar projects. These countries are taking advantage of their abundant sunlight resources to generate clean and sustainable electricity, reducing their dependence on fossil fuels and promoting energy independence. The growth of the solar energy sector in North Africa has the potential to create regional partnerships and collaborations, fostering economic integration and cooperation among neighboring countries. The development of the solar energy sector in North Africa can also have positive social impacts. Investment in large-scale solar projects can create job opportunities and contribute to poverty reduction, improving the overall quality of life for the local population. The shift toward renewable energy can also lead to a reduction in greenhouse gas emissions, helping combat climate change on a global scale. The use of solar energy can reduce dependence on fossil fuels and enhance energy security in the region. This can lead to a more sustainable and resilient energy infrastructure, ensuring a stable and reliable power supply for North African countries. The development of the solar energy sector can attract foreign investment and promote technological advancements, further driving economic growth and innovation in the region.

This high solar potential positions Africa as a prime candidate for a net zero energy transition. By harnessing this abundant solar resource, African countries can significantly reduce their reliance on fossil fuels and achieve sustainable development goals. The implementation of solar energy projects can provide clean and affordable electricity to remote areas, improve energy access, create job opportunities, and mitigate greenhouse gas emissions. The use of solar energy can also contribute to reducing energy poverty in Africa, as it provides a decentralized and reliable source of power that can reach even the most remote communities. The adoption of solar energy can enhance energy security for African countries by reducing their dependence on imported fossil fuels and volatile global oil prices. The utilization of solar energy can stimulate economic growth in Africa by attracting investments in the renewable energy sector and fostering local entrepreneurship. The deployment of solar energy systems can help diversify the energy mix and reduce the vulnerability of African countries to fluctuations in oil prices, thus promoting long-term stability and resilience in their economies.

In addition, the use of solar energy can also have positive environmental impacts by reducing greenhouse gas emissions and mitigating climate change. The widespread adoption of solar energy can improve access to electricity in remote areas, thereby improving the quality of life and opportunities for education and healthcare in these communities. By investing in solar infrastructure, local communities can develop a skilled workforce and attract investments from renewable energy companies, leading to sustainable development and poverty reduction. The use of solar energy can also reduce reliance on fossil fuels, which are finite resources and contribute to air pollution and global warming. This transition to renewable energy sources can also create new job opportunities in the clean energy sector, fostering economic growth and stability in these communities. The use of solar energy can enhance energy security by reducing dependence on imported fossil fuels, allowing African countries to become more self-sufficient and resilient in meeting their energy needs. Ultimately, the widespread adoption of solar energy can lead to a more sustainable and resilient future for both local communities and the planet as a whole.

11.2.2 Vast Potential for Wind Energy along Coastal Regions and High-Altitude Areas

According to a study conducted by the International Renewable Energy Agency (IRENA, 2022), Africa has the potential to generate over 1800 GW of wind energy, with coastal regions and high-altitude areas being particularly promising. The coastal regions benefit from strong and consistent winds due to their proximity to the ocean, while high-altitude areas experience higher wind speeds due to reduced surface friction. Harnessing wind energy along coastal regions and high-altitude areas in Africa can play a crucial role in achieving a net zero energy transition for the continent. Coastal regions, with their proximity to the ocean, offer a consistent and strong source of winds, making them ideal for wind power generation. Similarly, high-altitude areas experience higher wind speeds due to reduced surface friction, making them another promising location for harnessing wind energy. By strategically utilizing these resources, Africa can tap into its natural advantages and accelerate its transition toward sustainable and renewable energy sources.

For instance, the Eastern Cape province in South Africa has been identified as having significant wind energy potential along its coastline. With an average wind speed of around 8–10 m/s, this region alone could generate substantial amounts of clean electricity. Similarly, countries like Ethiopia and Kenya have identified their high-altitude regions, such as the Rift Valley and the Great Rift Valley, respectively, as prime locations for harnessing wind power. Other locations in Africa include the Atlas Mountains in Morocco and the Drakensberg Mountains in Lesotho, both of which have been recognized for their strong and consistent winds. These areas offer great potential for wind energy generation and could contribute significantly to the continent's renewable energy goals. Coastal regions like Cape Verde and Mauritius also have favorable wind conditions, making them attractive options for wind power development in Africa. The Nile River Valley in Egypt and the Namib Desert in Namibia have also been identified as promising areas for wind power due to their unique geographical features. The strong and consistent winds in these regions could provide a reliable source of renewable energy for their respective countries and contribute to the overall sustainability of the continent. The Sahel region, stretching across several countries, including Senegal, Mali, and Niger, has been identified as another promising area for wind power due to its consistent trade winds. Expanding wind energy projects in these regions could not only provide clean and sustainable electricity but also create job opportunities and boost economic growth in Africa.

In Nigeria, the coastal areas, such as Lagos, Port Harcourt, Akwa Ibom, Benue, Asaba, and Calabar, have the potential to harness wind energy due to their proximity to the Atlantic Ocean. The strong ocean breezes in these regions could be utilized to generate electricity and reduce the country's reliance on fossil fuels. Investing in wind power infrastructure in Nigeria could help diversify the energy sector and contribute to the country's goal of achieving sustainable development. The development of wind power in Nigeria could also attract foreign investment and technology transfer, leading to the creation of local manufacturing industries and job opportunities. This would not only boost economic growth but also enhance the country's energy security by reducing its dependence on imported fossil fuels. Harnessing wind energy in these coastal areas would have a minimal environmental impact compared to other forms of energy generation, aligning with Nigeria's commitment to reducing greenhouse gas emissions and mitigating climate change. The potential for wind power is particularly high in the northern states, such as Kano, Maiduguri, Katsina, and Sokoto, where the wind speeds are consistently strong. By harnessing this natural resource, Nigeria could reduce its reliance on fossil fuels and decrease greenhouse gas emissions, contributing to global efforts to combat climate change. Investing in wind energy infrastructure would diversify Nigeria's energy mix and improve energy security, ensuring a more stable and reliable electricity supply for its growing population. The development of wind power in Nigeria would create job opportunities and stimulate economic growth in the renewable energy sector. This would not only benefit the local communities but also attract foreign investments and enhance Nigeria's position as a candidate for sustainable development within Africa.

Other locations in Africa include the coastal areas of South Africa, such as Cape Town and Port Elizabeth, which experience strong and consistent winds from the Atlantic Ocean. These regions have the potential to harness wind power and reduce their reliance on fossil fuels, contributing to a greener and more sustainable energy mix. Countries like Kenya and Ethiopia have also shown great potential for wind energy development, with their high-altitude locations providing ideal conditions for wind farms. By tapping into these natural resources, Africa can take significant steps toward achieving its renewable energy goals and mitigating the impacts of climate change. Harnessing wind power in regions like Elizabeth, which experience strong and consistent winds from the Atlantic Ocean, can not only reduce reliance on fossil fuels but also create new job opportunities and stimulate economic growth. Investing in wind energy development in countries like Kenya and Ethiopia can not only address their energy needs but also contribute to regional energy security and foster international collaborations for a sustainable future.

Harnessing this vast potential for wind energy in coastal regions and high-altitude areas can play a crucial role in Africa's net zero energy transition. By investing in wind power infrastructure and utilizing these abundant resources, African countries can reduce their dependence on fossil fuels, mitigate greenhouse gas emissions, and contribute toward a sustainable future. The development of wind energy in Kenya and Ethiopia can also create job opportunities and stimulate economic growth in these countries. This can lead to improved living standards and increased social development, benefiting local communities and the overall region. By showcasing successful renewable energy projects, these countries can inspire other nations to follow suit and accelerate the global transition toward a greener and more sustainable energy system. The development of wind energy in Kenya and Ethiopia can help diversify their energy sources, reducing their vulnerability to fluctuations in fossil fuel prices. This can enhance energy security and stability in these countries, ensuring a reliable and affordable electricity supply for their populations. The expansion of wind energy can also contribute to the preservation of natural resources and ecosystems by reducing the need for environmentally harmful practices associated with fossil fuel extraction and combustion.

11.2.3 Availability of Geothermal Resources in Certain Volcanic Regions

Geothermal resources, which harness the heat from beneath the Earth's surface, have significant potential in volcanic regions of Africa. According to a study by the International Energy Agency and United Nations Environment Programme (IEA and UNEP, 2018), Africa has an estimated geothermal potential of 15,000 MW. This vast resource can play a crucial role in Africa's net zero energy transition. Volcanic regions such as the East African Rift System and the Great Rift Valley are known for their geothermal activity. For instance, Kenya has already made significant strides in tapping into its geothermal potential. The country's Olkaria Geothermal Complex is one of the largest in the world, generating over 800 MW of clean and renewable energy (IEA, 2023). Expanding geothermal power generation in these volcanic regions can contribute to Africa's net zero energy transition by reducing reliance on fossil fuels and decreasing greenhouse gas emissions. Geothermal energy provides a stable and reliable source of power that can support industrial growth and improve access to electricity in remote areas. With proper investment and technological advancements, harnessing geothermal resources can be a key driver toward achieving a sustainable and low-carbon future for Africa.

Geothermal power plants have a long lifespan and require minimal maintenance, making them a cost-effective solution for long-term energy production. The development of geothermal energy can create job opportunities and stimulate economic growth in these regions, further benefiting local communities and contributing to Africa's overall development goals. Geothermal energy is a reliable and consistent source of power, as it is not dependent on weather conditions like solar or wind energy. This makes it particularly suitable for meeting the energy needs of industries and businesses in remote areas. By reducing reliance on fossil fuels, geothermal energy can help Africa mitigate the impacts of climate change and reduce greenhouse gas emissions, aligning with global efforts to combat environmental degradation. Geothermal energy has the potential to create job opportunities and stimulate entrepreneurship in these regions. The development of geothermal power plants requires skilled labor and technical expertise, which can lead to the growth of local industries and the establishment of new businesses. This can contribute to poverty alleviation and improve the standard of living for communities in Africa. By diversifying its energy sources with geothermal power, Africa can enhance its energy security and reduce its vulnerability to fluctuations in global oil prices.

11.2.4 Potential for Hydropower Generation from Rivers and Waterfalls

According to a study conducted by the International Renewable Energy Agency (IRENA, 2022), Africa has an estimated hydropower potential of 1,750 TWh/year, which is equivalent to about 18% of the continent's current electricity consumption. This immense potential can significantly contribute to Africa's net zero energy transition. Hydropower is a reliable and consistent source of renewable energy that can provide baseload power, ensuring a stable electricity supply. This is particularly important for Africa, as it seeks to expand its energy access and meet the growing demand for electricity in a sustainable manner. By harnessing the power of rivers and waterfalls through hydropower generation, Africa can reduce its dependence on fossil fuels and make significant progress toward achieving a net zero energy future. Africa has enormous potential for hydropower generation from its abundant rivers and waterfalls. For instance, the Congo River in Central Africa is estimated to have the potential to generate up to 40% of the continent's total hydropower capacity (IEA, 2023). The Zambezi River in Southern Africa, with its famous Victoria Falls, holds significant untapped hydropower potential that could contribute significantly to Africa's net zero energy transition (World Bank, 2022).

The River Niger and Benue in Nigeria also offer significant opportunities for hydropower generation, with the potential to contribute to the country's renewable energy goals. Exploiting these water resources for hydropower could not only provide clean and reliable electricity but also create job opportunities and stimulate economic growth in the region. Investing in hydropower infrastructure would help diversify Nigeria's energy mix and reduce its dependence on fossil fuels, ultimately contributing to a more sustainable and environmentally friendly future. The development of hydropower infrastructure in Nigeria could create job opportunities in the construction, maintenance, and operation of these facilities. This would not only boost the local economy but also enhance skills and expertise in the renewable energy sector. By reducing the country's reliance on fossil fuels, Nigeria could potentially save significant amounts of money spent on importing these resources, allowing for more investment in other critical sectors such as healthcare and education. The utilization of hydropower in Nigeria would help reduce greenhouse gas emissions, mitigating the effects of climate change. This would not only benefit the country but also contribute to global efforts to combat environmental degradation. The availability of a reliable and sustainable energy source like hydropower would improve access to electricity in rural areas, promote economic development, and improve the quality of life for local communities.

Harnessing this vast hydropower potential would not only provide clean and renewable energy but also help address Africa's energy deficit and promote economic development. It could reduce reliance on fossil fuels and mitigate greenhouse gas emissions, aligning with global efforts to combat climate change. The development of hydropower infrastructure would create job opportunities and stimulate local economies. By investing in hydropower, African countries can establish themselves as leaders in renewable energy technology and attract foreign investment for further development projects. The revenue generated from hydropower projects could be reinvested into other sectors, such as education and healthcare, improving the overall quality of life for African citizens. The increased access to electricity would enhance productivity in industries and households, driving economic growth and poverty reduction. The development of hydropower infrastructure would also contribute to reducing reliance on fossil fuels, leading to a decrease in carbon emissions and a more sustainable future for Africa. The establishment of hydropower plants could provide a reliable and affordable source of electricity, which would not only benefit industries but also improve the living conditions of rural communities by enabling access to modern amenities such as lighting and communication devices.

11.2.5 Abundance of Biomass Resources for Bioenergy Production

Africa has a vast potential for biomass resources, which can play a crucial role in achieving a net zero energy transition. According to a study by the International Renewable Energy Agency (IRENA), Africa has an estimated biomass potential of over 1.2 billion tons per year, primarily consisting of agricultural residues, forestry waste, and dedicated energy crops (IRENA, 2022). This abundance of biomass resources provides a significant opportunity for bioenergy production, which can contribute to reducing greenhouse gas emissions and meeting the growing energy demands in the region. The utilization of biomass resources for bioenergy production aligns with Africa's commitment to sustainable development and climate change mitigation. The African Union's Agenda 2063 emphasizes the importance of renewable energy sources and calls for the promotion of bioenergy as a means to achieve sustainable industrialization and economic growth (African Union Commission, 2015). By harnessing its abundant biomass resources, Africa can reduce its reliance on fossil fuels and transition toward a more sustainable and climate-resilient energy system.

Most African countries in sub-Saharan Africa possessed a larger proportion of total biomass resources than their North African counterparts. This disparity in biomass resources highlights the potential for sub-Saharan African countries to lead the way in renewable energy development and climate change mitigation. With proper investment and support, these countries can not only achieve sustainable industrialization and economic growth but also contribute significantly to global efforts to combat climate change. For example, countries like Nigeria, Ethiopia, Tanzania, and the Democratic Republic of the Congo have significant biomass resources that can be utilized for bioenergy production. These resources include agricultural residues, forest biomass, and organic waste, which can be converted into biofuels, biogas, and biomass-based electricity. By tapping into these resources, African countries can not only reduce greenhouse gas emissions but also create new economic opportunities and improve energy access for their populations. The utilization of biomass resources can also contribute to sustainable development by promoting rural livelihoods and reducing dependence on fossil fuels. In addition, investing in bioenergy infrastructure and technologies can enhance energy security and resilience in African countries, ensuring a reliable and decentralized energy supply for both urban and rural areas. The development of bioenergy infrastructure can also stimulate job creation and drive economic growth in African countries. This can be particularly beneficial for rural communities, as it provides them with income-generating opportunities and helps to alleviate poverty. By diversifying their energy sources and reducing their reliance on fossil fuels, African countries can enhance their energy security and reduce their vulnerability to fluctuations in global oil prices.

In addition, the use of bioenergy can also contribute to mitigating climate change by reducing greenhouse gas emissions. This is crucial for African countries, as they are often highly vulnerable to the impacts of climate change, such as droughts and extreme weather events. By investing in bioenergy infrastructure, African countries can not only address their energy needs but also contribute to global efforts to combat climate change. The development of bioenergy can also have positive socio-economic impacts for African countries. It can create new job opportunities and stimulate economic growth in rural areas where bioenergy feedstocks are cultivated and processed. Promoting the use of bioenergy can enhance energy access for remote communities that are currently underserved by traditional energy sources, improving their quality of life and fostering inclusive development. The use of bioenergy can also reduce reliance on fossil fuels and decrease greenhouse gas emissions, aligning with global sustainability goals. The development of bioenergy technologies can promote technology transfer and knowledge sharing between African countries and the international community, fostering collaboration and innovation in the renewable energy sector. By investing in bioenergy, African countries have the potential to not only meet their energy needs but also contribute to a greener and more sustainable future for all.

11.2.6 Potential for Tidal and Wave Energy along Coastal Areas

According to recent studies, Africa has a vast potential for tidal and wave energy due to its extensive coastline spanning over 30,000 km (IRENA, 2020). For instance, the West African region alone has been estimated to possess a tidal energy potential of approximately 10 GW. The eastern coast of Africa is known for its strong and consistent wave energy resources, which could contribute significantly to the continent's renewable energy goals. In Nigeria, there have been initiatives to harness tidal and wave energy with the development of pilot projects and feasibility studies. The country's coastal areas, such as Lagos and Cross River, have shown promising potential for tidal and wave energy generation. These renewable energy sources have the potential to not only provide clean electricity but also create job opportunities and stimulate economic growth in the region. In addition, the utilization of tidal and wave energy can help reduce Nigeria's reliance on fossil fuels and decrease greenhouse gas emissions, contributing to global efforts to combat climate change. The development of a robust tidal and wave energy sector in Nigeria could also attract foreign investment and technology transfer, further bolstering the country's renewable energy industry.

In South Africa, the adoption of tidal and wave energy could also have significant benefits. By harnessing the power of the ocean, South Africa can diversify its energy mix and reduce its dependence on coal, which is a major contributor to air pollution and greenhouse gas emissions. The development of a thriving tidal and wave energy sector can create new job opportunities and stimulate economic growth, particularly in coastal regions where these resources are abundant. This transition toward renewable energy sources aligns with South Africa's commitment to sustainable development and can contribute to global efforts to mitigate climate change. Investing in ocean energy can enhance energy security for South Africa by reducing its reliance on imported fossil fuels. The development of this sector can position South Africa as a candidate for renewable energy technology and attract foreign investment and partnerships, fostering innovation and knowledge transfer. The growth of the ocean energy sector can create new job opportunities and stimulate economic diversification in South Africa. By harnessing the power of the ocean, the country can tap into a clean and inexhaustible source of energy, reducing its carbon footprint and promoting a greener economy. This transition can also serve as a model for other countries looking to transition to renewable energy sources and drive global efforts toward a more sustainable future.

In Egypt, the development of ocean energy can also have significant benefits. With its long coastline along the Mediterranean Sea and Red Sea, Egypt has immense potential for harnessing wave and tidal energy. By investing in this sector, Egypt can reduce its dependence on fossil fuels and contribute to its national goal of increasing renewable energy capacity. The growth of the ocean energy industry can create job opportunities and stimulate economic growth while also positioning Egypt as a candidate for clean energy technology within the region. The development of ocean energy in Egypt can also help address the issue of climate change by reducing greenhouse gas emissions. This can lead to improved air quality and a healthier environment for both current and future generations. By embracing ocean energy, Egypt can showcase its commitment to sustainability and inspire other countries in the region to follow suit in transitioning toward cleaner and more renewable sources of energy. The development of ocean energy in Egypt can also contribute to energy security by diversifying the country's energy sources and reducing its dependence on fossil fuels. This can help mitigate the risks associated with fluctuating oil prices and geopolitical tensions. The growth of the ocean energy sector can attract foreign investment and foster technological innovation, further boosting Egypt's economy and creating a competitive advantage in the global clean energy market.

Harnessing tidal and wave energy can play a crucial role in Africa's net zero energy transition. As these sources are renewable and predictable, they offer a reliable and consistent power supply that can complement other forms of renewable energy, such as solar and wind. Coastal communities can benefit from these technologies by creating job opportunities in the installation, operation, and maintenance of tidal and wave energy systems. By capitalizing on Africa's coastal potential for tidal and wave energy, the continent can make significant strides toward achieving its sustainable development goals while reducing its reliance on fossil fuels. In addition, the utilization of tidal and wave energy can also help address the issue of energy poverty in Africa. Many remote coastal communities currently lack access to reliable electricity, but by harnessing these abundant resources, they can have a more affordable and sustainable energy source. This not only improves their quality of life but also promotes economic growth and social development in these regions. The implementation of tidal and wave energy projects can create job opportunities in Africa. The development and maintenance of these renewable energy systems require skilled labor, providing employment for local communities and stimulating economic activity. The use of tidal and wave energy can contribute to reducing greenhouse gas emissions, mitigating the impacts of climate change, and preserving the continent's natural resources for future generations.

11.2.7 Possibility of Harnessing Ocean Thermal Energy

Ocean thermal energy is a promising renewable energy source that can contribute significantly to Africa's net zero energy transition. According to a study by the International Renewable Energy Agency (IRENA, 2020), the African continent has vast untapped potential for ocean thermal energy conversion (OTEC) due to its extensive coastline and proximity to the equator. The study estimates that Africa's OTEC resources could potentially generate up to 1170 gigawatts of electricity (IRENA, 2020), which is more than ten times the current installed capacity on the continent. Harnessing ocean thermal energy can not only provide a clean and sustainable source of power but also address Africa's energy challenges. With over 600 million people in sub-Saharan Africa lacking access to electricity (IRENA, 2020), utilizing OTEC can help bridge this energy gap and improve living conditions for millions. By reducing reliance on fossil fuels, OTEC can mitigate greenhouse gas emissions and contribute to combating climate change.

Investing in the research, development, and implementation of OTEC technology will be crucial for unlocking its full potential in Africa. Collaborative efforts between governments, international organizations, and private sector entities are needed to support the deployment of OTEC projects and create favorable policies and incentives. By capitalizing on its ocean resources, Africa can tap into a sustainable energy solution that aligns with its net zero ambitions while driving economic growth and social development. OTEC technology has the potential to provide a reliable and continuous source of clean energy, reducing Africa's dependence on fossil fuels. The implementation of OTEC projects can create job opportunities and stimulate local economies, contributing to poverty alleviation and improving living standards in coastal communities. In addition, OTEC projects can also contribute to the diversification of Africa's energy mix, reducing the risk of power shortages and enhancing energy security. The development of OTEC technology can position Africa as a leader in renewable energy innovation and attract foreign investment in the region's clean energy sector. This can lead to technology transfer and knowledge sharing, further strengthening Africa's renewable energy sector. OTEC projects have the potential to mitigate the impacts of climate change by reducing greenhouse gas emissions and promoting sustainable development in coastal regions. The implementation of OTEC technology can create job opportunities and stimulate economic growth in Africa. By harnessing the power of the ocean, countries can tap into a new source of renewable energy, reducing their dependence on fossil fuels and promoting a more sustainable future. OTEC projects can also have positive social impacts by providing electricity to remote coastal communities that may currently lack access to reliable power sources.

11.2.8 Potential for Hydrogen Production through Electrolysis Using Renewable Energy Sources

According to a study conducted by the International Renewable Energy Agency (IRENA, 2019), Africa has enormous potential for renewable energy generation, including hydrogen production through electrolysis. The study estimates that Africa's total technical potential for hydrogen production from renewable sources is around 2600 TWh/year (IRENA, 2019). This vast potential can play a crucial role in Africa's transition toward net zero energy, as hydrogen can be used as a clean and versatile energy carrier. It can be utilized in various sectors, such as transportation, industry, and power generation, thereby reducing greenhouse gas emissions and promoting sustainable development across the continent. In addition to its environmental benefits, the use of hydrogen as an energy carrier can also enhance energy security in Africa. By diversifying their energy sources and reducing their reliance on fossil fuels, African countries can become more resilient to fluctuations in global oil prices and geopolitical tensions.

The development of a hydrogen economy in Africa has the potential to create new job opportunities and stimulate economic growth, fostering a more sustainable and prosperous future for the continent. The adoption of hydrogen as an energy carrier can help African countries meet their commitments under international climate agreements, such as the Paris Agreement. By reducing greenhouse gas emissions, Africa can contribute to global efforts to mitigate climate change and protect the planet for future generations. The development of a hydrogen economy can attract foreign investment and technology transfer, driving innovation and technological advancements in various sectors of the African economy. The utilization of hydrogen as an energy carrier can also enhance energy security for African countries by diversifying their energy sources and reducing their dependence on fossil fuels. The establishment of a hydrogen economy can create new job opportunities and stimulate economic diversification, further strengthening the continent's overall resilience and sustainability.

11.2.9 Opportunity for Decentralized Off-Grid Solutions to Reach Remote Communities

In a report published in 2019, the International Renewable Energy Agency (IRENA, 2018) stated that over 70% of rural Africans lack access to electricity currently; however, decentralized off-grid solutions have the potential to change that (IRENA, 2018). This is particularly crucial in remote communities where extending the central grid infrastructure may not be economically viable. A study conducted by the World Bank found that decentralized off-grid solutions can contribute significantly to Africa's net zero energy transition. By harnessing renewable energy sources such as solar and wind power at a local level, these solutions can reduce greenhouse gas emissions and promote sustainable development in remote areas of Africa (World Bank, 2022). Decentralized off-grid solutions can also enhance energy access and improve the quality of life for people living in these remote communities. By providing reliable and affordable electricity, these solutions can support economic activities, education, healthcare, and other essential services that are crucial for community development. The implementation of decentralized off-grid solutions can create local job opportunities and empower communities to become more self-sufficient in meeting their energy needs. Decentralized off-grid solutions can also have positive environmental impacts by reducing reliance on fossil fuels and promoting the use of renewable energy sources. This not only helps combat climate change but also contributes to the overall sustainability of these remote areas in Africa. By involving local communities in the planning and implementation process, decentralized off-grid solutions can foster a sense of ownership and pride among community members, leading to long-term commitment and sustainability.

11.2.10 Africa's Renewable Energy Potential Is Further Enhanced by Advancements in Technology and Decreasing Costs of Renewable Energy Systems

Africa has the capacity to produce over 310 gigawatts of renewable energy by 2030, which is equal to nearly 22% of the continent's total electricity consumption, according to the International Renewable Energy Agency (IRENA, 2018). This potential is further bolstered by advancements in technology, such as improved solar panel efficiency and storage systems, as well as decreasing costs of renewable energy systems. For instance, the cost of solar photovoltaic (PV) panels has decreased by around 80% since 2010 (IRENA, 2018), making them a more affordable option for African countries. Innovations in battery storage technology have made it possible to store excess renewable energy generated during peak production hours for use during low production periods or at night.

These advancements and decreasing costs contribute to Africa's propensity for a net zero energy transition, enabling countries to rely more on clean and sustainable sources of power. By harnessing its abundant renewable energy potential and leveraging technological advancements, Africa can not only meet its growing energy demands but also reduce greenhouse gas emissions and combat climate change. The adoption of battery storage technology in Africa can also enhance energy access in remote areas that are not connected to the grid. This can greatly improve the quality of life for communities by providing them with reliable and sustainable power sources, leading to economic growth and development. The implementation of these innovations can attract investments in the renewable energy sector, create job opportunities, and drive economic prosperity across the continent. The use of battery storage technology can also help stabilize the grid and reduce reliance on fossil fuels, making Africa less vulnerable to fluctuations in global energy prices. By embracing renewable energy solutions, African countries can position themselves as leaders in sustainability and attract international partnerships and collaborations to further accelerate their energy transition efforts. The adoption of renewable energy solutions can also contribute to reducing greenhouse gas emissions and mitigating the impacts of climate change. This is particularly important for African countries, as they are often disproportionately affected by the adverse effects of climate change, such as droughts and extreme weather events. By prioritizing renewable energy development, Africa can play a crucial role in global efforts to combat climate change and ensure a sustainable future for generations to come.

11.2.11 Africa's Vast Landmass Provides Ample Space for Large-Scale Solar and Wind Farms

According to a study conducted by the International Renewable Energy Agency (IRENA, 2018), Africa has the potential to generate 310 GW of wind power and 280 GW of solar power by 2030. This is due to its vast landmass, which offers abundant open spaces suitable for the installation of large-scale solar and wind farms. The continent's favorable climate conditions, with high levels of solar radiation and consistent wind patterns, further enhance its renewable energy potential. These factors make Africa an ideal region for harnessing renewable energy sources, such as wind and solar power. The development of these clean energy technologies can not only address the continent's growing energy demand but also create job opportunities and stimulate economic growth.

11.2.12 The Continent's Growing Population and Increasing Energy Demand

According to the United Nations (2022), Africa's population is projected to double by 2050, reaching approximately 2.5 billion people. This rapid population growth will undoubtedly lead to an increased demand for energy across various sectors, such as residential, commercial, and industrial. The International Renewable Energy Agency (IRENA, 2019) estimates that Africa has the potential to generate more than 310 gigawatts of renewable energy by 2030, highlighting the vast opportunities for investment and development in this sector. These factors combined make Africa an attractive market for renewable energy investments, not only to meet the growing energy demand but also to contribute toward achieving a net zero energy transition on the continent. In addition, the abundance of natural resources in Africa, such as solar and wind power, further strengthens its potential for renewable energy development. Investing in renewable energy in Africa can also bring socio-economic benefits by creating job opportunities and promoting sustainable development in the region. The increasing political will and commitment from African governments to prioritize renewable energy initiatives have created a favorable regulatory environment for investors. This, coupled with the continent's rapidly growing population and urbanization, presents a significant opportunity for renewable energy companies to establish a strong market presence and achieve long-term profitability.

11.3 ADDRESSING CLIMATE CHANGE BY REDUCING FOSSIL FUEL DEPENDENCY AND MITIGATING ITS IMPACTS

Africa's inclination toward a net zero energy transition is evidence of its determination to combat climate change by lowering its reliance on fossil fuels and lessening its effects inside the continent. This cutting-edge approach not only demonstrates Africa's determination to combat climate change but also highlights its recognition of the urgent need for sustainable and renewable energy sources. By embracing innovative technologies and fostering international collaborations, Africa is paving the way for a greener future while simultaneously addressing the pressing challenges posed by fossil fuel dependency and climate change impacts on the continent. Through initiatives such as investing in renewable energy infrastructure and promoting sustainable practices, Africa is positioning itself as a global player in the fight against climate change. These efforts not only contribute to the overall reduction of greenhouse gas emissions but also create new opportunities for economic growth and job creation within the continent. With its proactive stance, Africa is setting an inspiring example for other regions to follow in their pursuit of a more sustainable and resilient future. By embracing renewable energy sources such as solar and wind power, Africa is not only reducing its reliance on fossil fuels but also mitigating the adverse effects of climate change. These initiatives are empowering local communities and fostering technological innovation, paving the way for a greener and more prosperous Africa. As the continent continues to prioritize sustainability, it is demonstrating that economic development and environmental preservation can go hand in hand.

According to the International Energy Agency (IEA, 2023), Africa is particularly vulnerable to the impacts of climate change due to its high dependence on fossil fuels and limited adaptive capacity. The burning of fossil fuels contributes significantly to greenhouse gas emissions, exacerbating global warming and leading to various climate-related challenges in Africa, such as droughts, floods, and food insecurity. To address this issue, it is crucial for African countries to reduce their reliance on fossil fuels and transition toward renewable energy sources. For instance, investing in solar power can not only help mitigate climate change but also provide access to clean and affordable energy for communities across the continent. Implementing sustainable land management practices and promoting reforestation efforts can contribute to carbon sequestration and enhance resilience against climate impacts.

By adopting these measures, African nations can make significant progress in reducing their carbon footprint while simultaneously building a more sustainable and climate-resilient future for their citizens. It is essential for international cooperation and financial support to be provided to assist African countries in their efforts toward addressing climate change effectively. International cooperation and financial support are crucial in assisting African countries to implement these measures effectively. This support can help African nations access the necessary resources and technologies to transition to low-carbon economies and adapt to the changing climate. By providing international cooperation and financial support, African countries can receive the necessary expertise and funding to develop renewable energy sources and implement sustainable practices. This will not only reduce their carbon footprint but also create opportunities for economic growth and job creation in these nations. Such support can enable African countries to strengthen their resilience to climate change impacts, such as extreme weather events and food security challenges, ultimately improving the well-being of their citizens. Investing in renewable energy and sustainable practices can help African countries diversify their energy sources and reduce their dependence on fossil fuels. This can contribute to a more stable and secure energy supply, reducing vulnerability to price fluctuations in the global oil market. By embracing clean technologies, African nations can position themselves as leaders in the global fight against climate change, attracting investments and partnerships that can further drive economic development.

In South Africa, the government has implemented policies to promote renewable energy, such as the Renewable Energy Independent Power Producer Procurement Program (REIPPPP). This program has successfully attracted private investments in renewable energy projects, leading to job creation and economic growth. The country has set ambitious targets for renewable energy generation, aiming to increase the share of renewables in its energy mix to 18% by 2030 (IRENA, 2022). These efforts not only contribute to a more sustainable future but also enhance South Africa's energy security and resilience. The increased use of renewable energy sources reduces South Africa's dependence on fossil fuels, mitigating the negative environmental impacts associated with their extraction and combustion. By diversifying its energy sources, South Africa is better prepared to handle potential disruptions in the global oil and gas markets, ensuring a stable and reliable energy supply for its citizens. Investing in renewable energy technologies creates job opportunities and stimulates economic growth in South Africa. The development of a robust renewable energy sector attracts both domestic and foreign investments, fostering innovation and driving the country's transition toward a low-carbon economy. By reducing greenhouse gas emissions, South Africa can play a significant role in global efforts to combat climate change and meet international climate targets.

In Egypt, the government has implemented a feed-in tariff program to incentivize the development of renewable energy projects. This has led to an increase in solar and wind power installations, created job opportunities, and boosted the country's energy independence. The shift toward renewable energy sources has also improved air quality and reduced pollution levels, benefiting the health and well-being of Egyptian citizens. The government's commitment to renewable energy has positioned Egypt as a role model for other countries in the region, inspiring them to take similar actions. This has not only strengthened Egypt's global reputation but also fostered collaboration and knowledge sharing among nations in tackling climate change. The increased use of renewable energy has contributed to a more stable and reliable electricity grid in Egypt. With a diversified energy mix, the country is less dependent on volatile fossil fuel prices and is better equipped to handle potential disruptions in the energy supply. The transition to renewable energy has stimulated innovation and technological advancements in the sector, attracting investment and driving economic growth in Egypt. The shift toward renewable energy has also led to a reduction in greenhouse gas emissions, contributing to Egypt's efforts to mitigate climate change. This has not only improved the overall environmental sustainability of the country but has also positioned Egypt as a leader in sustainable development on the global stage. The success of Egypt's renewable energy initiatives has inspired other nations to follow suit and explore similar strategies, fostering international cooperation in combating climate change.

11.4 THE ECONOMIC BENEFITS OF INVESTING IN RENEWABLE ENERGY INDUSTRIES

Investment in Africa's renewable energy sector has the potential to yield significant economic returns, according to a report published by the International Renewable Energy Agency (IRENA, 2023). The report estimates that by 2030, such investments could generate around 2 million jobs and contribute to an increase in GDP of up to 1.1% in the region. Transitioning toward renewable energy sources can reduce dependence on fossil fuel imports, leading to improved energy security and reduced trade deficits.

These economic benefits are not limited to job creation and GDP growth alone. The transition to renewable energy can also attract foreign direct investment (FDI) and stimulate innovation and technological advancements in the region. By leveraging Africa's vast renewable energy potential, countries can position themselves as attractive investment destinations for both domestic and international investors, fostering sustainable economic development. Investing in renewable energy industries aligns with Africa's pursuit of a net zero energy transition while simultaneously unlocking significant economic opportunities for the continent. The development of renewable energy industries can contribute to job creation and poverty reduction, providing opportunities for local communities to participate in the green economy. This can lead to improved living standards and social well-being, ultimately promoting inclusive growth across Africa. By reducing reliance on fossil fuels and mitigating the impacts of climate change, investing in renewable energy can help African countries meet their international commitments under the Paris Agreement and contribute to global efforts to combat climate change. The shift toward renewable energy can also enhance energy security in Africa by diversifying the energy mix and reducing dependence on imported fossil fuels. This can help stabilize energy prices and ensure a more sustainable and reliable supply of electricity for both urban and rural areas. Investing in renewable energy can attract foreign direct investment and stimulate economic growth, creating a ripple effect that benefits various sectors of the economy.

The adoption of renewable energy in Africa can also help mitigate the impacts of climate change by reducing greenhouse gas emissions. This transition can also promote sustainable development and improve access to electricity in remote and underserved areas, thus enhancing social equity and inclusivity. The shift toward renewable energy sources can reduce dependence on fossil fuels, reducing the vulnerability of African countries to fluctuations in global oil prices. This can lead to greater energy security and stability in the region. The development of renewable energy infrastructure can create job opportunities and foster local entrepreneurship, contributing to poverty reduction and social empowerment. The adoption of renewable energy can contribute to mitigating climate change and reducing greenhouse gas emissions, which is crucial for the long-term sustainability of African countries. Investing in renewable energy technologies can attract foreign investments, support economic growth, diversify the energy sector, and reduce reliance on traditional industries.

11.5 UNLOCKING ENERGY POTENTIAL: EMPOWERING REMOTE AND UNDERSERVED COMMUNITIES WITH ENHANCED ACCESS AND RELIABILITY

Recent research indicates that Africa has enormous unrealized potential for renewable energy sources like hydroelectric, solar, and wind power. For instance, the continent receives an average of 325 days of sunshine per year, making solar energy a highly viable option. Africa is home to some of the strongest and most consistent wind resources globally. Unlocking this energy potential can play a crucial role in Africa's net zero energy transition. A report by the International Renewable Energy Agency (IRENA, 2022) states that by harnessing just a fraction of its renewable energy resources, Africa could not only meet its own energy demands but also become a major exporter of clean energy to other regions. Investing in enhanced access and reliability of energy in remote and underserved communities is vital for achieving sustainable development goals in Africa. Improved access to reliable electricity can uplift these communities by enabling economic growth, enhancing education and healthcare services, and reducing poverty levels.

Therefore, recognizing and leveraging Africa's vast renewable energy potential while prioritizing access and reliability in remote areas will be crucial steps toward achieving a net zero energy transition on the continent. By harnessing Africa's renewable energy potential, such as solar and wind power, the continent can reduce its dependence on fossil fuels and mitigate the harmful effects of climate change. Prioritizing access and reliability in remote areas will help bridge the energy gap and ensure that no community is left behind in the transition to clean energy. In addition, investing in renewable energy infrastructure will also create job opportunities and stimulate economic growth in Africa. By developing a skilled workforce in the renewable energy sector, the continent can not only address its energy needs but also contribute to sustainable development and poverty reduction. Promoting research and innovation in renewable energy technologies will enable Africa to become a global player in clean energy solutions, attract investment, and foster technological advancements for the benefit of its people and the planet.

11.6 EMBRACING MODERN TECHNOLOGIES COULD ENABLE THE LEAPFROGGING OF OUTDATED AND INEFFICIENT ENERGY INFRASTRUCTURE

Renewable energy resources, such as solar, wind, and hydropower, are abundant in Africa, according to a report published by the International Renewable Energy Agency (IRENA, 2022). Embracing modern technologies such as advanced solar panels, efficient wind turbines, and smart grid systems could tap into these resources and enable the leapfrogging of outdated energy infrastructure. For instance, the falling costs of solar PV panels have made them increasingly affordable and accessible. By harnessing solar power through the widespread adoption of PV systems, Africa could reduce its reliance on fossil fuels and achieve a significant shift toward clean energy. Implementing smart grid systems would enhance the efficiency of energy distribution and management, reducing losses in transmission and ensuring optimal utilization of renewable energy sources.

By leveraging these modern technologies in a strategic manner, Africa can bypass the need for costly investments in traditional fossil fuel–based infrastructure. This leapfrogging approach would not only help accelerate the continent's transition toward net zero energy but also contribute to economic growth, job creation, and improved access to electricity for its population. The adoption of smart grid systems can also lead to improved monitoring and control of energy consumption, allowing for better demand response and load management. This can help address issues of power outages and blackouts, ensuring a more reliable and stable electricity supply for African communities. Ultimately, embracing smart grid technologies presents a unique opportunity for Africa to leapfrog into a sustainable energy future while simultaneously addressing pressing social and economic challenges.

11.7 PROMOTING ENERGY INDEPENDENCE THROUGH DIVERSIFICATION AND REDUCED RELIANCE ON IMPORTS

Promoting energy independence through diversification and reduced reliance on imports is crucial for Africa's net zero energy transition. According to a report by the International Renewable Energy Agency (IRENA, 2022), Africa has abundant renewable energy resources, including solar, wind, hydro, and geothermal power. By tapping into these resources and developing domestic renewable energy capacities, Africa can significantly reduce its dependence on imported fossil fuels. Diversifying the energy mix will not only enhance energy security but also contribute to economic growth and job creation. The African Development Bank (2022) estimates that achieving universal access to electricity in Africa by 2030 would require an investment of around $120–140 billion per year. This investment could stimulate the development of local renewable energy industries, attract private investments, and create employment opportunities across the continent.

By prioritizing energy diversification and reducing reliance on imports, Africa can pave the way for a sustainable net zero energy transition while unlocking economic potential and improving access to clean and affordable energy for its population. Investing in renewable energy infrastructure would not only reduce greenhouse gas emissions but also mitigate the impacts of climate change in Africa. This would contribute to the continent's resilience and ability to adapt to a changing climate, ensuring a sustainable future for generations to come. Investing in renewable energy infrastructure would create job opportunities and stimulate economic growth in Africa. By developing a robust renewable energy sector, the continent can attract foreign investments and foster technological innovation, driving sustainable development and reducing poverty. This transition to clean energy sources would also improve air quality and public health, addressing the pressing issue of pollution in many African cities.

11.8 ENVIRONMENTAL BENEFITS OF RENEWABLE ENERGY GAIN RECOGNITION

According to a report by the International Renewable Energy Agency (IRENA, 2022), renewable energy sources have the potential to provide Africa with numerous environmental benefits. For instance, transitioning to renewable energy can significantly reduce greenhouse gas emissions, contributing to mitigating climate change. This is crucial for Africa, as the continent is highly vulnerable to the impacts of climate change, such as extreme weather events and rising sea levels. Renewable energy technologies offer opportunities for improved air quality and public health. Traditional sources of energy like coal and diesel generators often contribute to air pollution, leading to respiratory diseases and other health issues. By embracing renewable energy sources such as solar and wind power, Africa can reduce its reliance on fossil fuels and improve the overall well-being of its population.

Recognizing these environmental benefits, African countries are increasingly prioritizing renewable energy in their national policies and strategies. For example, South Africa has set a target of generating 18 gigawatts of renewable energy by 2030, while Kenya aims to achieve 100% clean energy access by 2022 (IEA, 2023). Morocco has invested heavily in solar power with the construction of the world's largest concentrated solar power plant. These efforts not only help to combat climate change but also create job opportunities and stimulate economic growth in these countries. Egypt has also made significant strides in renewable energy with the construction of the Benban Solar Park, one of the largest solar installations in the world. The country aims to generate 20% of its electricity from renewable sources by 2022 (IEA, 2023), attracting investments and boosting its green economy. These initiatives demonstrate a growing commitment among African nations to transition toward sustainable energy sources and reap the multiple benefits they offer. Algeria is another African country that has recognized the potential of renewable energy. It has set a target to generate 27% of its electricity from renewable sources by 2030 (IEA, 2023). The government has implemented various policies and incentives to attract investment in solar and wind energy projects, create job opportunities, and promote economic growth. This transition toward sustainable energy not only helps combat climate change but also reduces dependence on fossil fuels, making Algeria more resilient to fluctuations in global oil prices. Kenya and Ethiopia are also leading the way in renewable energy adoption in Africa. Kenya has made significant progress in harnessing geothermal and wind energy, with plans to achieve 100% renewable energy generation by 2030 (IEA, 2023). Similarly, Ethiopia has invested heavily in hydropower and aims to become a major exporter of clean energy to neighboring countries. These efforts not only contribute to reducing greenhouse gas emissions but also provide access to electricity for remote communities, improving quality of life, and driving economic development. Nigeria and Ghana are also making strides in renewable energy adoption in Africa. Nigeria has been investing in solar power projects and aims to increase its renewable energy capacity by 30% by 2030. Ghana, on the other hand, has been focusing on expanding its bioenergy sector and has set a target of generating 10% of its electricity from renewable sources by 2020 (IEA, 2023). These countries recognize the importance of transitioning to clean energy sources for sustainable development and are actively working toward achieving their goals.

These commitments not only demonstrate Africa's recognition of its environmental advantages but also highlight its determination to transition toward a net zero energy future. By embracing renewable energy sources such as solar, wind, and hydropower, African countries can also stimulate economic growth and create job opportunities in the clean energy sector. Transitioning to renewable energy can help Africa mitigate the impacts of climate change and improve energy security by reducing its dependence on imported fossil fuels. Investing in renewable energy can also lead to improved air quality and public health outcomes, as it reduces the reliance on polluting fossil fuel power plants. Transitioning to renewable energy can attract foreign investment and partnerships, fostering technological advancements and knowledge sharing in the region. In addition, transitioning to renewable energy can also provide a sustainable solution to Africa's energy access challenges, particularly in rural areas where traditional electricity grids may not reach them. This can improve the quality of life for millions of people by enabling access to reliable and affordable electricity for lighting, cooking, and powering essential appliances. The development of renewable energy infrastructure can create a ripple effect by driving innovation and job creation in related industries such as manufacturing, installation, and maintenance of renewable energy systems. The shift toward renewable energy can reduce Africa's dependence on fossil fuels, mitigating the negative impacts of climate change and improving air quality. By harnessing the continent's abundant renewable resources, such as solar, wind, and hydropower, Africa can become more self-sufficient in meeting its energy needs and reduce its reliance on costly imports.

11.9 OPPORTUNITIES FOR GLOBAL COLLABORATION IN THE RENEWABLE ENERGY SECTOR

Africa is estimated to have 310 GW of untapped renewable energy capacity, according to a report by the International Renewable Energy Agency (IRENA, 2023). This indicates that Africa has enormous potential for the development of renewable energy. This presents a significant opportunity for global collaboration in the renewable energy sector, as Africa's transition toward net zero energy aligns with the global goal of reducing greenhouse gas emissions and combating climate change.

Collaborations between African countries and international partners can bring in expertise, technology transfer, and investment to accelerate the deployment of renewable energy projects. For instance, initiatives like the Africa Renewable Energy Initiative (AREI) aim to mobilize investments from both African governments and international stakeholders to achieve universal access to sustainable energy in Africa by 2030 (IEA, 2023). These collaborations can also foster knowledge sharing and capacity building, enabling African countries to develop their own renewable energy industries and create local job opportunities. The shift toward renewable energy in Africa can contribute to reducing reliance on fossil fuels, improving energy security, and promoting sustainable economic growth in the region. The transition to renewable energy can also have positive environmental impacts by reducing greenhouse gas emissions and mitigating the effects of climate change. Investing in renewable energy infrastructure can help African countries diversify their energy sources and reduce their vulnerability to fluctuations in global fuel prices.

By leveraging these opportunities for global collaboration, Africa can not only meet its own growing energy demands sustainably but also contribute significantly to global efforts to achieve a clean and green future. The adoption of renewable energy technologies can create new job opportunities and stimulate local economies. This can lead to increased income and improved living standards for communities across Africa. By embracing renewable energy, African countries can position themselves as leaders in sustainable development and attract foreign investment in the clean energy sector. The shift toward renewable energy can also help African countries reduce their dependence on fossil fuels and mitigate the adverse effects of climate change. By investing in renewable energy infrastructure, African nations can improve energy access and reliability, particularly in rural areas where electricity is scarce. This can have a transformative impact on education, healthcare, and the overall quality of life for millions of people in Africa.

11.10 PRIORITIZING ALIGNMENT WITH GLOBAL SUSTAINABLE DEVELOPMENT GOALS AND CLIMATE CHANGE COMMITMENTS

According to the International Energy Agency (IEA, 2023), Africa is expected to experience significant growth in energy demand in the coming decades. This presents a unique opportunity for the continent to prioritize alignment with global sustainable development goals and climate change commitments as it transitions toward a net zero energy future. The IEA projects that Africa's energy demand will increase by 60% by 2040 (IEA, 2023), making it crucial for the continent to adopt sustainable and climate-friendly energy solutions. By prioritizing alignment with global sustainable development goals and climate change commitments, Africa can not only meet its growing energy needs but also contribute to global efforts to mitigate climate change. This transition would require significant investments in renewable energy sources such as solar and wind power, as well as implementing energy efficiency measures across various sectors to ensure a sustainable and low-carbon future for Africa.

By harnessing these resources and investing in clean energy infrastructure, the continent can not only meet its growing energy needs but also reduce its carbon emissions. Aligning with global sustainable development goals and climate change commitments can attract international investments and partnerships. The United Nations Sustainable Development Goals (SDGs) provide a framework for addressing poverty, inequality, and environmental challenges. By integrating these goals into its energy transition plans, Africa can create sustainable and inclusive growth while mitigating climate change impacts. By promoting renewable energy sources such as solar and wind power, Africa can tap into its abundant natural resources and decrease its reliance on fossil fuels. This transition to clean energy can also create job opportunities and stimulate economic development in the region. By prioritizing energy efficiency measures and implementing innovative technologies, Africa can optimize its energy use and further reduce its carbon footprint.

11.11 UNLOCKING THE PROMISE OF RENEWABLE ENERGY: EMPOWERING CONSUMERS WITH AFFORDABLE ENERGY CHOICES

Unlocking the Promise of Renewable Energy: Empowering Consumers with Affordable Energy Choices is a crucial step toward Africa's net zero energy transition. The report by IRENA (2022) reveals that the continent has enormous potential to generate over 310 GW of renewable energy by 2030, which is equivalent to two-thirds of its current power capacity. This demonstrates the significant opportunity for

Africa to embrace clean and sustainable energy sources, paving the way for a greener and more sustainable future. Harnessing this potential will not only contribute to reducing carbon emissions but also provide affordable energy choices for consumers, empowering them to actively participate in the transition toward a net zero economy.

Unlocking the promise of renewable energy is crucial to empowering consumers with affordable energy choices in Africa. By harnessing renewable sources such as solar, wind, and hydropower, African countries can reduce their dependence on expensive fossil fuels and improve access to clean and affordable energy for their citizens. This transition will not only contribute to mitigating climate change but also foster economic growth and create job opportunities in the renewable energy sector. In addition, the transition to renewable energy will also enhance energy security in Africa as it reduces reliance on imported fossil fuels. This will enable African countries to have more control over their energy resources and reduce their vulnerability to fluctuations in global fuel prices. The development of a robust renewable energy sector will attract investments and stimulate innovation, leading to technological advancements and economic diversification in the region.

Investing in renewable energy will contribute to reducing air pollution and improving public health in Africa. By shifting away from fossil fuels, countries can mitigate the harmful effects of greenhouse gas emissions and improve the overall quality of the air. This will have a positive impact on the well-being of communities and reduce healthcare costs associated with respiratory illnesses and other pollution-related diseases. The growth of the renewable energy sector will create a ripple effect in other industries, fostering a more sustainable and resilient economy for Africa. Investing in renewable energy will also help Africa diversify its energy sources and reduce its dependence on imported fossil fuels. This will enhance energy security and promote economic stability by reducing vulnerability to fluctuations in global oil prices. The development of renewable energy infrastructure will create job opportunities and stimulate local economies, leading to increased prosperity and social development in the region. The shift toward renewable energy will also contribute to mitigating the adverse effects of climate change by reducing greenhouse gas emissions. This will not only benefit Africa but also contribute to global efforts to combat climate change and achieve sustainable development goals. Investing in renewable energy will foster innovation and technological advancements, positioning Africa as a candidate for clean energy solutions and attracting further investments and partnerships from international stakeholders.

11.12 ENSURING A RELIABLE ENERGY MARKET FOR SUSTAINABLE ECONOMIC GROWTH

Africa's energy demand is expected to rise dramatically over the next several decades due to population growth and economic development, according to a report published by the International Energy Agency (IEA, 2023). To ensure a reliable energy market for sustainable economic growth, it is crucial for Africa to embrace a net zero energy transition. The IEA estimates that Africa will need to invest around \$2.5 trillion in energy infrastructure by 2040 to meet its growing energy demand (IEA, 2023). This presents an opportunity for the continent to shift toward renewable energy sources and reduce its reliance on fossil fuels. Transitioning to a net zero energy system can not only address climate change concerns but also foster economic growth through job creation and technological innovation. By investing in renewable energy infrastructure, Africa can tap into its vast natural resources, such as solar and wind power, to meet its energy needs sustainably. This transition can also help reduce air pollution and improve public health by decreasing reliance on fossil fuels. Embracing a net zero energy system can attract foreign investment and create a competitive advantage for African countries in the global clean energy market. The development of a net zero energy system can enhance energy security for African countries by reducing their dependence on imported fossil fuels. This would not only save valuable foreign exchange but also provide a stable and reliable energy supply for domestic consumption. The transition to renewable energy can stimulate local industries and create new business opportunities, leading to economic diversification and increased resilience in the face of global energy price fluctuations. The adoption of clean energy technologies can also contribute to mitigating the impacts of climate change, as African countries can reduce their greenhouse gas emissions by shifting away from fossil fuels. This would not only help in meeting international climate commitments but also position African nations as candidates for sustainable development and environmental stewardship.

11.13 AFRICA'S ABUNDANT NATURAL RESOURCES FUELING CLEAN TECHNOLOGY REVOLUTION

Africa's abundance of natural resources presents a huge potential for the production of renewable energy, according to a report published by the International Renewable Energy Agency (IRENA, 2023). The continent is blessed with vast solar, wind, hydro, and geothermal resources that can be harnessed to fuel the clean technology revolution. For instance, Africa receives the highest solar radiation levels globally, making it a prime location for solar power generation. In addition to solar power, Africa also has strong wind resources, particularly along its coastlines and in certain inland regions. These wind resources can be tapped into to further expand the clean technology revolution on the continent. Africa's rich hydroelectric potential, with its numerous rivers and waterfalls, offers another significant opportunity for renewable energy generation and sustainable development.

The African Union's Agenda 2063 emphasizes the need for a sustainable and low-carbon energy transition. This commitment has led to increased investments in renewable energy projects across the continent. For example, countries like Morocco and South Africa have made significant strides in deploying large-scale renewable energy projects. Morocco has constructed one of the world's largest solar power plants, the Noor Ouarzazate Solar Complex, which has a capacity of 580 MW (IRENA, 2022). South Africa, on the other hand, has successfully implemented a renewable energy procurement program that has attracted substantial private sector investment in wind and solar projects. These advancements demonstrate Africa's potential to become a leader in clean energy production and contribute significantly to global efforts to combat climate change. The combination of Africa's rich natural resources and its commitment to a net zero energy transition creates a favorable environment for the growth of clean technology industries. These industries not only contribute to reducing greenhouse gas emissions but also provide opportunities for job creation and economic development in the region.

The development of clean technology industries in Africa can also help to address energy poverty and improve access to electricity for underserved communities. By investing in renewable energy projects, countries in Africa can reduce their reliance on fossil fuels and increase their energy independence. This shift toward clean energy sources not only benefits the environment but also promotes sustainable development and resilience in the face of climate change. The transition to clean energy in Africa can attract foreign investments and promote technology transfer, fostering collaboration between African nations and international partners. This collaboration can lead to knowledge sharing and capacity building, empowering local communities, and creating a skilled workforce for the growing clean energy sector. Ultimately, investing in clean technology industries in Africa can pave the way for a more sustainable and prosperous future for the continent.

11.14 LEVERAGING THE CONTINENT'S ECOSYSTEMS FOR SUSTAINABLE CARBON SEQUESTRATION AND GREENHOUSE GAS MITIGATION

Recent research indicates that Africa's extensive ecosystems may be essential for mitigating greenhouse gas emissions and sequestering carbon. For instance, the continent's forests alone are estimated to absorb around 1.2 billion metric tons of CO_2 annually (IEA, 2023). Wetlands and grasslands in Africa have also been identified as significant carbon sinks, contributing to the overall reduction of greenhouse gas emissions (IEA and UNEP, 2018). These natural ecosystems not only have the capacity to store carbon, but they also provide numerous co-benefits such as biodiversity conservation, water regulation, and soil fertility enhancement. Expanding efforts to protect and restore these ecosystems can contribute to Africa's net zero energy transition while simultaneously addressing other pressing environmental challenges. For example, by preserving wetlands and grasslands, countries in Africa can enhance their water security by regulating water flow and reducing the risk of floods. These ecosystems play a crucial role in supporting local communities by providing opportunities for sustainable livelihoods through activities such as fishing, agriculture, and tourism. Therefore, investing in the protection and restoration of wetlands and grasslands can have wide-ranging benefits for both the environment and the well-being of African societies.

By leveraging these ecosystems for sustainable carbon sequestration and incorporating them into Africa's net zero energy transition strategy, the continent can make substantial progress toward achieving its climate goals. This approach not only offers an opportunity to combat climate change but also provides co-benefits such as biodiversity conservation, improved water quality, and enhanced resilience to natural disasters (IRENA, 2019). Therefore, harnessing Africa's ecosystems for carbon sequestration presents a win–win solution that aligns with global sustainability objectives while addressing the specific challenges faced by the continent. By leveraging Africa's rich natural resources and ecosystems, such as forests, wetlands, and grasslands, for carbon sequestration, the continent can not only reduce its greenhouse gas emissions but also contribute to global efforts to mitigate climate change. This approach can create new economic opportunities and promote sustainable development in Africa by attracting investments in renewable energy projects and nature-based solutions. By investing in the protection and restoration of these natural resources, Africa can enhance its resilience to climate change impacts such as extreme weather events and water scarcity. This will not only benefit local communities but also contribute to the preservation of biodiversity and ecosystem services that are crucial for the continent's long-term sustainability.

REFERENCES

African Development Bank. (2022). African Development Bank – Building today, a better Africa tomorrow. In Http://Www.Afdb.Org/En/

African Union Commission. (2015). The African Union Commission: AGENDA 2063, The Africa We Want. First Ten-Year Implementation Plan. African Union.

Agency, I. R. E. (2020). Renewable Power Generation Costs in 2019. International Renewable Energy Agency.

IEA. (2023). Global EV Outlook 2023. *Geo*, *Geo*.

IEA and UNEP. (2018). International Energy Agency and the United Nations Environment Programme - Global Status Report 2018: Towards a zero-emission, efficient and resilient buildings and construction sector. Global Status Report.

IRENA. (2018). International Renewable Energy Agency. Renewable Power Generation Costs in 2017. International Renewable Energy Agency.

IRENA. (2019). Future of Solar Photovoltaic: Deployment, investment, grid integration and socio-economic aspects. International Renewable Energy Agency (IRENA).

IRENA. (2020). Innovation landscape brief: Peer-to-peer electricity trading. *Innovation Landscape Brief: Peer-to-Peer Electricity Trading*.

IRENA. (2022). World Energy Transitions Outlook 2023: 1.5°C Pathway. World Energy Transitions.

IRENA. (2023). World Energy Transitions Outlook 2023: 1.5°C Pathway. In *World Energy Transitions*.

Ituen, E. E., Esen, N. U., Nwokolo, S. C., & Udo, E. G. (2012). Prediction of global solar radiation using relative humidity, maximum temperature and sunshine hours in Uyo, in the Niger Delta Region, Nigeria. *Advances in Applied Science Research*, *3*(4), 1923–1937.

Laura Cozzi et al. (2023). World Energy Outlook. In *World Energy Outlook*. International Energy Agency. www.iea.org/weo.

Nwokolo, S. C., Echeng, E., Obiwulu, A., Meyer, E. L., Christian, C., Ogbulezie, J. C., & Proutsos, N. (2024). A multi-model approach based on CARIMA-SARIMA-GPM for assessing the impacts of climate change on concentrated photovoltaic (CPV) potential. *Physics and Chemistry of the Earth*, *134*(January), 103560. https://doi.org/10.1016/j.pce.2024.103560

Nwokolo, S. C., Meyer, E. L., & Ahia, C. C. (2023). Credible pathways to catching up with climate goals in Nigeria. *Climate*, *11*(9), 1–32. https://doi.org/org/10.3390/cli11090196

Nwokolo, S. C., Meyer, E. L., & Ahia, C. C. (2024). Exploring the interactive influences of climate change and urban development on the fraction of absorbed photosynthetically active radiation. *Atmosphere*, 1–37.

Nwokolo, S. C., Obiwulu, A. U., & Ogbulezie, J. C. (2023). Machine learning and analytical model hybridization to assess the impact of climate. *Physics and Chemistry of the Earth*, 130(February 2023), 103389. https://doi.org/10.1016/j.pce.2023.103389

Nwokolo, S. C., & Ogbulezie, J. C. (2018a). A qualitative review of empirical models for estimating diffuse solar radiation from experimental data in Africa. *Renewable and Sustainable Energy Reviews*, *92*(18), 353–393. https://doi.org/10.1016/j.rser.2018.04.118

Nwokolo, S. C., & Ogbulezie, J. C. (2018b). A quantitative review and classification of empirical models for predicting global solar radiation in West Africa. *Beni-Suef University Journal of Basic and Applied Sciences*, *7*(4), 367–396. https://doi.org/10.1016/j.bjbas.2017.05.001

Nwokolo, S. C., Ogbulezie, J. C., & Ushie, O. J. (2023). A multi-model ensemble-based CMIP6 assessment of future solar radiation and PV potential under various climate warming scenarios. *Optik*, *285*(May), 170956. https://doi.org/10.1016/j.ijleo.2023.170956

Nwokolo, S. C., & Otse, C. (2019). Impact of sunshine duration and clearness index on diffuse solar radiation estimation in mountainous climate. *Trends in Renewable Energy*, *5*(3), 307–332. https://doi.org/10.17737/tre.2019.5.3.00107

Nwokolo, S. C., Proutsos, N., & Meyer, E. L. (2023). Machine Learning and Physics-Based Hybridization Models for Evaluation of the Effects of Climate Change and Urban Expansion on Photosynthetically Active Radiation. *Atmosphere*, *14*(687), 1–36. https://doi.org/doi.org/10.3390/atmos14040687

Obiwulu, A. U., Chendo, M. A. C., Erusiafe, N., & Nwokolo, S. C. (2020). Implicit meteorological parameter-based empirical models for estimating back temperature solar modules under varying tilt-angles in Lagos, Nigeria. *Renewable Energy*, *145*, 442–457. https://doi.org/10.1016/j.renene.2019.05.136

Obiwulu, A. U., Erusiafe, N., Olopade, M. A., & Nwokolo, S. C. (2020). Modeling and optimization of back temperature models of mono-crystalline silicon modules with special focus on the effect of meteorological and geographical parameters on PV performance. *Renewable Energy*. https://doi.org/10.1016/j.renene.2020.02.103

Obiwulu, A. U., Erusiafe, N., Olopade, M. A., & Nwokolo, S. C. (2022). Modeling and estimation of the optimal tilt angle, maximum incident solar radiation, and global radiation index of the photovoltaic system. *Heliyon*, *8*(6), e09598. https://doi.org/10.1016/j.heliyon.2022.e09598

The World Bank Group. (2022). *Risk of Global Recession in 2023 Rises Amid Simultaneous Rate Hikes*. The World Bank Group.

United Nations. (2022). United Nations World Population Prospects 2022. In *Department of Economic and Social Affairs, Population Division* (Issue 9).

Index